Two Manuals Combined into One Book

Cavalry Troop

And

Browning Machine Gun .50 HB M2

Department of Defense

Pentagon
Publishing

***FM 17-97**

Field Manual
No 17-97

HEADQUARTERS
DEPARTMENT OF THE ARMY
Washington, DC 3 October 1995

CAVALRY TROOP

TABLE OF CONTENTS

		Page
Preface		iv
Chapter 1	INTRODUCTION	1-1
	Organizations	1-2
	Missions, Capabilities, and Limitations	1-16
	Responsibilities	1-18
Chapter 2	BATTLE COMMAND	2-1
	Troop Leading Procedures	2-2
	Situational Awareness	2-13
	Communications	2-17
	Techniques of Tactical Control	2-22
	Command Guidance and Organizational Control	2-27
	Tactical Movement Formations	2-34

*This publication supersedes FM 17-97, September 1988.

Chapter 3 RECONNAISSANCE 3-1
Purpose, Fundamentals, and Capabilities 3-2
Planning Considerations, Methods, and Techniques 3-5
Route Reconnaissance....................... 3-29
Zone Reconnaissance 3-41
Area Reconnaissance......................... 3-53

Chapter 4 SECURITY... 4-1
Purpose, Fundamentals, and Capabilities 4-2
Screen... 4-5
Guard 4-31
Cover 4-32
Route Security.................................... 4-33
Area Security...................................... 4-33
Convoy Security 4-34

Chapter 5 OFFENSE.. 5-1
Purpose and Fundamentals................ 5-1
Movement to Contact.......................... 5-2
Hasty Attack....................................... 5-11
Deliberate Attack................................ 5-22
Raid.. 5-24

Chapter 6 DEFENSE.. 6-1
Purpose, Fundamentals, and Schemes of Maneuver...................... 6-2
Defend From a Troop

Battle Position 6-6
Defend in Troop Sector....................... 6-10
Delay in Troop Sector 6-22

Chapter 7 OTHER TACTICAL OPERATIONS 7-1
Tactical Road Marches 7-1
Assembly Areas 7-9
Relief in Place 7-15
Battle Handover/Passage of Lines...... 7-21
Hasty Water Crossing.......................... 7-31
In-Stride Breach 7-35
NBC Defensive Operations................. 7-42

Chapter 8 COMBAT SUPPORT 8-1
The Fire Support Team........................ 8-1
Mortar Support.................................... 8-9
Artillery Support................................... 8-12
Engineer Support 8-14
Ground Surveillance Radar Support... 8-20
Air Defense 8-23
Army Aviation Support 8-26
Close Air Support................................ 8-28

Chapter 9 COMBAT SERVICE SUPPORT 9-1
Organization.. 9-1
Logistics 9-6
Personnel Service Support 9-16
Prisoners of War 9-19

Appendix A FRATRICIDE PREVENTION.............. A-1

Appendix B STANDING OPERATING PROCEDURES.................................. B-1

Appendix C OPERATIONS ORDER C-1

Appendix D OPERATIONS OTHER THAN WAR .. D-1

Glossary Glossary-1

References References-1

Preface

This publication serves as a doctrinal guide for commanders and subordinate leaders of both light (HMMWV equipped) and heavy (M1 and M3 equipped) cavalry troops. Heavy cavalry troops are found in Armored Cavalry Regiments (ACR), in Armor/Mechanized Divisions, and in Separate Armor/Mechanized Brigades. Light cavalry troops are found in Light Armored Cavalry Regiments (LACR), in Light Infantry Divisions, and in Separate Light Infantry Brigades.

Although FM 17-97 does not specifically address the TOE of the Light Division Cavalry Troop or the troops of a Separate Heavy or Light Brigade, the tactics, techniques, and procedures outlined in this publication still apply.

This publication lays out the organization, command and control, tactical employment, and service support of the cavalry troop in combat. It establishes the responsibilities and duties of key personnel in the troop during combat.

FM 17-97 is one of three publications in the cavalry troop family. It provides troop leaders with a doctrinal reference to link training and fighting. ARTEP 17-487-30-MTP, *Mission Training Plan for the Regimental Armored Cavalry Troop*, outlines how to train the troop. FKSM 17-97-3, *Cavalry Troop Common SOP*, describes troop operating procedures. This family of publications supports FM 17-15, *Tank Platoon*; FM 7-91, *Tactical Employment of Antiarmor Platoons, Companies, and Battalions*; and FM 17-98, *Scout Platoon*.

The information in this manual is based on the objective table of organization and equipment (TOE) for the ACR and LACR troops. Differences between actual unit organizations and equipment and those described in this manual are caused by differences in the modified table of equipment and organization (MTOE).

The proponent for this publication is HQ TRADOC. Submit changes for improving this publication on DA Form 2028 (Recommended Changes to Publications and Blank Forms), and forward it to Commandant, USAARMS, ATTN: ATSB-SBA-F, Fort Knox, KY 40121-5211.

Although this manual does not implement any particular international agreement, the material presented herein is in accordance with related international agreements. A list of related international agreements and other references can be found in the References section. The paragraph in Chapter 2, Section I, entitled Issue a Warning Order is in compliance with STANAG 2014. Chapter 7, Section III, Relief in Place, and Section IV, Battle Handover and Passage of Lines, are in compliance with STANAG 2082. Tables 7-2 and 7-4 are in compliance with STANAG 2083.

Unless this publication states otherwise, masculine nouns and pronouns do not refer exclusively to men.

Article I. Chapter 1

Section 1.01 Introduction

The role of the cavalry troop in Army operations remains unchanged from the traditional role of cavalry throughout the history of warfare. The troop is organized, equipped, and trained to protect and preserve the fighting ability of other combined arms forces. While its primary missions are reconnaissance and security, the cavalry troop may be called upon to execute attack, defend, and delay missions as part of squadron and regimental missions. *The troop accomplishes its missions by communicating, moving, and shooting in that order.*

The purpose of this chapter is—

- To depict organizations of the cavalry troop in—

  Heavy Troop:

  - The Armored Cavalry Regiment (ACR) (M1 and M3 equipped).
  - The Heavy Division Cavalry Squadron (M1 and M3 equipped).

  Light Troop:

  - The Light Armored Cavalry Regiment (LACR) (HMMWV equipped).

Contents

Page

Section I. Organizations 1-2
Section II. Missions, Capabilities, and Limitations ... 1-16
Section III. Responsibilities 1-18

- To outline missions each troop performs and to highlight each troop's capabilities and limitations.

- To establish responsibilities of key personnel in combat.

(a) Section I. Organizations

(I) THE HEAVY CAVALRY TROOP

The heavy cavalry troop consists of 6 officers and 126 enlisted soldiers. The troop is organized into a headquarters section, two scout platoons, two tank platoons, a mortar section, and a maintenance section (see Figure 1-1).

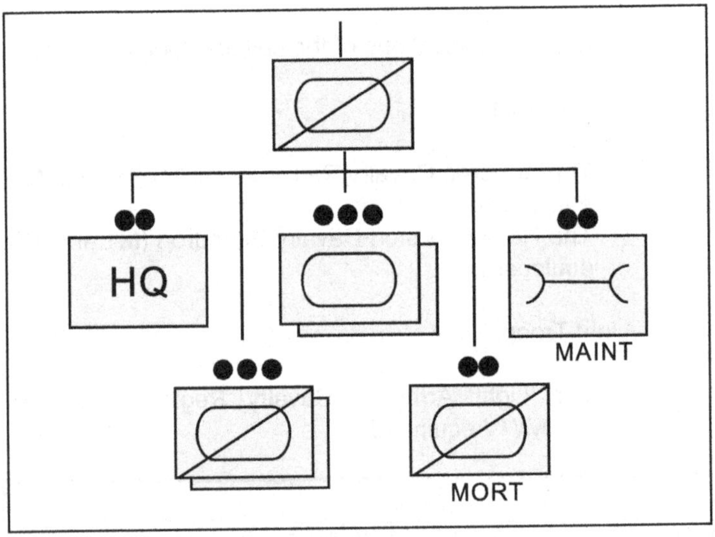

a) Figure 1-1. Heavy cavalry troop organization.

2) The Heavy Troop Scout Platoon

The scout platoon is organized and equipped to conduct reconnaissance and screening in support of its parent troop. However, when the parent unit is performing missions within an economy-of-force role, the scout platoon may conduct offensive, defensive, and retrograde operations in support of the troop mission. The platoon consists of 1 officer and 29 enlisted soldiers, and is equipped with six M3 cavalry fighting vehicles (CFV) organized into three scout sections (see Figure 1-2).

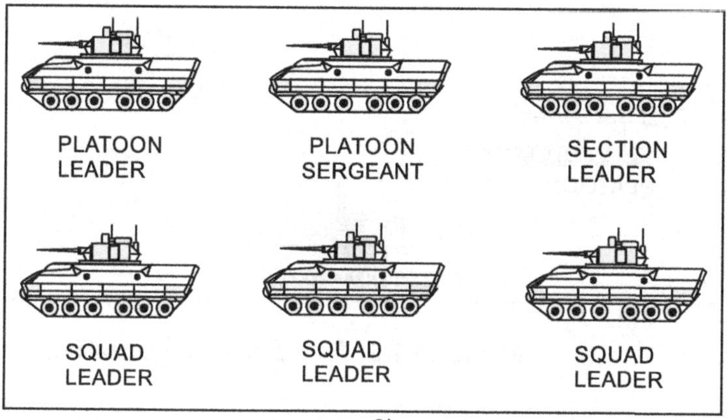

3)

a) Figure 1-2. Scout platoon organization.

4)

5) The Heavy Troop Tank Platoon

The tank platoon is organized and equipped to perform its three primary missions—attack, defend, and move. The platoon consists of 1 officer and 15 enlisted soldiers manning four M1A1 Abrams main battle tanks organized into two sections (see Figure 1-3).

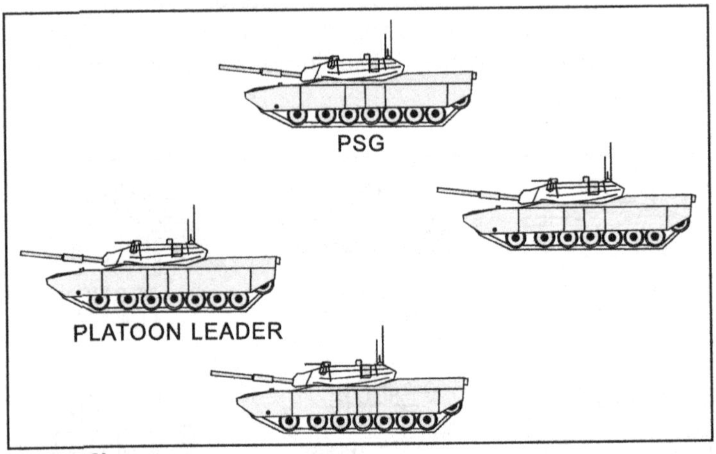

6)

a) Figure 1-3. Tank platoon organization.

7) The Heavy Troop Mortar Section

The mortar section is organized and equipped to provide immediate indirect fires in support of troop operations. Such supporting fires are usually suppression, screening, obscuration, or illumination. The section consists of nine enlisted soldiers. It is equipped with two 107-mm mortars mounted in two self-propelled mortar carriers (see Figure 1-4).

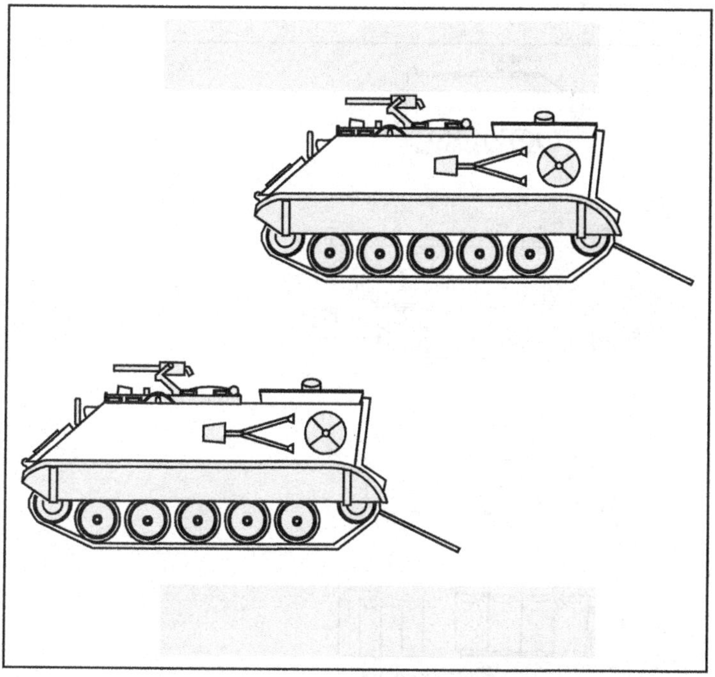

Figure 1-4. Mortar section organization.

8) The Heavy Troop Maintenance Section

This section is organized and equipped to diagnose and repair most equipment faults at troop level. It has the capability to recover all troop vehicles and to maintain the troop's equipment records. The section consists of 18 enlisted soldiers equipped with one armored personnel carrier (APC), one heavy recovery vehicle, one utility truck with cargo trailer, and two cargo trucks with cargo trailers (see Figure 1-5).

a) Figure 1-5. Maintenance section organization.

9) The Heavy Troop Headquarters Section

a)

The troop headquarters section is organized and equipped to perform command and control and logistical support functions for the troop. The section consists of 2 officers and 11 enlisted soldiers. It is equipped with one main battle tank, one command post (CP) carrier, one APC, one M3 CFV, one cargo truck with a 400-gallon water trailer, and two utility trucks with one cargo trailer (see Figure 1-6).

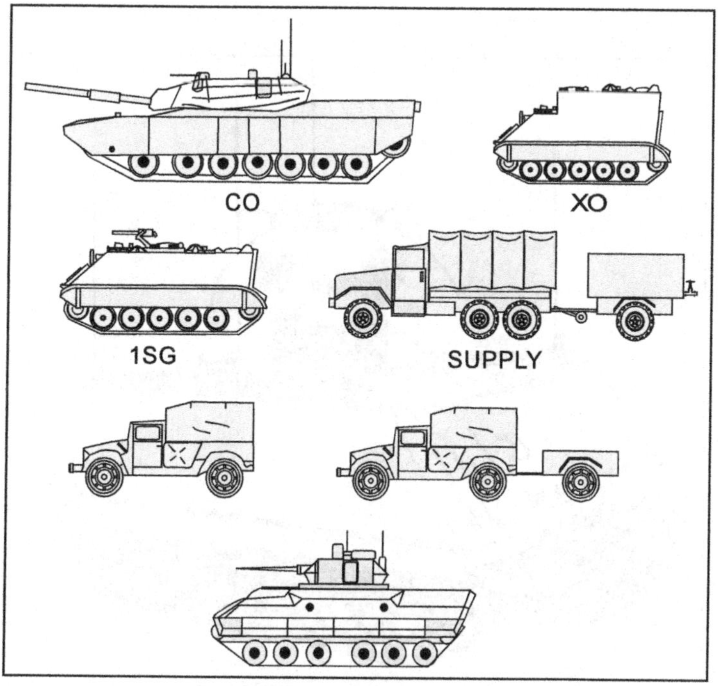

b) Figure 1-6. Headquarters section organization.

10) **The Fire Support Team (FIST)**

The FIST comes from the squadron howitzer battery in the heavy cavalry regiment and the division artillery in the division cavalry squadron. The FIST consists of one fire support officer (FSO), one NCO (team chief), one enlisted fire support specialist, and one enlisted radio operator. The FIST is responsible for coordinating indirect fires for the troop. In the heavy troop the team is mounted on an M981 FIST vehicle (see Figure 1-7).

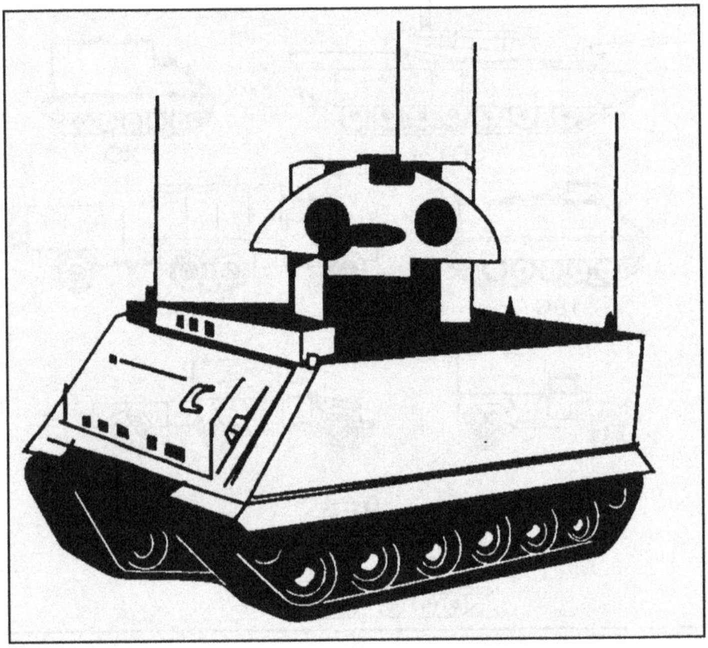

11)

a) Figure 1-7. FIST organization.

(II) THE LIGHT CAVALRY TROOP

The light cavalry troop consists of 6 officers and 107 enlisted soldiers. The troop is organized into a headquarters section, two scout platoons, two antitank (AT) platoons, a mortar section, and a maintenance section (see Figure 1-8).

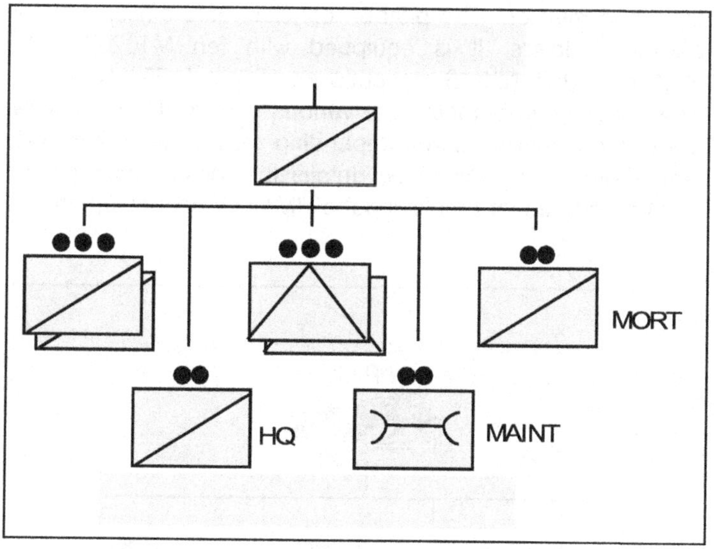

a) Figure 1-8. Light cavalry troop organization.

2) The Light Troop Scout Platoon

The scout platoon is organized and equipped to conduct reconnaissance and screening in support of its parent troop. However, when the parent unit is performing missions within an economy-of-force role, the scout platoon may conduct offensive, defensive, and retrograde operations in support of the troop mission. The platoon consists of 1 officer and 29 enlisted soldiers. It is equipped with ten M1025/M1026 HMMWVs (five MK-19 equipped and five cal .50 equipped). The platoon can organize into various configurations, usually from two to five sections, depending on factors of METT-T (see Figure 1-9). FM 17-98 details the tactics, techniques, and procedures for employing the HMMWV scout platoon.

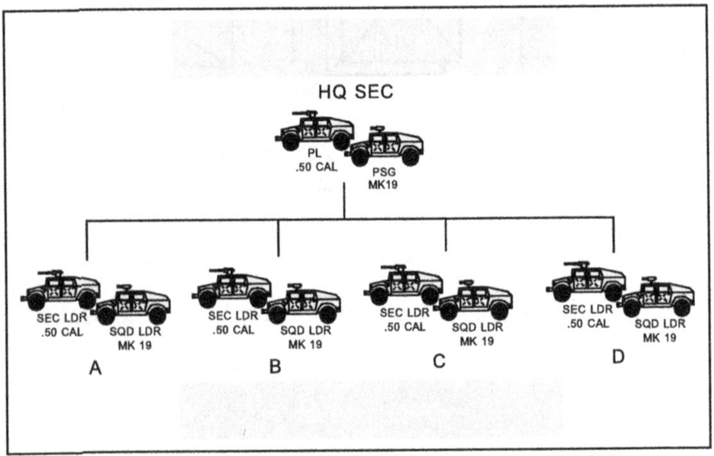

a) Figure 1-9. Light troop scout platoon organization.

3) The Light Troop Antitank Platoon

The AT platoon is organized and equipped to perform its three primary missions—attack, defend, and move. The platoon consists of 1 officer and 11 enlisted soldiers manning four M996 HMMWVs (TOW carriers) organized into two sections (see Figure 1-10).

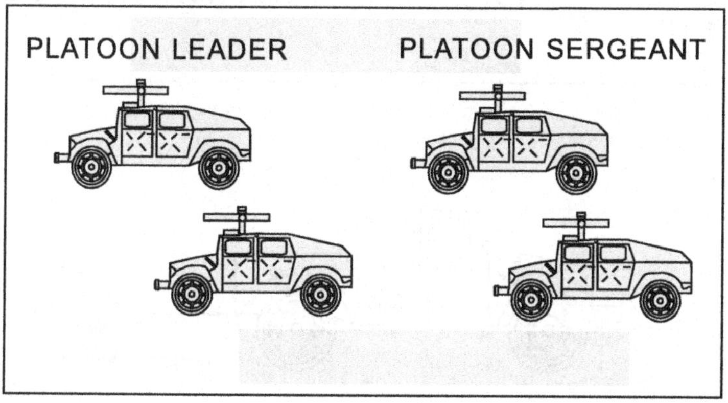

a) Figure 1-10. Light troop AT platoon organization.

4) The Light Troop Mortar Section

The mortar section is organized and equipped to provide immediate indirect fires to support troop operations. Such supporting fires are usually suppression, screening, obscuration, or illumination. The section consists of nine enlisted soldiers. It is equipped with two 120-mm mortars trailered by three M1025/M1026 HMMWVs (see Figure 1-11).

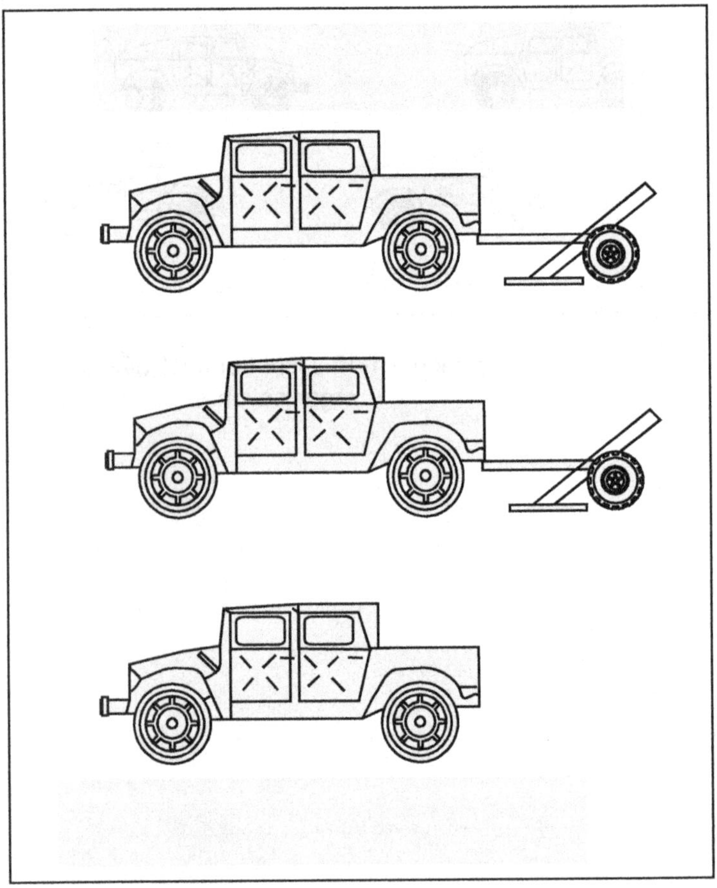

a) Figure 1-11. Light troop mortar section organization.

5) The Light Troop Maintenance Section

This section is organized and equipped to diagnose and repair most equipment faults at troop level. It has the capability to recover all troop vehicles and to maintain the troop's equipment records. The section consists of eight enlisted soldiers equipped with one M998 cargo/troop carrier HMMWV, one cargo truck with cargo trailer, and one 5-ton wrecker (see Figure 1-12).

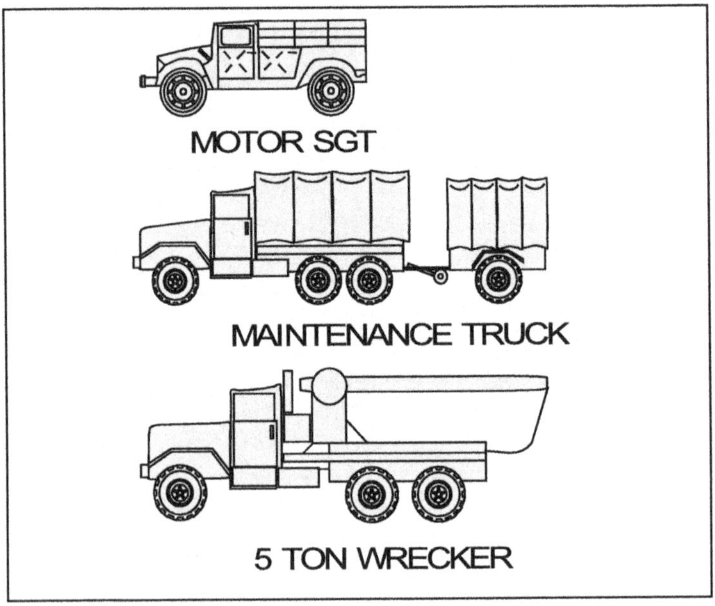

6)

a) Figure 1-12. Light troop maintenance section organization.

7) The Light Troop Headquarters Section

The troop headquarters section is organized and equipped to perform command and control and logistical support functions for the troop. The section consists of two officers and ten enlisted soldiers. It is equipped with one M1025 HMMWV, one M1037 HMMWV shelter carrier with S250 shelter, one M998 HMMWV with cargo trailer, and one 2 1/2-ton cargo truck with 400-gallon water trailer (see Figure 1-13).

a) Figure 1-13. Light troop headquarters section organization.

8) The Fire Support Team (FIST)

The FIST comes from the squadron howitzer battery in the light cavalry regiment. The FIST consists of one officer (FSO), one NCO (team chief), one enlisted fire support specialist, and one enlisted radio operator. The FIST is responsible for coordinating indirect fires for the troop. In the light troop the team is mounted on a HMMWV (see Figure 1-14).

(b)

a) Figure 1-14. Light troop FIST organization.

(c) Section II. Missions, Capabilities, and Limitations

(I) MISSIONS

Heavy and light cavalry troops perform reconnaissance and security missions to protect and preserve the fighting ability of the units to which they are assigned or attached. Both troops also conduct offensive, defensive, and retrograde operations in an economy-of-force role. Cavalry troops in general have limitations and capabilities associated with their TOEs and METT-T that must be considered when employing them in a specific mission role (see Figure 1-15).

	Hvy Trp	Lt Trp
Reconnaissance Missions		
Route Recon	✓	✓
Zone Recon	✓	✓
Area Recon	✓	✓
Security Missions		
Screen	✓	✓
Area Security	✸	✸
Convoy Security	x	x
Economy-of-Force Role		
Offensive Missions		
Hasty Attack	✓	✸
Attack	✓	✸
Movement to Contact	✓	✸
Defensive Missions		
Defend a Battle Position	✓	✸
Defend in Sector	✓	✸
Retrograde Missions		
Delay	✓	✸

✓ = fully capable
x = capable when reinforced
✸ = capable under permissive METT-T

a) Figure 1-15. Cavalry troop mission profiles.

(II) CAPABILITIES

a) The heavy cavalry troop is a unit that can be deployed by rail, sea, or both into a theater of operations. It can perform its missions under all visibility conditions and in any terrain that supports heavy armor movement. The integrated thermal sights on the M1A1 tank and M3 CFV along with ancillary passive night sights and night vision devices authorized to the troop provide an outstanding nighttime/reduced visibility acquisition and fighting capability. Additionally, the firepower and survivability organic to the heavy cavalry troop allow it to execute missions aggressively across the spectrum of warfare.

The light cavalry troop is a unit that can be rapidly deployed by air, rail, or sea into a theater of operations. It can perform its missions under all visibility conditions and in any terrain that supports wheeled vehicle movement. The ancillary thermal and night vision devices authorized to the troop provide an excellent reduced visibility acquisition capability. However, the light cavalry troop's ability to fight under reduced visibility conditions is limited because of its lack of integrated passive and thermal sights on stabilized weapons platforms.

(III)

(IV) LIMITATIONS

a) Given suitable terrain conditions, the ability of heavy and light cavalry troops to accomplish their assigned missions is mainly limited by the size and strength of threat forces encountered. Both troops are limited in close terrain, especially urban environments, due to their limited number of dismounted scouts. Also, the troops are dependent on their parent squadrons and regiments for additional combat support (CS) and combat service support (CSS) assets.

(d)

(e)

(f) Section III. Responsibilities

(I) THE TROOP COMMANDER

The troop commander is responsible to the squadron commander (SCO) for the discipline, combat readiness, and training of the troop, and for the maintenance of its equipment. He must be proficient in the tactical employment of the troop and its assigned and attached CS elements. He must also know the capabilities and limitations of the troop's personnel and equipment as well as those of CS elements attached to him.

a) The troop commander's responsibility in combat is twofold. He will—

- Accomplish all missions assigned to the troop in accordance with the SCO's intent.
- Preserve the fighting capability of the troop.

(II) THE EXECUTIVE OFFICER

In combat, the troop executive officer (XO) is second in command. He supervises the troop tactical operations center (TOC), where he stays abreast of the tactical situation in the troop's area of operations (AO). He manages the flow of combat information between the troop and the squadron from the troop TOC. With the assistance of the troop first sergeant (1SG), he plans and coordinates CSS for the troop.

(III) THE FIRST SERGEANT

The primary responsibility of the troop 1SG is sustaining the troop's ability to fight. He supervises the procurement and distribution of fuel, ammunition, food, water, clothing, equipment, replacements, and repair parts. He receives incoming personnel and assigns them to subordinate elements as needed. He is responsible for the medical evacuation of sick, injured, and wounded soldiers to the supporting medical treatment facility. He is also responsible for the evacuation of soldiers killed in action to the supporting graves registration collection point. He is also responsible for the recovery and evacuation of damaged combat equipment.

(IV) THE TROOP FIRE SUPPORT OFFICER

The troop FSO is responsible for the planning and coordination of the troop fire support plan. He relays troop requests for indirect fire and advises the commander on the employment of indirect-fire weapons in support of the troop's maneuver. The troop FSO may control and position the mortars during combat operations.

(V) THE PLATOON LEADER

The platoon leader is responsible to the troop commander for the discipline, combat readiness, and training of the platoon, and for the maintenance of its equipment. He must be proficient in the tactical employment of the platoon and know the capabilities and limitations of the platoon's personnel and equipment.

The platoon leader's responsibility in combat is twofold. He must—

- Accomplish all missions assigned to the platoon in accordance with the troop commander's intent.

- Preserve the fighting capability of the platoon.

(VI) THE PLATOON SERGEANT

The platoon sergeant (PSG) leads elements of the platoon as directed by the platoon leader, and assumes command of the platoon in his absence. The PSG assists the platoon leader in maintaining discipline, conducting training, and exercising control. He supervises platoon CSS, which includes supply and equipment maintenance.

(VII) THE MORTAR SECTION SERGEANT

The mortar section sergeant is responsible for providing responsive indirect fires to support the troop commander's concept of the operation.

(VIII) THE SUPPLY SERGEANT

The supply sergeant picks up, transports, and issues supplies and equipment to the troop. He works closely with the 1SG to accomplish these tasks. He also evacuates enemy prisoners of war and assists in the evacuation of soldiers who are killed in action to the graves registration collection point.

(IX) THE MAINTENANCE SERGEANT

The maintenance sergeant is responsible for the prompt battlefield repair or recovery of damaged or inoperable equipment. He works closely with the 1SG to accomplish these tasks.

(X) THE COMMUNICATIONS SERGEANT

The communications sergeant ensures the troop TOC and its crew are prepared for combat operations, and assists the XO in the TOC during combat operations. Within his capability, he repairs the communications equipment of subordinate elements.

(XI) THE NUCLEAR, BIOLOGICAL, CHEMICAL (NBC) NCO

The troop NBC NCO is responsible for troop NBC defense activities. He supervises radiological monitoring, chemical detection, and decontamination operations. He assists in maintaining NBC equipment and training NBC equipment operators and decontamination teams.

Chapter 2

Battle Command

Battle command is the art of battle decision making, leading and motivating soldiers and their organizations into action to accomplish missions. Battle command includes visualizing the current state and the future state, then formulating concepts of operations to get from one state to the other at least cost. Other functions of battle command include assigning missions; prioritizing and allocating resources; selecting the critical time and place to act; and knowing how and when to make adjustments during the fight.

Instantaneous response to orders is the hallmark of effective battle command in any cavalry operation. Leaders must quickly analyze information, make tactical decisions, and turn those decisions into successful battlefield actions.

Battle command of cavalry units is typically decentralized due to the size of the area of operations, vagueness of the enemy situation, and terrain unknowns. This places the burden of sound, timely decision making at the lowest levels. Leaders must develop a keen sense of situational awareness and constantly track the actions of subordinate units as well as those to the front, flank, and rear.

Contents

Page

Section I. Troop Leading Procedures 2-2
Section II. Situational Awareness 2-13
Section III. Communications 2-17
Section IV. Techniques of Tactical Control 2-22
Section V. Command Guidance and Organizational Control 2-27
Section VI. Tactical Movement Formations 2-34

Effective battle command begins in the planning phase and continues through the execution phase of each mission. This chapter outlines the tools and techniques a troop commander needs to effectively command and control a cavalry troop in combat.

Section I. Troop Leading Procedures

The nine troop leading procedures are—

- Receive and analyze the mission.
- Issue a warning order.
- Make a tentative plan.
- Start necessary movement.
- Conduct a reconnaissance.
- Make final decisions and complete the plan.
- Issue an operations order.
- Rehearse.
- Supervise and refine.

RECEIVE AND ANALYZE THE MISSION

The troop commander will receive missions from squadron in the form of written, briefed, or radioed operation orders (OPORD) and fragmentary orders (FRAGO). Upon receipt of a mission, he will conduct a mission analysis to determine the *who*, *what*, *when*, *where*, and *why* elements of the mission and how much time is available until mission execution.

Mission. The mission analysis must identify the following:

- Specified tasks.
- Implied tasks.
- Essential tasks.
- Intent of the higher commander.
- Any constraints or limitations.

The items for analysis are derived from the squadron order as noted below.

- Paragraph 2, MISSION.
  - Mission Statement.

- Paragraph 3, EXECUTION.
  - Concept of the Operation.
  - Commander's Intent.
  - Maneuver.
  - Specific Instructions.
  - Coordinating Instructions.

- Execution matrixes.

- Thorough map reconnaissance.

List all specified and implied tasks not covered by unit SOP in a mission list. Identify those tasks that are vital to the squadron accomplishing its mission. These are essential tasks. Put the essential tasks in the form of a restated mission for the troop. The restated mission statement should be short and simple, and should cover *who, what, when, where,* and *why.*

One final note on this—keep the mission list at hand. It makes the job easier when preparing a course of action and assigning missions or tasks to platoons and sections.

Time. Because time is usually the most limited resource available, it must be used as efficiently as possible. Figure out what has to be done and what can be done in the time available. Use the backwards planning process to determine when critical events in the planning process must occur and stick to the timeline developed. Get the timeline out to subordinates in the warning order. Strive to use no more than one-third of the time available in the planning process at troop level. The remaining two-thirds is for subordinate leaders to plan and prepare.

Consider the following in terms of time when conducting mission analysis and preparing the troop warning order:

- Combat service support.

- Precombat inspections.

- Route reconnaissance (time/distance factors):
  - Assembly area to route start point.
  - Start point to command posts.
  - Command posts to release point.
  - Release point to line of departure.

- Subordinate's planning time.

- Platoon rehearsals:
  - Battle drill rehearsals.
  - Actions on contact.
  - Obstacle breaching/emplacement.
  - Movement formations.

Note. The preparatory actions above are triggered by receipt of a warning order at platoon level.

- Time to issue troop OPORD.

Note. One-third of the available time includes the time it takes to issue the troop OPORD.

- Troop rehearsals:
  - Backbriefs.
  - Rock drills.
  - Radio rehearsals.

- Time to move from—
  - Assembly area to line of departure.
  - Phase line to phase line.
  - Line of departure to objective.

- Effect of weather on movement.

Once the mission and time available have been analyzed, move to the next step in the trooping leading procedures.

ISSUE A WARNING ORDER

Issue the warning order promptly. Elements of the warning order include—

- Situation.
- Mission.
- Earliest time of movement.
- Time and place of OPORD.
- Attachments and detachments.
- Actions triggered:
  - Security.
  - Reconnaissance.
  - Combat service support actions.
  - Precombat inspections.
  - Sleep plans.
  - Rehearsals.

MAKE A TENTATIVE PLAN

Develop a course of action (COA). Consider the factors of METT-T when developing two, or better yet, three separate COAs.

- **Mission.** See *Receive and Analyze the Mission* paragraph on page 2-2.

- **Time.** In addition to the factors mentioned in the *Receive and Analyze the Mission* paragraph on page 2-2, the commander should consider the following details in relation to time when developing his tentative plan.

  - Time to move from—

    Line of departure to phase lines.
    Phase line to phase line.
    Line of departure to objective.

- Time for enemy to—
  - Attack.
  - Counterattack.
  - Reinforce.
- Effect of weather on movement and identification of friend or foe.
- Effect of NBC operations on movement.

• **Enemy.** Conduct intelligence preparation of the battlefield (IPB). IPB integrates enemy doctrine with weather and terrain to determine and evaluate enemy capabilities, vulnerabilities, and probable courses of action. IPB is a key part of preparing for battle. Much of the IPB workup can be found in Paragraph 1 of the squadron OPORD.

• **Terrain.** Analyze terrain in the area of operations. Consider the factors of OCOKA. (Arranged here in a more logical, systematic approach, based on IPB procedures and preparation of the modified combined obstacle overlay [MCOO].)

■ ***Obstacles.*** Identify all obstacles and restricted and severely restricted terrain throughout the sector or zone (a terrain sketch may be beneficial). Consider the following as obstacles:

- Water.
- Power lines.
- Bridges.
- Defiles.
- Slopes.
- Towns.
- Embankments.
- Railroads.
- Existing engineer obstacles.
- Wooded areas.

Terrain that flows around, or in some instances through or over, the obstacles identified is the primary maneuver corridor, or avenue of approach.

■ ***Avenues of Approach.*** Consider the following for each avenue of approach:

- What size force does it support? In what formation?
- Does it support long range observation and fields of fire? From where?
- Does it provide any cover and concealment? Where?
- Where does it lead?
  Through sector/zone?
  Enters/exits a flank?

The answers to these questions will help determine what terrain, if any, is key terrain.

■ ***Key Terrain.*** Key terrain is that terrain which control over would provide a clear advantage over another force. Seizing, securing, or even avoiding key terrain in sector/zone will figure prominently in the final plan. Identify what terrain is key and evaluate it in terms of observation and fields of fire along each avenue of approach.

■ ***Observation/Fields of Fire.*** Evaluate all terrain along the avenues of approach in terms of observation and fields of fire. Determine the following:

- What terrain provides the best observation of critical areas or along avenues of approach?
  At night?
  When visibility is obscured by smoke or fog?
- What terrain offers effective range of friendly and enemy weapon systems?
- What terrain provides the best fields of fire? For what size force?

- ***Cover and Concealment.*** Evaluate all terrain along the avenues of approach in terms of cover and concealment. Determine the following:

  - What terrain affords the best cover and concealment to the enemy and to the friendly forces?
  - Where are likely hide positions that support potential enemy and friendly battle positions?
  - What routes would a force use to move from hide positions to battle positions?

- **Troops Available.** Analyze the troops available to execute the mission. Determine the following:

  - ***Soldiers.***

    - What is the strength of the platoons in the troop?
    - Can all weapon systems be manned?
    - How is the morale of the troop?
    - What is the experience level of the troop?
    - What training and skills are required? Are specialists such as engineers needed?
    - What is the general health status and fatigue level of the soldiers?

  - ***Equipment.***

    - How many weapon systems are operational and employable?
    - Is special equipment or additional equipment needed to accomplish the mission?

START NECESSARY MOVEMENT

While developing his tentative plan, the commander may decide to reposition some of his forces before the operation starts so that he is prepared to meet mission requirements. If the entire troop needs to move, he will often execute the movement in accordance with a squadron plan. The amount of time he has to reposition his forces is determined by the readiness condition (REDCON) in effect.

Note. See Chapter 7 for a discussion of REDCON.

CONDUCT A RECONNAISSANCE

If sufficient time is available after determining his tentative course of action, the commander should make every effort to get out into his AO and look at the terrain. Ground reconnaissance is the norm; however, air transportation may be available. Staying on the ground gives a better feel for the terrain than does reconnoitering by helicopter, but it is slower. Although faster, aerial reconnaissance does not provide an appreciation of the terrain from a vehicle commander's point of view in terms of fields of fire and cover and concealment. If circumstances permit, subordinate leaders should accompany the commander to help perform the reconnaissance. It gives them an opportunity to become familiar with the area.

MAKE FINAL DECISIONS AND COMPLETE THE PLAN

Based on the new information gathered by his personal reconnaissance, the commander makes final adjustments to his plan and nails down the details. Once the plan is complete, he prepares the order. Troop orders need not be written and handed to subordinate leaders. Speak face to face with them if possible. Put the plan in the form of a standard five-paragraph field order. Jot it down in a notebook so every detail can be recalled.

ISSUE THE OPERATIONS ORDER

Assemble the orders group. Make sure everyone in the orders group is present. Issue necessary graphics to subordinate leaders and give them time to prepare their maps. Ensure the overlays are neat and accurate. Messy overlays with broad pen strokes cause confusion and waste valuable time. Once the subordinate's overlays are prepared, check them for accuracy. (A technique for ensuring accuracy is to include in the order a list of grid coordinates for key locations that might be ambiguous in the operational graphics. Even neat graphics, once copied from squadron down to platoon, may often deviate by as much as 500 meters.) Arrange the orders group in a semicircle from left to right in the order in which they will be addressed when giving specific instructions.

Issue the order. Use notes when issuing the order; do not rely on memory. If possible, issue the order while overlooking the AO. Keep in mind that the order will be translated a couple of times before the privates receive it, and it must make sense to them. Talk in their language. The order should be short and simple, and must be logical and easy to follow; be clear, organized, and concise so that no one becomes confused. Misunderstanding the commander's intent most often occurs in the transmission of orders to the platoon leader and his subordinates.

Note. See Appendix C for the operations order format and sample OPORD.

Conduct the confirmation brief. Immediately after the order is issued, call for questions from the subordinate leaders. After all questions concerning the order have been answered, begin the confirmation brief by each of the troop's subordinate leaders.

The confirmation brief is used by the commander to confirm that his intent and guidance for the conduct of the operation are clearly understood by everyone in the orders group before they are dismissed to begin their planning. The confirmation brief adjourns only when the commander is confident his subordinates understand their mission, his and the higher commander's intent, his concept of the operation, the scheme of maneuver, the time plan, and the type and location of the rehearsal.

REHEARSE

Rehearsals are of paramount importance before executing any plan. Rehearsals help in the following ways.

- Clarify the commander's intent.
- Expose combat, combat support, combat service support, or disconnected activities in the plan.
- Reinforce the scheme of maneuver and fire support plan.
- Focus on actions and decision points critical to mission accomplishment.
- Ensure subordinates explicitly understand their missions, how their missions relate to each other, and how each mission relates to the commander's plan.
- Outline conditions that, when present, would necessitate execution of branch plans.
- Provide feedback to the commander.

Rehearsals instill confidence in participants by—

- Giving participants faith in the success of their own plan as well as in their commander's plan.
- Providing subordinates with purpose, direction, and motivation.
- Enabling leaders to execute missions with speed, flexibility, and audacity.

There are generally seven types of rehearsals—full, key leader, terrain model, sketch map, map, radio, and backbrief. Preparation time and resources for each range from extensive to minimal. At troop level, the most common rehearsal types are backbrief, map, sketch map, terrain model, and radio.

Backbrief. During the backbrief, each subordinate leader briefs the commander of how he intends to accomplish his mission before he issues his OPORD to his respective unit. By having subordinates explain their intent and concept of operation, the commander can ensure their plans support his own. Flaws or potential problems with the plan may be revealed at this time.

Do not confuse the backbrief rehearsal with the confirmation brief the commander uses immediately after he issues an OPORD to determine how well his subordinate leaders understand the order.

Map/Sketch Map/Terrain Model. Using either a map and overlay of the same scale as used to plan and execute the operation, a large scale sketch map, or a terrain model, the commander and subordinate leaders move unit markers in a sequential (either by phase, event, or time), interactive, verbal execution of the operation. These type rehearsals are essentially conducted like a war game to show the planned sequence of action-reaction-counteraction to critical events or phases of the operation. This is a good opportunity to coordinate not only actions of the unit but also critical locations such as contact points, checkpoints, boundaries, battle positions, and hide positions among subordinate leaders.

Radio. Using existing communications networks (either FM or wire), the commander and subordinate leaders verbally and interactively execute critical portions of the operation. This technique can have obvious communications security disadvantages; if so, then only the essential, most critical portions of the operation are rehearsed. Radio rehearsals require few resources and little time and are best used in conjunction with other methods or to further refine the plan.

SUPERVISE AND REFINE

The last step in troop leading procedures must not be neglected. Failure to supervise the execution of the order, or to refine the plan as the situation changes, is the road to ruin. The importance of this step cannot be overemphasized. Actions during the final step may include, but are not limited to, higher commander and his staff being present in the subordinate unit staff planning process.

Section II. Situational Awareness

Situational awareness is always keeping a clear picture of the tactical situation, both mentally and graphically. This picture includes both the friendly and enemy situation and an understanding of the relevant terrain. Since the troop normally operates dispersed over wide frontages, it is essential that all leaders maintain situational awareness so they can make sound, quick tactical decisions. Situational awareness also permits the leaders to anticipate events and relate separate pieces of information to form logical conclusions. One of the critical outcomes of situational awareness observed by all leaders is a reduction of fratricide incidents.

BATTLEFIELD FRAMEWORK

How the commander has structured the battlefield impacts the troop commander's ability to maintain good situational awareness. A commander will structure the battlefield based on the conditions of METT-T. The framework of the battlefield can vary from a very rigid extreme with an obvious front and rear boundary and closely tied adjacent units, to a very dispersed and decentralized structure with no secure areas, unit boundaries, or definable front. Between these extremes is an unlimited number of possible variations. Maintaining situational awareness will become more difficult the more unstructured the battlefield is. Modern, highly mobile operations with small forces lend themselves to a less rigid framework which will challenge the ability to maintain a good picture of the battlefield.

PICTURING THE BATTLEFIELD

To have a good picture of the battlefield, all leaders must have virtually perfect knowledge of the friendly situation one level higher than their own. This means that all principal subordinate leaders must know the troop situation and the troop commander must know the squadron situation. All leaders must have accurate knowledge of the terrain, and they must know as much as possible about the enemy. The requirement to maintain a real-time picture of the battlefield one level higher does not relieve leaders of the requirement to understand the situation two levels higher. The difference is that a leader's understanding of the situation two levels higher than his own does not have to be as specific or in real time.

Most of the information passed between elements is in the form of reports over FM radio. The troop commander receives many reports as a result of his troop graphics. Good graphics require that the subordinate elements report periodically as they accomplish tasks. The troop commander must be aware of when subordinates report so he will know how current his visualization of the situation is. If an element does not report in a timely manner, the commander must quickly determine the situation of the overdue element.

Although many reports may not be addressed specifically to him, particularly on the squadron net, the commander must monitor them by eavesdropping on the nets as traffic is sent. How effectively he can accomplish this is, to some degree, experience dependent; however, there are techniques he can use to relate the information to his map and thereby track the tactical situation. The troop XO plays an important role in assisting the troop commander and platoon leaders in maintaining situational awareness. The troop XO provides leaders with periodic updates of friendly units to the front, flanks, and rear, based on traffic from the squadron operations and intelligence (OI) net.

The commander's map is the key to maintaining situational awareness (see Figure 2-1). He should plot all friendly position reports up to one level higher than his own. Information from spot reports should also be plotted. Using different colors for friendly and enemy elements allows quick distinction. To avoid cluttering the map, he should place a dot or symbol with a number on his map where the element is located. The same number is then written in the map margin (or beyond the area of operations) with the complete spot report or unit identification next to it. This notation should also include the time. As positions or reports are updated, the old symbol is crossed off and a new one with a corresponding notation is added. This system allows the commander to easily track and monitor the tactical situation. This system is augmented by a formal operations log kept in the troop TOC by the XO or NCOIC.

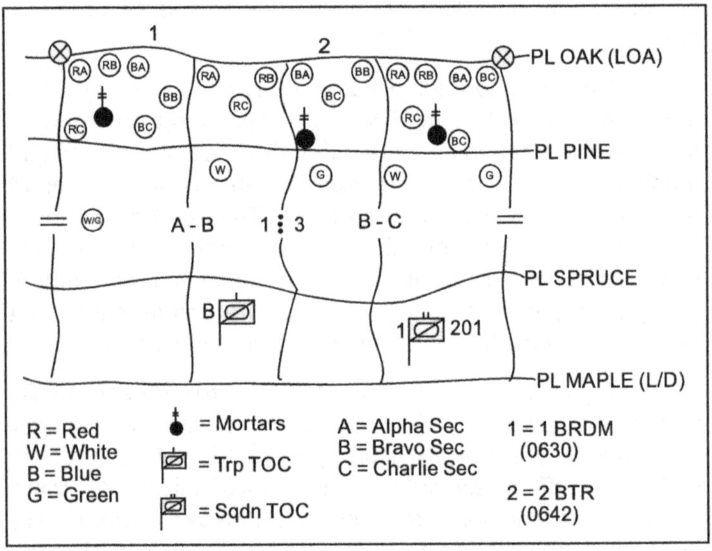

Figure 2-1. Sample situation map (troop commander).

BATTLE SPACE

Situational awareness, as previously discussed, is critical to successful reconnaissance and security missions; however, the troop commander's primary attention must be on his battle space. Battle space for the cavalry troop is determined by the location of individual platoons and sections, the range of direct-fire weapons, observation, sensors, and the terrain on which they are applied. This space is the immediate area over which the troop has influence. The troop commander must be aware of the general situation, but he must actively direct and manage all activity within his battle space. Most of the troop commander's command and control efforts focus on what is happening within his battle space and shaping his battle space to make it more efficient.

FRATRICIDE

Fratricide is a significant danger to all forces operating on a mobile battlefield where weapon system lethality is significantly greater than friend or foe identification ability. For this reason, situational awareness on the part of all leaders, particularly the commander, is critical not only to mission success but also to survival.

Under these types of conditions, it is critical that the commander know where other friendly elements are operating. With this knowledge of the situation, he must anticipate dangerous situations and take steps to either avoid or mitigate them. The commander must constantly be vigilant to changes and developments in the situation that may place his elements in danger. He must also ensure all subordinate unit positions are constantly sent to higher headquarters so all other friendly elements are aware of where they are and what they are doing. When the commander perceives a potentially dangerous situation, he must personally use the squadron command net to coordinate directly with the friendly element involved.

Section III. Communications

NETS

Several troop radio nets are used to circulate the volume of information in any combat operation. A troop command net links the troop commander with his subordinate units (see Figure 2-2). A troop fire support net links subordinate units with the fire support team (FIST) and mortar section (see Figure 2-3). Each net has an assigned radio frequency, but not necessarily a dedicated radio.

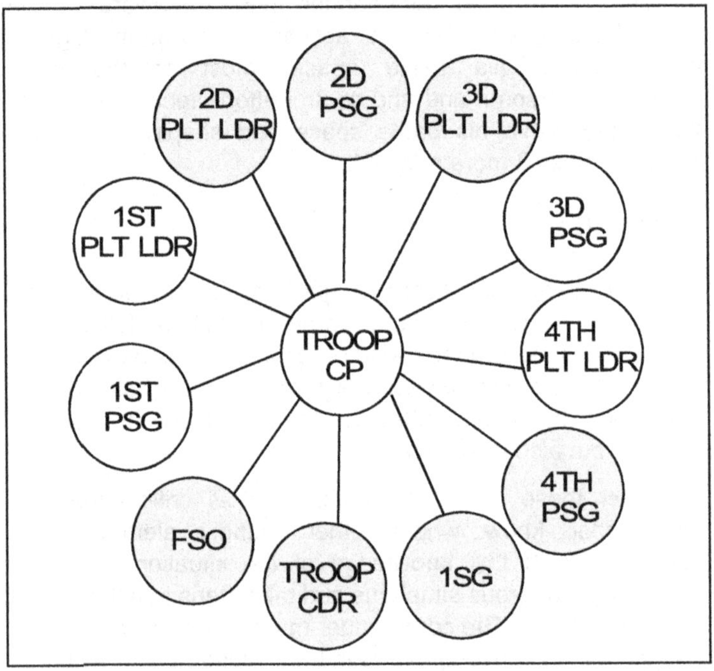

Figure 2-2. Troop command net.

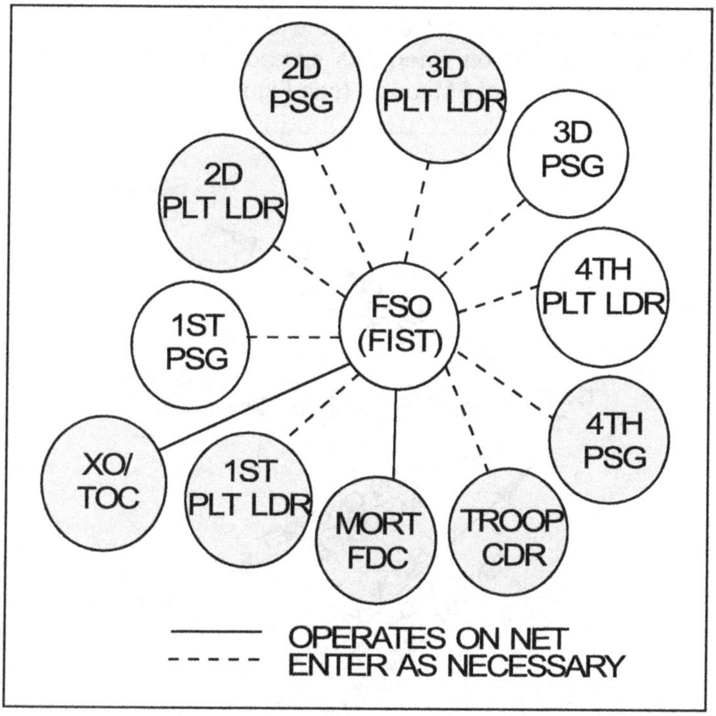

Figure 2-3. Troop fire support net.

In the troop CP, three additional radio nets are established to link the troop with the SCO and his headquarters. The squadron operations and intelligence (OI) net is used primarily to transfer enemy and friendly information by FM radio. All routine and recurring reports from the troop to squadron are transmitted on this net. Squadron headquarters uses this net to transmit information about current enemy and friendly situations within the squadron's AO. It is often used to issue FRAGOs. The troop commander and the XO are linked to the SCO by an FM squadron command net. This net is reserved strictly for use by the SCO, XO, S3, and CPs; and by subordinate troop, company, and battery commanders. In addition, an AM squadron command net allows the SCO to command and control the squadron when it is spread over distances that exceed the range of FM radios (see Figure 2-4).

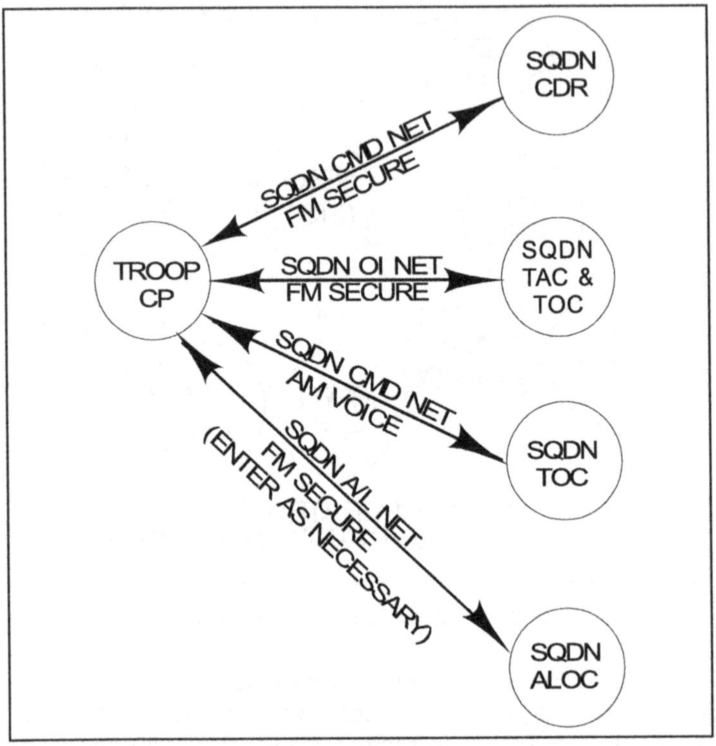

Figure 2-4. Troop command post external nets.

EAVESDROP

This tool is used by all cavalrymen. It is designed to give leaders time to think before they act. Most leaders have a hard time making decisions while they are talking on the radio. Decisions are usually better if the commander listens closely to what is said, and if he has uninterrupted time to consider information received. Eavesdropping is a method of communicating based on that principle. Everyone simply listens to the information sent from one leader to another, and unless necessary, stays off the radio net. In the troop, for example, reports from scout platoons are transmitted to the troop CP, where the XO reviews and records them. All others listen in and record the information themselves. Everyone stays off the net yet remains informed. If all or part of a transmission is missed, enter the net and quickly get the needed information from the troop CP.

FIXED CALL SIGNS

Fixed call signs are tools that establish immediate identity of subunits and leaders, thereby reducing the length of transmissions on the radio. These call signs eliminate the confusion often caused by the ever-changing signal operation instructions (SOI). Each platoon is referred to by a color, and leaders are referred to by numbers. These call signs do not change. An example of fixed call signs for a cavalry troop is outlined below.

Fixed Call Signs

Element	Call Sign
TROOP HEADQUARTERS	BLACK
FIST	Black 1
Medics	Black 2
TOC	Black 3
Supply	Black 4
XO	Black 5
Commander	Black 6
1SG	Black 7
Maintenance	Black 8
Mortars	Black 9
NBC	Black 10

1ST PLATOON (Scout)	RED
Plt Ldr 11	Red 1
Sec Ldr 12	Red 2
Sqd Ldr 13	Red 3
PSG 14	Red 4
Sec Ldr 15	Red 5
Sqd Ldr 16	Red 6
*Sec Ldr 17	Red 7
*Sqd Ldr 18	Red 8
*Sec Ldr 19	Red 9
*Sqd Ldr 10	Red 0
2D PLATOON (Tank or AT)	WHITE
Plt Ldr 21	White 1
TC 22	White 2
TC 23	White 3
PSG 24	White 4
3D PLATOON (Scout)	BLUE
Plt Ldr 31	Blue 1
Sec Ldr 32	Blue 2
Sqd Ldr 33	Blue 3
PSG 34	Blue 4
Sec Ldr 35	Blue 5
Sqd Ldr 36	Blue 6
*Sec Ldr 37	Blue 7
*Sqd Ldr 38	Blue 8
*Sec Ldr 39	Blue 9
*Sqd Ldr 30	Blue 0
4TH PLATOON (Tank or AT)	GREEN
Plt Ldr 41	Green 1
TC 42	Green 2
TC 43	Green 3
PSG 44	Green 4

*Light troop only

Section IV. Techniques of Tactical Control

CHAIN OF COMMAND

The troop commander cannot personally supervise the execution of all combat tasks he assigns. He must depend on a trusted team of subordinate leaders. The chain of command in the troop extends from the commander to the platoon leaders and section leaders. Within the platoons, it extends from platoon leaders and their platoon sergeants to section sergeants to vehicle/crew commanders. The troop XO and the first sergeant serve as the commander's principal assistants. They help coordinate the actions of subordinate units to ensure unity of effort. They also ensure the troop has what it needs to accomplish its assigned tasks.

ORDERS GROUP

The troop orders group consists of all subordinate leaders required to receive and execute the troop commander's orders. As a minimum, the orders group should include the XO, the first sergeant, the platoon leaders, the mortar section sergeant, the fire support officer (FSO), and the leaders of any units that are attached or under the troop's operational control (OPCON).

STANDING OPERATING PROCEDURES

A troop SOP is a written collection of standing orders. It establishes procedures for performing routine combat tasks. For example, the SOP tells subordinates what procedures to follow when the troop occupies an assembly area, breaches an obstacle, conducts a passage of lines, or reacts to air attack. It lists reports that are routinely sent to troop and squadron headquarters, and shows their correct formats. It also states procedures for communicating within the troop, obtaining indirect fire, or getting additional fuel or ammunition. The SOP is a valuable time saver for troop leaders. Most important, the SOP keeps everyone in step, reduces confusion, and provides the backbone of combat proficiency. Disciplined execution of the troop SOP is the hallmark of outstanding units.

TERRAIN INDEX REFERENCE SYSTEM

Terrain index reference system (TIRS) is used to maneuver the troop, rapidly report friendly locations without fear of compromise, and pass out control measures. TIRS is not used to report enemy locations.

Description

Each TIRS point is shown by a tick mark located on a grid line intersection. Each point is given a designator of one letter and two numbers, such as X56, placed in the upper right quadrant of the tick mark. TIRS point designation is SOP, and units determine which letters they will use. They may designate specific letters for specific unit sectors or AO. The TIRS point is then recorded as X56, located at PA2450 (four-digit grid).

TIRS point-to-point references designate kilometers. For example, 500 meters is given as "POINT FIVE," 1,000 meters as "ONE," and 3,500 meters as "THREE POINT FIVE."

For shifts from the TIRS point, use cardinal directions instead of left, right, up, or down. Using TIRS, "FROM X-RAY FIVE SIX—EAST ONE POINT SEVEN—NORTH POINT SEVEN" translates to "From the tick mark for TIRS point X56, shift east 1,700 meters and north 700 meters." When each TIRS point is placed on a four-digit grid intersection, the use of shifts makes the TIRS as accurate as the six-digit grid system. For an example of a TIRS point in use, refer to Figure 2-5.

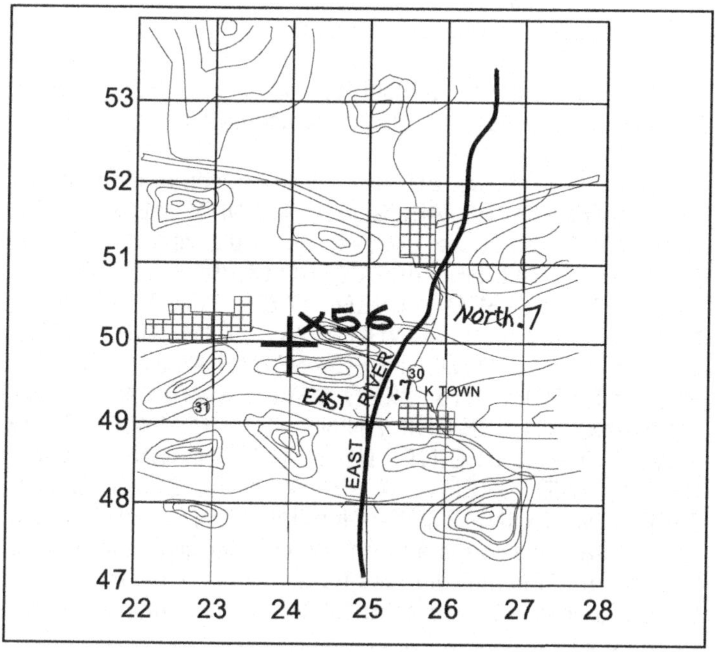

Figure 2-5. Placing TIRS on the map.

Application

The squadron normally issues the TIRS to be used for the operation as early as possible, perhaps with the warning order. The TIRS list should be issued to elements as an annex to a written OPORD.

The squadron should designate four to six TIRS points in each 10-kilometer square. The squadron TIRS should normally be sufficient for the troop to operate; however, if the troop designates additional TIRS, it should always ensure only squadron TIRS are referenced in its communications with higher headquarters.

TIRS should be used routinely to control combat operations. Use them—

- To identify BPs and to pass out control measures (such as LDs, PLs, and boundaries) quickly.
- To report friendly unit locations.

Passing control measures, such as BPs, sector boundaries, and PLs, are quick and accurate using TIRS.

Figure 2-6 shows BP B22. A troop commander could report "TROOP ESTABLISHED VICINITY ALFA ZERO SEVEN." To be more precise, he could report "TROOP BRAVO ESTABLISHED ALFA ZERO SEVEN WITH RED AT FROM ALFA ZERO SEVEN EAST POINT THREE—NORTH POINT FIVE—WHITE AT FROM ALFA ZERO SEVEN WEST POINT THREE—NORTH POINT EIGHT—BLUE AT FROM ALFA ZERO SEVEN EAST ONE POINT NINE—NORTH ONE POINT NINE—TO FROM ALFA ZERO SEVEN WEST POINT NINE—NORTH ONE POINT SEVEN—AND GREEN AT FROM ALFA ZERO SEVEN EAST ONE POINT ZERO—NORTH POINT FIVE." In the second transmission, the troop commander gives the precise location of the center of mass for all his platoons: red, white, blue, and green represent the 1st, 2d, 3d, and 4th platoons.

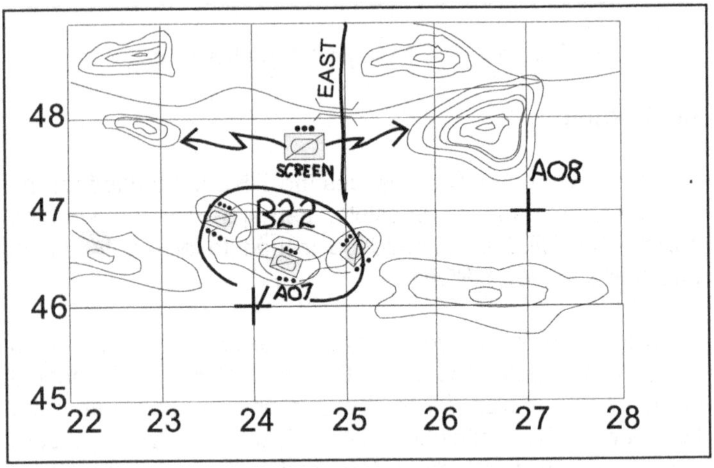

Figure 2-6. Reporting location using TIRS.

Using this transmission, the SCO orders Troop B to occupy a BP that is not on the troop commander's map: "BRAVO SIX SIX OCCUPY A BP EXTENDING FROM ALFA ZERO EIGHT WEST POINT SEVEN—SOUTH ONE POINT ZERO—TO FROM ALFA ZERO EIGHT EAST POINT SIX—SOUTH POINT FIVE—ORIENT NORTHEAST."

Figure 2-7 shows how the SCO could quickly establish new phase lines and boundaries between troops.

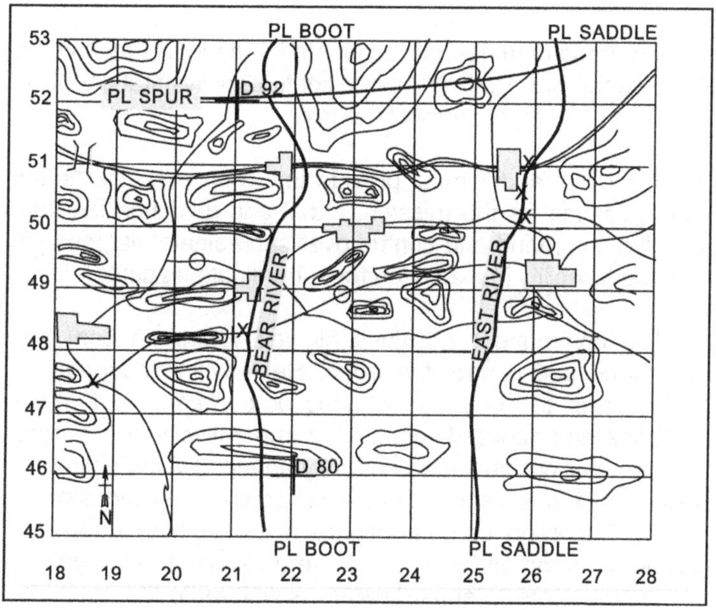

Figure 2-7. Establish graphic control measures using TIRS.

The SCO could transmit a new boundary between troops B and C by radio: "BRAVO SIX SIX AND CHARLIE SIX SIX BOUNDARY IS FROM DELTA EIGHT ZERO EAST POINT NINE—NORTH POINT EIGHT—TO FROM DELTA NINE TWO EAST TWO—SOUTH THREE—TO FROM DELTA NINE TWO EAST ONE POINT NINE—SOUTH POINT ONE."

PL SPUR could be sent as "PL SPUR IS FROM DELTA NINE TWO WEST POINT NINE—TO FROM DELTA NINE TWO EAST FIVE POINT THREE—NORTH POINT SEVEN."

Section V. Command Guidance and Organizational Control

TROOP COMMANDER

The troop commander is mounted in an M1A1 tank in the heavy troop or a HMMWV in the light troop. His first priority is the command and control of the troop.

The commander must provide the troop with clear intent and guidance for the mission. Intent and guidance facilitate positive command and control over subordinate platoons yet provide flexibility in execution for subordinate leaders.

The commander positions himself where he can best command and control the troop. He may maneuver along with one of the tank/AT platoons or between the tank/AT platoons and scout platoons. It is imperative that the troop commander not become decisively engaged with fighting his vehicle. He must retain the ability to control the actions of his troop. The troop commander operates on both troop and squadron command radio nets. He must have the capability to communicate to subordinates as well as to the squadron commander.

TROOP TOC

The purpose of the troop tactical operations center (TOC) is to report information to the higher headquarters. The troop TOC is the critical link between squadron and troop. One of the primary purposes of the cavalry troop is to report, and the troop TOC is central to that function. The troop TOC operates under the direction of the troop XO who is responsible for assisting the commander in the command

and control of the troop. The troop TOC takes reports from subordinate units in the troop, processes the information, and reports the information to the higher headquarters. The TOC tracks the battle at the troop and squadron level and relays information to the commander and subordinate platoons pertaining to the friendly/enemy situation. The TOC is the net control station for the troop command net and operates on the squadron command and OI net. The troop TOC also monitors and operates as necessary on the squadron A/L and the troop fire direction net.

In the heavy troop the TOC vehicle is an M577. An M3 CFV moves with the heavy troop TOC to provide local security for the M577 and to serve as a command and control vehicle for the XO if the commander is out of action. The XO should not conduct TOC operations out of the CFV because his experience is needed in the TOC vehicle to handle the information flow between troop and squadron over the OI and command nets. If the commander is out of action, the XO should avoid, if possible, rushing forward to assume command of the troop because of the importance of the activities in the TOC. The XO should assume command of the troop from the TOC until the battle is over. However, if the XO does need to move forward to command and control the troop, a qualified replacement must be placed in the TOC to assume his duties. This individual may be the FIST chief, one of the platoon leaders, or the first sergeant. The correct replacement for the XO in the TOC should be based on the depth available at any one of the above mentioned positions.

The light troop TOC is a cargo HMMWV with a shelter.

The primary concern when positioning the troop TOC is based on the TOC's ability to communicate with the squadron and the subordinate elements of the troop. The TOC position should also be an area that provides good cover and concealment from enemy and limited access to civilian population. During reconnaissance or offensive operations, maneuver the TOC at least one terrain feature behind troop combat elements. During security or defensive operations, the TOC should be positioned in sufficient depth to avoid contact with the enemy yet maintain communications with the forward scout platoons. The TOC and the troop trains may collocate for increased local security.

FIST

The FIST is the critical link with the supporting artillery and responsible for coordinating indirect fires (both mortar and FA) for the troop. The team processes calls for fire from the platoons and allocates the appropriate indirect-fire system based on the commander's guidance for fire support. The FIST can also assist the squadron ALO with the employment of close air support.

The FIST operates on three radio nets: troop command, troop fire direction, and squadron FSE digital/voice. The FIST monitors at least one of the following nets: squadron command, squadron OI, and howitzer battery (supporting artillery headquarters in the heavy and light division).

The FIST vehicle also may serve as the alternate troop TOC. The FSO has ready access to the higher level situation and the radio systems to replicate the troop TOC if it becomes damaged or destroyed.

Command guidance to the FIST should include the following:

- **Purpose of indirect fires.** How does the commander intend to use field artillery and mortar fires to support his maneuver?
  - Screening.
  - Suppression.
  - Development of the situation.
  - Disengagement.

- **Engagement/attack criteria.** How many rounds and of what type and mix will be fired at a particular target? Which targets will be engaged with artillery and which with mortars?

- **Control of troop mortars.** If the FIST controls movement of troop mortars, how far forward of the scouts will the mortars be able to range? Where are the mortars going to move? When are the mortars going to move?

- **FIST movement.** The primary considerations when positioning the FIST are security of the team and the ability to communicate with the squadron FSE, howitzer battery, or the DS artillery. The FIST is not the forward observer team for the troop; the troop has 19Ds who act as forward observers. The five techniques to maneuvering the FIST are—

  - Maneuvers with the commander.

  - Maneuvers with or near the mortar section.

  - Maneuvers with the TOC.

  - Maneuvers alone to maintain communications.

  - Maneuvers with the scouts or tanks to directly control fires or to use the ground laser designator (GLD).

Note. See Chapter 8 for a more in-depth discussion of troop fire support command and control techniques.

TROOP MORTARS

The troop mortars provide responsive indirect-fire support to the troop. In the light troop they are equipped with three cargo HMMWVs and two 120-mm towed mortars. The heavy troop is equipped with two M106 mortar carriers with 107-mm mortars. The mortars operate on the troop command and fire direction nets.

Command guidance for the mortars should include the following:

- **Purpose of mortar fires.** How does the commander intend to use mortar fires to support his maneuver?
  - Screening.
  - Suppression.
  - Development of the situation.
  - Disengagement.

- **Engagement/attack criteria.** How many rounds and of what type and mix will be fired at a particular target?

- **Positioning.** How far forward of the scouts will the mortars be able to range?

- **Movement.** Where will the mortars move and what will the mortars orient their movement on, for example, checkpoints, a route, mortar firing positions? Who will control the movement of mortars—the mortar section sergeant, the FIST, or the troop XO?

SCOUT PLATOONS

The scout platoons, HMMWV or Bradley equipped, perform reconnaissance and security missions for the troop. They may also perform missions within an offensive or defensive role in certain METT-T situations. To enhance command and control and to provide scout platoons the freedom of action required in cavalry operations, the troop commander should give the following guidance to scouts, regardless of what missions they are performing.

- Engagement, disengagement, destruction, and bypass criteria.

- Tempo of the operation. Tempo is dictated by the higher commander's intent for the operation, and is a function of the enemy situation as it relates to the survivability and firepower of the systems conducting the operations (METT-T). Tempo for the scout platoon determines the rate at which they conduct their mission and develop the situation. High tempo operations, for example, could be characterized by very aggressive reconnaissance focusing on mounted avenues of approach and likely enemy defensive positions. Engagement criteria in high tempo operations would be unrestricted, allowing the scouts to rapidly develop the situation by fire and maneuver. Low tempo operations could be characterized by scouts conducting a very detailed reconnaissance and maximizing the use of stealth to develop the situation.

The scout platoon leader and platoon sergeant operate on both the troop command and platoon command nets. As required, they will operate on the troop fire direction net. For platoon organization for combat, see FM 17-98, Chapter 2, Section VI.

TANK/AT PLATOONS

The tank/AT platoons are the principal killers in the troop. They assist the troop commander in the rapid development of the situation during operations by fire and maneuver. The troop commander should give the platoons the following guidance to facilitate their command and control during battle.

- Location in the troop formation and distance behind the scout platoon.

- Engagement, disengagement, and destruction criteria. When and where will they be employed and against what size force? Will they be used to suppress in support of another maneuver force, or destroy by fire and maneuver?

- Tempo of the operation. (Same as for the scout platoons above.)

The platoon leader and platoon sergeant operate on troop command and platoon command nets. As required, they will operate on the troop fire direction net.

TROOP TRAINS

The troop trains provide logistical support and casualty evacuation to the troop. The trains are under the direction of the first sergeant and consist of the troop combat and field trains. The first sergeant monitors the troop command and squadron A/L nets. He directs resupply, treatment and casualty evacuation, and maintenance/recovery operations for the troop.

The troop trains may be consolidated or echeloned. Either way, the trains should move one terrain feature behind the trail combat elements of the troop. If echeloned, the field trains move with the squadron field trains under the direction of the squadron HHT commander. The troop combat trains consist of the first sergeant in his vehicle, a maintenance contact team with recovery vehicle, and a medical evacuation vehicle. The field trains are made up of the supply sergeant and the remaining members of the maintenance team.

Command guidance to the troop trains should include the following:

- Trains organization. Should they move consolidated or echeloned?

- Movement guidance. Where do the trains move in the troop formation and what is their orientation for movement? Use checkpoints, routes, or trains locations to orient the movement of the trains.

Note. See Chapter 9 for a more in-depth discussion of the TTP for combat service support.

Section VI. Tactical Movement Formations

During tactical operations, the troop normally moves in one of four movement formations: troop column, troop line, troop vee, or troop split-vee.

TROOP COLUMN

Use the troop column when moving on a designated route, when speed is essential, and when contact with the enemy is not expected. This formation moves the troop quickly and efficiently from one place to another. Control of the troop while moving in column is eased, but the troop is vulnerable to enemy air or ground attack.

The troop usually moves at a designated speed and with a set distance between vehicles when moving in column formation. The march speed and distance between vehicles are designated by the troop SOP, based on the situation.

Note. See Chapter 7 for a discussion of column formation.

TROOP LINE

The troop line formation can be used when conducting defensive or retrograde operations in the heavy and light troops (see Figure 2-8). The formation may also be used by the light troop during reconnaissance operations when the troop is scrambled.

Figure 2-8. Troop line formation.

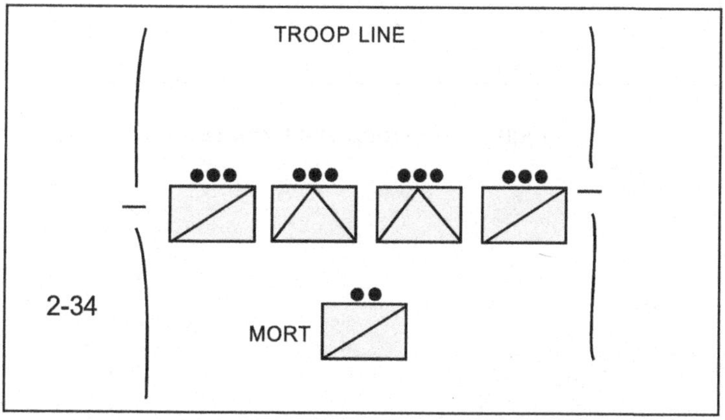

TROOP VEE

The troop vee formation is used to conduct reconnaissance or security missions (see Figure 2-9). The purpose of the vee formation is to keep maximum reconnaissance forward; the two tank or AT platoons are positioned in depth, ready to attack or defend any where in the troop zone/sector. This formation allows the troop to mass combat power at the decisive point (see Figure 2-10).

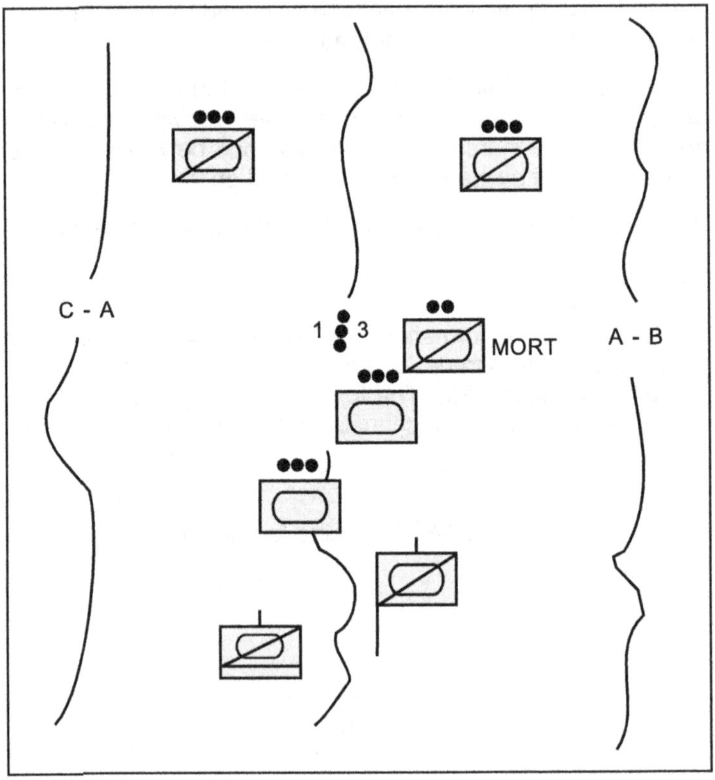

Figure 2-9. Troop vee formation.

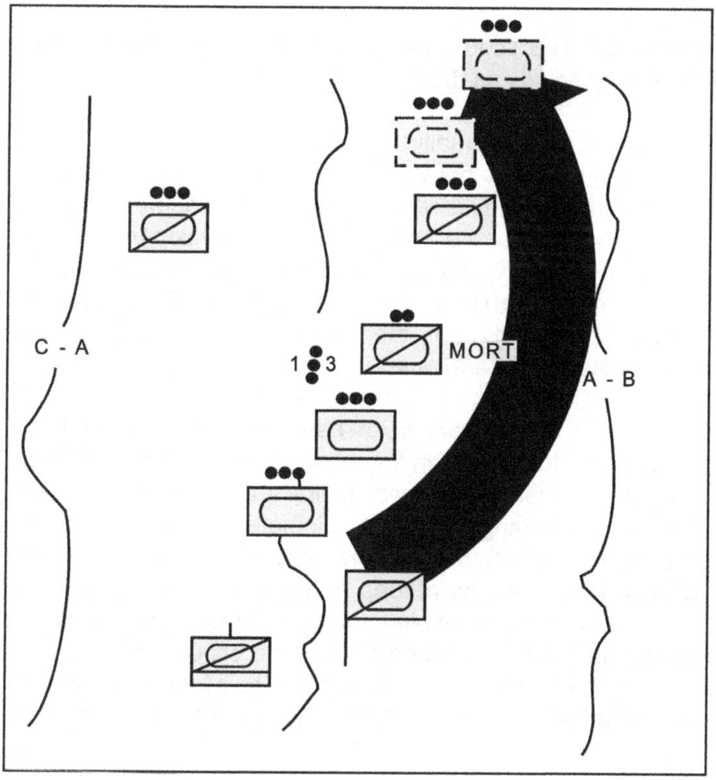

Figure 2-10. Troop commander maneuvers his tank platoons (troop vee).

In the troop vee formation, the two scout platoons work abreast in their respective zone/sector. The two tank/AT platoons are collocated in depth of the scouts, and they maneuver in the troop zone/sector.

Heavy Troop. Normally the distance between the scouts and tanks varies from 1 kilometer to 4 kilometers. The tanks should be close enough to respond quickly, but far enough back to respond to enemy contact based on the commander's guidance for engagement. Generally, do not position the tank platoons to overwatch the scout platoons. When the scouts make contact with the enemy, the tanks should be far enough in depth to maintain their ability to maneuver. The scouts are usually responsible for providing their own security as they move.

Light Troop. Normally the distance between the scouts and TOWs varies between 500 meters to 1,500 meters. TOW vehicles should respond quickly to enemy contact in order to assist in the scout platoon's development of the situation. Because of the reduced fire power in the HMMWV scout platoon, the TOW systems may provide the scouts with additional security in the form of overwatch during reconnaissance operations.

Position the mortars in or near the center of the troop zone/sector to provide indirect-fire support across the troop front. Keep them positioned to fire about one third to two-thirds of their range (about 3 kilometers to 3.5 kilometers, terrain dependent) beyond the scouts, so the scouts can engage enemy forces at long range with indirect fires. (Line of sight is the determining factor in range forward of the scouts. In densely wooded terrain, the scout's visibility may be only 100 to 200 meters forward of their front line trace, so mortar range need not always be 3,000 meters forward of the scouts.)

TROOP SPLIT-VEE

The troop split-vee is a variation of the troop vee, and can be used in reconnaissance and security missions (see Figure 2-11). In the split-vee formation, the scout platoons work abreast and forward of the troop. The two tank/AT platoons work in depth of the scout platoons, but each tank/AT platoon initially follows a single scout platoon in its zone/sector. The scout platoons conduct their reconnaissance or screening mission within their zone/sector, and the tank/AT platoons key their movement on the progress of the scout platoons. Give the tank/AT platoon leaders command guidance on where they should move in relation to the scout platoons. Tank/AT platoon leaders eavesdrop on the troop command net to monitor the progress of the scouts and to move their platoons to maintain their orientation with the scouts based on the troop commander's guidance.

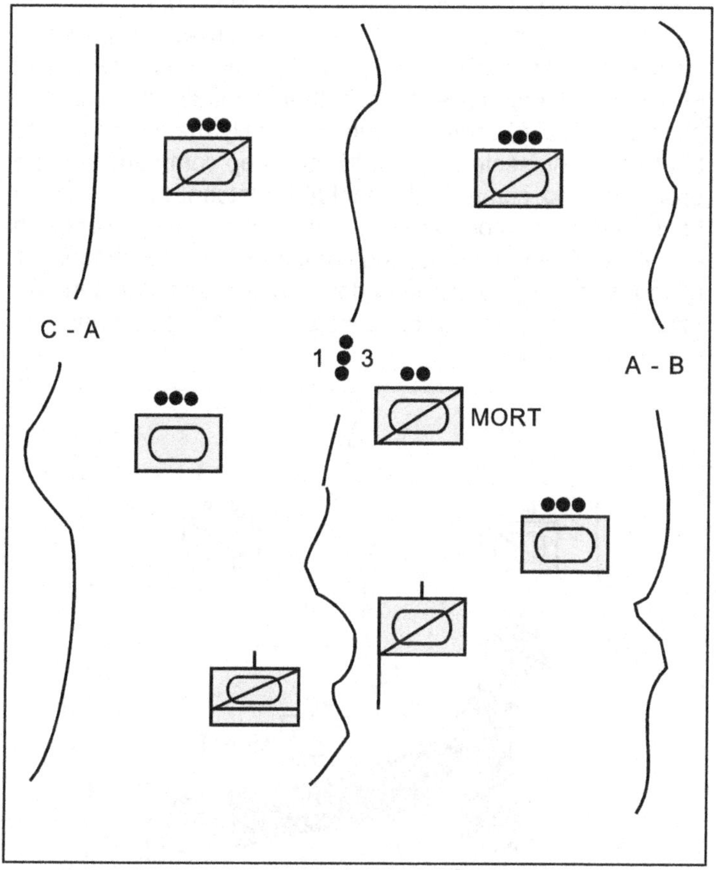

Figure 2-11. Troop split-vee formation.

Develop a habitual relationship between a scout platoon and a tank/AT platoon. First and second platoons work together, and third and fourth platoons work together. This builds teamwork between the scout and tank/AT units, and gives them confidence in each other.

All four platoons still work for the troop commander in this formation. The purposes of having a tank/AT platoon work with a scout platoon are to have combat power spread across the entire troop front, to keep the tanks/TOWs in a position to respond rapidly in support of a scout platoon, and to have scouts available to reconnoiter routes and positions for use by the tanks/TOWs. The troop split-vee is often used in close or restrictive terrain that reduces the tank/AT platoon's mobility so that it cannot quickly maneuver across the troop zone/sector. Generally the troop commander is the only one who commits the tank/AT platoons to conduct a hasty attack, to counterattack, or to move into overwatch positions. If he must mass the combat power of both tank/AT platoons, the troop commander can still shift the tank/AT platoons across the troop zone/sector during the mission (see Figure 2-12).

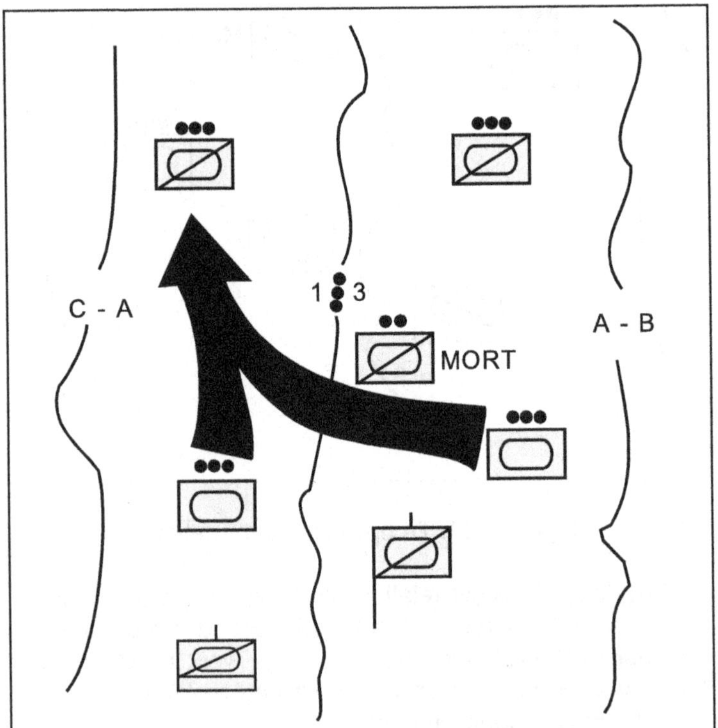

Figure 2-12. Troop commander maneuvers his tank platoons (troop split-vee).

LIGHT TROOP SCRAMBLE

The light troop scramble is a task organization that allows the troop commander to use a mix of weapon systems to enhance acquisition and destruction of the enemy under reduced visibility conditions and in restrictive terrain, where massed TOW fires are not possible. This task organization can also be used to respond to multiple but separate taskings of a troop usually associated with operations other than war, for example, manning checkpoints and area security missions. The troop commander may maneuver his troop in a vee, split-vee, column, or line formation when task organized into a scramble.

The troop is task organized into four like platoons of seven vehicles each. Five scouts and two AT vehicles are in each platoon (see Figure 2-13). The platoon may task organize and operate two ways:

- TOWs integrated with the scout sections (see Figure 2-14).
- TOWs maneuvering as a section to overwatch the movement of the scouts (see Figure 2-15).

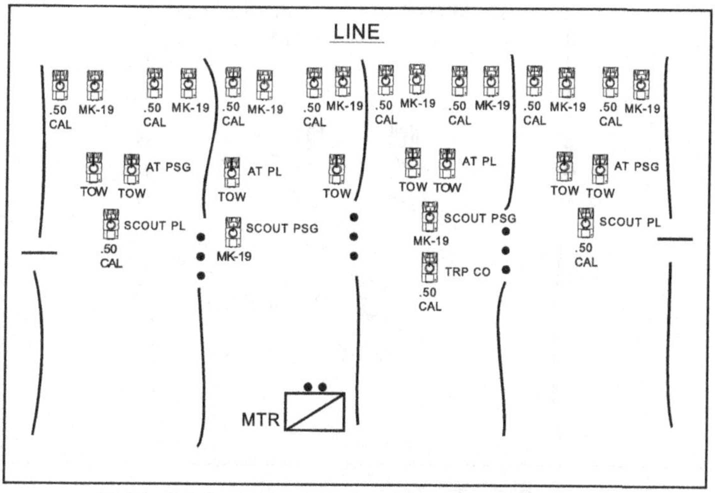

Figure 2-13. Light troop scramble.

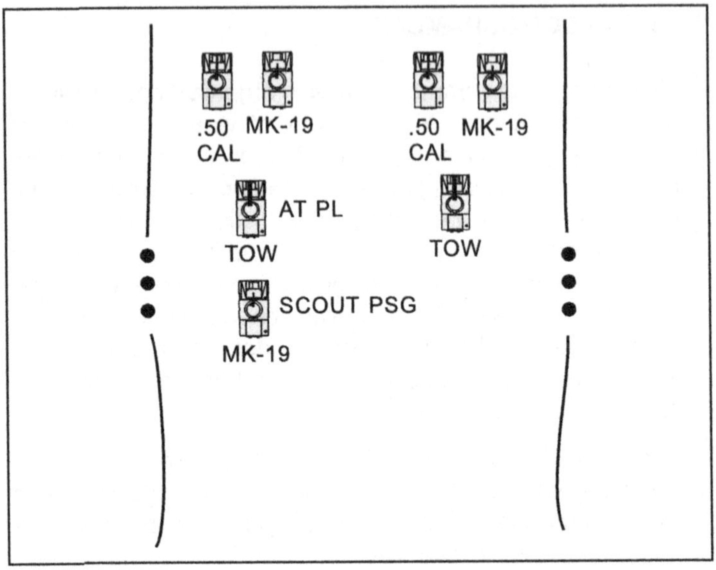

Figure 2-14. TOWs integrated with scout section.

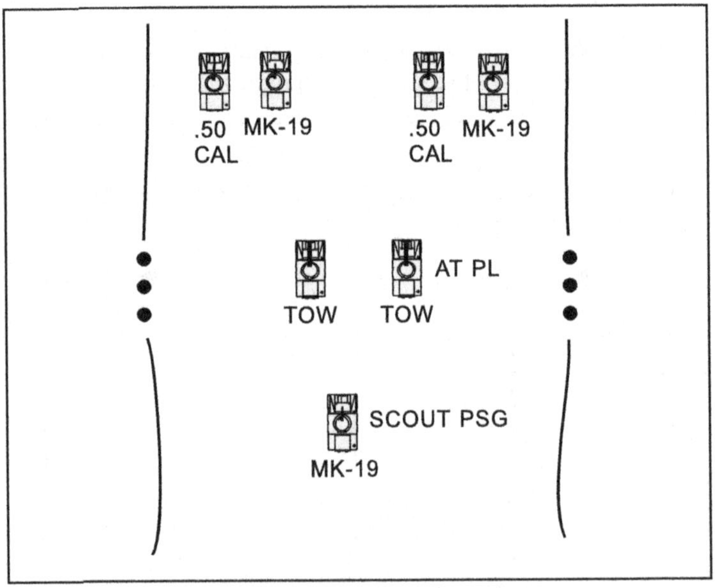

Figure 2-15. TOWs maneuvering as a section
to provide overwatch.

Chapter 3

Reconnaissance

Reconnaissance is a mission undertaken to obtain information about the activities and resources of an enemy or about the meteorologic, hydrographic, or geographic characteristics of a particular area. Reconnaissance produces combat information. Combat information is a by-product of all operations acquired as they are in progress. Reconnaissance, however, is a focused collection effort. It is performed before and during other combat operations to provide information used by the commander to confirm or modify his concept.

Cavalry is the corps or division commander's principal reconnaissance organization. Cavalry troops perform reconnaissance using a combination of mounted and dismounted techniques. The cavalry troop is uniquely organized, trained, and equipped to perform the crucial task of reconnaissance for other combined arms forces. Reconnaissance is the cavalryman's specialty.

CONTENTS

		Page
Section I.	Purpose, Fundamentals and Capabilities	3-2
Section II.	Planning Considerations, Methods, and Techniques	3-5
Section III.	Route Reconnaissance	3-29
Section IV.	Zone Reconnaissance	3-41
Section V.	Area Reconnaissance	3-53

Section I. Purpose, Fundamentals, and Capabilities

PURPOSE

Cavalry troops conduct reconnaissance forward of another friendly force to provide information about the terrain and enemy within the area. The reconnaissance allows the follow-on force an opportunity to maneuver freely and rapidly to its objective. Reconnaissance keeps the follow-on force from being surprised or interrupted, and protects it against losing soldiers and equipment on the way to the objective. The cavalry troop performs three types of reconnaissance—route, zone, and area.

FUNDAMENTALS

Successful reconnaissance operations are planned and performed with six fundamentals in mind.

- Maximum reconnaissance force forward.
- Orient on the location or movement of the reconnaissance objective.
- Report all information rapidly and accurately.
- Retain freedom to maneuver.
- Gain and maintain enemy contact.
- Develop the situation rapidly.

Maximum Reconnaissance Force Forward. In reconnaissance, every pair of eyes makes a difference. Do not keep scouts in reserve. This does not mean that every scout should be forward in a strictly linear sense, but actively employed in the conduct of the reconnaissance.

Orient on the Location or Movement of the Reconnaissance Objective. From the IPB (intelligence preparation of the battlefield) the S2 will identify gaps in the squadron's knowledge of the enemy and terrain it will be operating in. These gaps together with the commander's guidance form the PIR (priority intelligence requirements) and direct the reconnaissance efforts of the troop. Focus the efforts of the troop in selecting a course of action and a scheme of maneuver that continuously orients the troop on its reconnaissance objective based on the commander's guidance. The troop's focus may be a terrain feature, a specific area, or an enemy force.

Report All Information Rapidly and Accurately. Higher commanders base their decisions and plans on the battlefield information cavalry troops gather during reconnaissance. Combat information loses value as it ages, so the fresher the better. Scouts report exactly what they see. Troop commanders and executive officers may summarize or clarify reports to the squadron, but must never delete raw information reported by the scouts. Information that seems unimportant to the troop may be extremely valuable at higher echelons.

Retain Freedom to Maneuver. Cavalry troops must be able to maneuver on the battlefield to continue the reconnaissance mission. If the troop becomes decisively engaged, the reconnaissance stops. Use of proper movement techniques and overwatch helps prevent decisive engagement. IPB provides the troop commander information that allows him to anticipate battlefield events and to retain freedom of maneuver.

Gain and Maintain Enemy Contact. Contact is any condition ranging from a surveillance sighting to engaging in close combat. Surveillance is often sufficient and is the preferred method of maintaining contact. When necessary or required, troops will use fire and maneuver to maintain contact with an enemy force. Once contact is gained, however, it is not lost unless ordered by higher headquarters.

Develop the Situation Rapidly. During reconnaissance operations cavalry troops frequently and repeatedly encounter situations that require action to determine what the troop must face. These situations may be terrain oriented, obstacles, or enemy. Terrain or obstacle situations require close reconnaissance, bypass, hasty breach, if necessary, and marking. If an enemy force is encountered, the troop will determine the size, composition (What is the enemy force made up of—tanks, personnel carriers?), disposition (Is the enemy force in an offensive or defensive posture? Is he dug in? What is his orientation?), and activity. IPB provides the threat situational information that guides the effort. Reconnaissance techniques, often in the form of drills, are used while developing the situation.

Troops must develop the situation rapidly to get inside the enemy's decision cycle and force him to react to the actions of the troop rather than the troop reacting to the enemy. This requires the troop to rapidly execute development of the situation at the platoon level through the execution of rehearsed battle drills.

CAPABILITIES

The ability of a cavalry troop to conduct reconnaissance is a function of the enemy situation, the terrain the troop is operating in, and the type of troop conducting the reconnaissance.

- Heavy Troop
  - Can reconnoiter up to a 10-kilometer wide zone.
  - Can reconnoiter up to two routes simultaneously.
  - Can conduct reconnaissance at the rate of about 1 kilometer per hour, depending on the terrain.
  - When faced with a heavy-equipped threat, will conduct either aggressive or stealthy reconnaissance, depending on the higher commander's guidance.
- Light Troop
  - Can reconnoiter up to a 10-kilometer wide zone.
  - Can reconnoiter up to two routes simultaneously.
  - Can conduct reconnaissance at the rate of about 1 kilometer per hour, depending on the terrain.

- When faced with a heavy-equipped threat, will conduct stealthy reconnaissance and aggressive reconnaissance when reinforced with heavy forces.
- When faced with a light-equipped threat, will conduct either aggressive or stealthy reconnaissance, dependent on the commander's guidance.

Section II. Planning Considerations, Methods, and Techniques

The purpose of this section is to outline the planning considerations, methods, and procedures a cavalry troop uses to execute reconnaissance missions.

PLANNING CONSIDERATIONS

When planning a reconnaissance mission, the troop commander must take into account his unit's capabilities and limitations and consider the following:

- Time available from mission receipt to completion.
- Threat size, composition, disposition, and will to fight.
- Terrain and weather effects on the troop's ability to maneuver.
- Tempo of the operation.
- The squadron commander's (SCO) intent and guidance, as follows:
  - What are the SCO's focus and desired endstate of the reconnaissance?
  - What triggers the squadron's employment of the tank/AT company?
  - What does the SCO want the troop to destroy, fix, and bypass?
- Task organization or reinforcements.
- Critical tasks to be accomplished by the troop. Specifically, identify which critical tasks may be deleted during the reconnaissance.

Based on the considerations above, the troop commander determines the following:

- What is the focus of the reconnaissance? What critical tasks must be accomplished within the constraints of time and terrain?
- What specified or implied missions are associated with the squadron endstate?

Note. If given a limit of advance (LOA) for the reconnaissance mission, the troop commander should plan to screen along the LOA (see Chapter 4).

- How will the troop deal with enemy contact? What are the troop's criteria for engagement, destruction, and bypass?
- Under what situations does the commander see employment of his tank (heavy troop) or antitank (AT) (light troop) platoons?
- How will the commander use indirect fires from mortars and artillery to support his maneuver?
- Who controls the troop's attachments and how are they integrated into the reconnaissance?

The troop may receive attachments from higher headquarters. These assets may be maintained under troop control or tasked down to platoons for their use in the execution of the platoon's specified tasks. Examples include attached engineers, ground surveillance radar (GSR), or a chemical reconnaissance element.

- **Engineers.** If an engineer platoon is attached to the troop, the commander may elect to keep them under his control and treat them as a maneuver platoon. He may assign them to conduct the reconnaissance of the route, while the scout platoons move just ahead and reconnoiter terrain on either side. If the troop receives a squad or section, the troop commander may elect to task it to the scout platoon conducting the reconnaissance of the route.

- **GSR**. If a GSR squad or section is attached, the troop commander may elect to task organize them to a platoon or keep them under troop control. During reconnaissance operations, GSRs may be focused on flank avenues of approach into the troop zone with their movement controlled by the scout platoon leader or commander. In either case the commander should give clear guidance for positioning and orientation of the reconnaissance.
- **Chemical reconnaissance element.** If a chemical reconnaissance element (squad, section, or platoon) is task organized to the troop, the commander may maintain the element in reserve to reinforce a scout platoon that comes in contact with a contaminated area, or he may task organize the element down to one of the scout platoons.

METHODS OF RECONNAISSANCE

There are three methods of conducting reconnaissance at the cavalry troop level: *dismounted, mounted, and reconnaissance by fire.* The troop commander may use any method or combination of methods to accomplish the reconnaissance mission under the restrictions placed on him by METT-T (mission, enemy, terrain, troops, and time available) and the higher commander's intent and guidance. A fourth method of reconnaissance, *aerial reconnaissance*, may be executed by an air cavalry troop conducting a coordinated reconnaissance forward of the ground troop.

Dismounted Reconnaissance

The troop commander may direct scouts to conduct dismounted reconnaissance when—

- Time is not a limiting factor.
- Detailed information is required.
- Stealth is required.
- Enemy contact is expected or has been achieved through visual means.

- Scout vehicles cannot move through an area because of terrain.
- Security is the primary concern.

Dismounted reconnaissance permits the cavalry troop to collect the most detailed information about the terrain and enemy within a given zone, area, or route. However, dismounted reconnaissance is also the most time-consuming of all the reconnaissance methods. The cavalry troop, heavy or light, is limited in the number of dismounted scouts it can employ at any time.

Mounted Reconnaissance

The troop commander directs scouts to conduct mounted reconnaissance when—

- Time is limited.
- Detailed reconnaissance is not required, or mounted method affords the same opportunity to collect information as the dismounted method.
- Enemy locations are known.
- Enemy contact is not likely.
- An air cavalry troop is conducting a coordinated air reconnaissance.

Reconnaissance by Fire

When conducting reconnaissance by fire, the troop places direct and/or indirect fire on positions the enemy is suspected of occupying. This action causes the enemy to disclose his presence by moving or by returning fire. The troop commander may use reconnaissance by fire when—

- Time is critical.
- Natural or man-made obstacles that could be overwatched by an enemy force are encountered.
- A suspected enemy position fits the situational template.
- Bunker complexes that may or may not be occupied are encountered.
- Enemy locations are known.

The disadvantage of the reconnaissance-by-fire method is that the troop will lose any element of surprise it may have had. However, reconnaissance by fire may reduce the chance of some portion of the troop being caught in an enemy kill zone. Reconnaissance by fire may not always provide the desired effect. A well-disciplined force will resist the inclination to move when probed by weapon fires.

When indirect-fire situations exist, the troop commander ensures scouts are in a position to observe the target area. Once the decision is made to use reconnaissance by fire, weapons should be used in the following priority.

- Indirect-fire systems.
- Machine gun.
- 25-mm chain gun or MK-19.
- TOW or tank cannon fire.

Reconnaissance by fire does not mean the indiscriminate use of direct and indirect fires at all woodlines and hilltops in the hopes of causing the enemy to react. Not only will the enemy recognize this ploy for what it is and not react to it, but also it wastes valuable ammunition.

Aerial Reconnaissance

Air cavalry troops perform aerial reconnaissance when time is critical. Aerial reconnaissance is often coordinated closely with a ground reconnaissance troop. The air and ground forces complement each other. The air element can move forward of the ground unit and reconnoiter key pieces of terrain or restrictive terrain, allowing the ground troop to concentrate its efforts in other areas or to increase the tempo of its reconnaissance. The air troop provides the ground troop with added security by clearing the ground forward of the ground unit, thereby facilitating movement of the ground force and quickening the pace of the operation. The ground troop can move rapidly mounted to the areas of interest within its area or zone of operations and have the time to dismount and collect detailed information (see Figure 3-1).

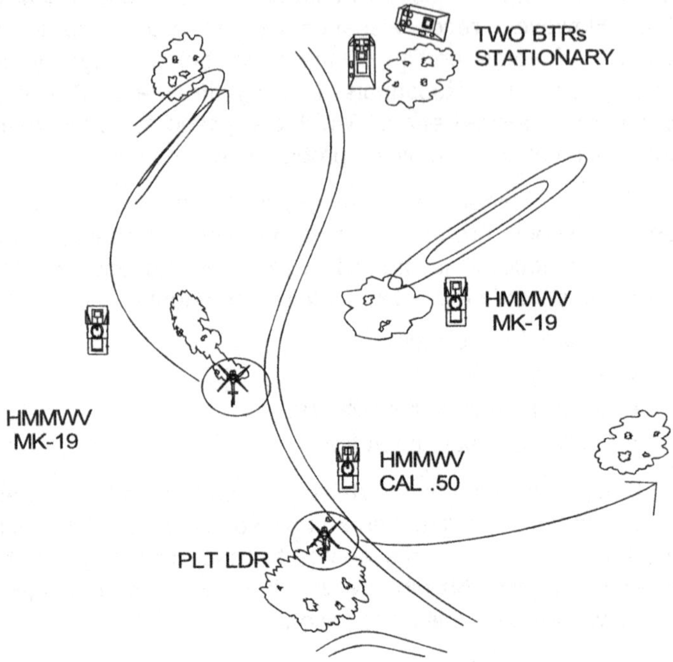

- Air cavalry platoon identifies two stationary BTRs along the woodline.
- Air cavalry platoon leader sends spot report to air cavalry troop commander. Troop commander tells platoon leader to develop situation and get in contact with ground platoon leader in zone.
- Air cavalry platoon leader sends information to ground platoon leader and assists ground scout section in developing situation.
- Air cavalry troop commander and ground troop commander choose a course of action.
- This coordination of air-ground supports squadron commander's intent, which increases troop survivability and squadron operating tempo of reconnaissance.

Figure 3-1. Air scouts conduct coordinated reconnaissance of zone with ground troop.

Aggressive Versus Stealthy Reconnaissance

The cavalry troop uses various combinations of reconnaissance methods. The method or combination of methods chosen by the troop and its scout platoons is based on the higher commander's guidance and METT-T. The method or methods chosen will characterize the reconnaissance as either aggressive or stealthy.

The stealthy approach to reconnaissance is more time-consuming. It emphasizes avoiding detection and engagement by the enemy. To be effective, a stealthy approach must rely primarily on dismounted reconnaissance and maximum use of covered and concealed terrain.

Aggressive reconnaissance is characterized by the speed and manner in which the reconnaissance force develops the situation once contact is made with an enemy force. A troop conducting aggressive reconnaissance uses firepower from direct- or indirect-fire systems and maneuvers to rapidly develop the situation. The troop will primarily use mounted reconnaissance and reconnaissance by fire when conducting aggressive reconnaissance.

Cavalry troops must be trained to conduct both aggressive and stealthy reconnaissance. The troop commander, scout platoon leader, and squad leaders will most likely use many different methods over the course of a reconnaissance operation. The constraints of commander's guidance, the opposing threat, the capabilities of the troop, the terrain the troop is operating in, and the time available to complete the mission (METT-T) all have an impact on the method or methods chosen by the troop leadership.

TECHNIQUES

Actions on Contact

Actions on enemy contact are a series of steps the troop takes when it encounters an enemy force or situation that warrants/demands action. Actions on contact are important because they allow the troop to maintain its tempo of operation by rapidly developing the situation and taking action before the enemy can gain the initiative and force the troop to react. At platoon level, actions on enemy contact consist of four steps.

- Deploy and report.
- Develop the situation.
- Choose a course of action.
- Recommend or execute a course of action.

Note. See FM 17-98 for an in-depth discussion of platoon actions on contact during reconnaissance.

While the platoon that makes contact executes actions on contact, the commander must continue to maneuver the remainder of the troop to ensure a clear picture of the enemy situation across the entire troop front. The following steps demonstrate the actions taken by the platoon in contact and the corresponding actions at the troop level.

Deploy and Report

Platoon Action. The elements of the scout platoon that make initial contact with the enemy immediately deploy to terrain that affords them both cover and good observation. If necessary, scouts return fire to suppress the enemy, and then deploy to their positions. The scout making contact sends a contact report to his platoon leader. The platoon leader forwards the report to the troop commander. Once the scout in contact is in a good covered and concealed position, he sends his initial spot report, which is forwarded through the platoon leader to the troop commander (see Figure 3-2).

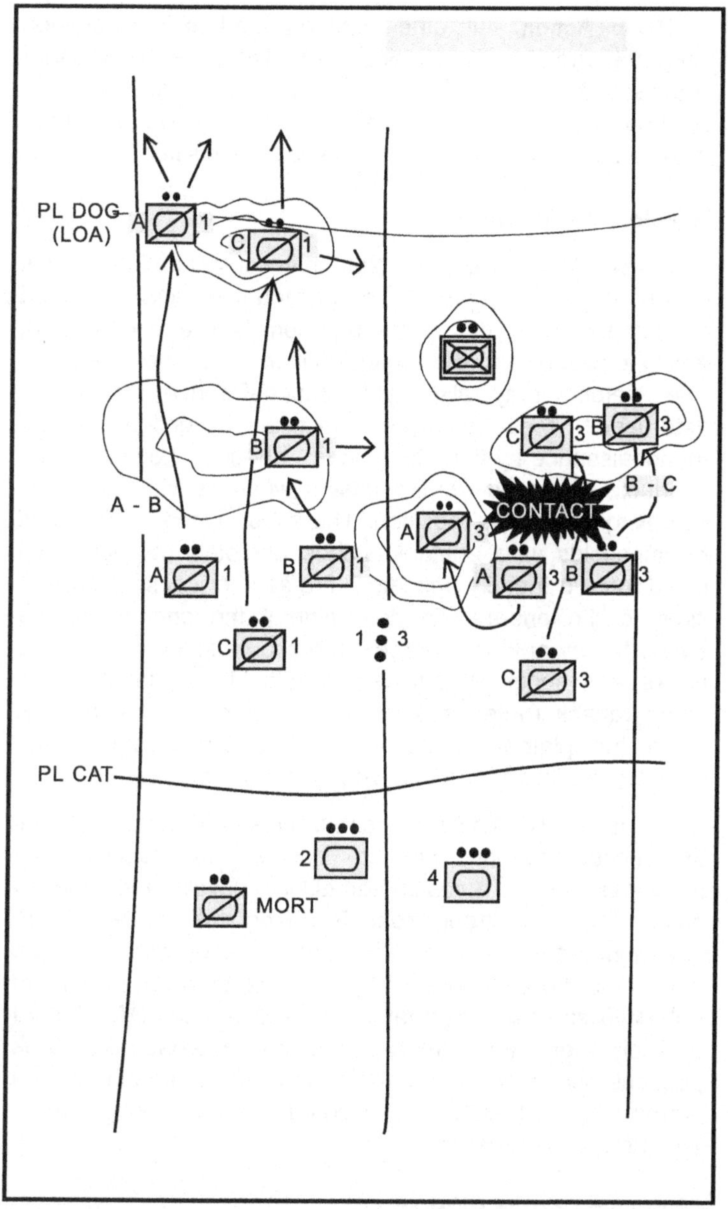

Figure 3-2. Scouts deploy and report then begin to develop the situation.

Troop Action. All other platoon leaders and platoon sergeants monitor the contact report. The troop commander assesses the information and moves, if necessary, to a position where he can observe the action. However, the troop must not lose focus of the reconnaissance mission.

Develop the Situation

Platoon Action. Next, the platoon in contact defines what it is up against. Scouts use dismounted and mounted reconnaissance to determine the enemy's size, composition, and orientation, and the exact location of weapon systems. The platoon may also use reconnaissance by fire to determine the enemy's tactical intentions. The reconnaissance-by-fire technique should, however, be conducted with indirect-fire assets when possible to avoid revealing the scouts' position. The scout platoon will search for minefields, wire, antitank ditches, and other obstacles that could force a friendly unit into a fire sack or cause it to flank itself to the enemy. To determine if the enemy can be supported by any other forces, the scouts search for enemy flanks and scour all adjacent terrain. They identify good counterattack routes into the flanks or rear of the enemy. Once the platoon leader determines the extent of the situation, he forwards a follow-up spot report.

Troop Action. The troop commander will most likely tell the scout platoon not in contact to continue its reconnaissance to a designated LOA to develop the situation across the entire troop front. By doing this, the troop can determine if there are any other enemy forces with which the troop must be concerned. The scout platoon not in contact will establish hasty observation posts along the LOA oriented on likely enemy locations or avenues of approach. The tank platoons will continue to monitor the troop command net and prepare to act based on the commander's intent/guidance and the enemy situation.

Choose a Course of Action

Platoon Action. Now that the platoon leader knows what he is up against, he considers two or three possible courses of action and selects the one that best meets the commander's intent/concept of the operation, is within his capabilities, and allows the troop to resume its reconnaissance mission as soon as possible. The possible courses of action open to him, based on the commander's intent/concept, might be hasty attack, bypass, hasty defense/screen, or support of a hasty attack by another platoon(s). Some courses of action will quickly be ruled out because they do not meet the commander's intent and guidance for the operation.

- **Hasty Attack.** The platoon leader can conduct a hasty attack if he has enough combat power to defeat the enemy quickly. In most cases, the scout platoon does not have the capability to mass enough combat power to defeat an enemy in prepared positions. In addition, the scout platoon leader may not want to risk battle losses that would reduce his effectiveness and his ability to complete the mission.

- **Bypass.** If the platoon does not have enough combat power to conduct a hasty attack, or if it wants to remain undetected and continue the reconnaissance mission, the scout platoon can bypass the enemy. The platoon leader must receive the troop commander's permission to bypass. If he has permission to bypass, the platoon leader must leave scouts in contact with the enemy force unless ordered to break contact by the troop commander. The platoon leader will know from the troop commander's guidance in the operations order if this is a viable course of action.

- **Hasty Defense/Screen.** If the platoon cannot conduct a hasty attack and cannot bypass, it establishes a hasty defense or screen. The platoon will conduct a hasty defense if it can defend against an enemy force. If the enemy contact exceeds the platoon's capability to conduct a hasty defense, it may elect to establish a screen and maintain contact through observation. The platoon concentrates on maintaining contact with the enemy and fixing it in place with indirect or possibly direct fire until additional support comes from the troop.

- **Support a Hasty Attack by Another Platoon(s).** The platoon in contact may become the support element for a hasty attack by a tank/AT platoon(s).

Troop Action. The troop commander continues to maneuver the troop and assists the platoon leader as required. He maintains situational awareness.

Recommend/Execute a Course of Action

Platoon Action. If the course of action the platoon leader chooses meets the commander's guidance for actions on contact, the platoon leader executes the course of action. If the situation dictates that the platoon execute a hasty defense/screen, or support a hasty attack by another platoon, the platoon leader reports this information to the troop. The report is brief. He updates his spot report (Blue 1) with any additional information, tells the troop commander what he is doing about the situation, and recommends the course of action that he thinks best suits the situation. The platoon in contact has then completed its actions on contact.

Troop Action. The commander must approve or disapprove the recommended course of action, based on its effect on the troop and squadron mission.

If the scout platoon is required to establish a hasty defense/screen, the troop commander assumes responsibility for continuing to develop the situation. The commander can do this best in a position from which he can influence troop actions. Based on what the scouts have found and reported and what the commander can see, he must weigh the advantages and disadvantages of each course of action, and choose the one that best meets the troop's mission requirements.

- **Hasty Attack**. The troop can conduct a hasty attack, which throws the weight of one or both tank/AT platoons against the enemy and destroys him quickly, then continue its reconnaissance. The commander may coordinate the efforts of the scout platoon and tank/AT platoon(s) in the hasty attack or delegate the responsibility to the scout platoon in contact. The hasty attack at troop level should have two elements: the *assault/attack-by-fire* element and the *support* element.

  Steps to the execution of a hasty attack are as follows:

  - The scouts determine that the enemy force encountered cannot be bypassed based on the enemy disposition and composition. The enemy encountered meets the commander's intent for destruction.

  - The scout platoon leader recommends hasty attack to the troop commander and identifies a good attack-by-fire or assault position for the tank/AT platoons.

  - The troop commander approves the scout platoon leader's recommended course of action and issues orders to execute a hasty attack.

  - The troop commander decides to use either attack by fire or assault as his method of destruction. Heavy troops equipped with tanks can perform attack by fire or assault. Because of the defensive nature of the TOW, light troops tend to use attack by fire when conducting hasty attack.

  - The scout squad or section moves to link up with the tank/AT platoons and guides them to an assault/attack position. The assault element reports when set in assault positions, and if time is available, reconnoiters the attack-by-fire position or the attack axis.

- The troop commander moves to the scouts who are positioned in overwatch of the enemy contact. He also positions the FIST in overwatch to assist in controlling indirect fires.

- The troop commander designates the scouts in overwatch as the support element. He establishes direction of fire and methods of control for direct-fire weapon systems; for example, left and right limits for fires, and signals for the initiation of direct fires and the shifting of fires as the assault element begins its attack.

- The troop commander directs the FIST to fire for effect on enemy positions when given the signal (AT MY COMMAND). The FIST reports when guns are ready.

- The troop commander issues the fire command for the support force and indirect-fire systems and directs the attack-by-fire or the assault element to execute its movement. The troop commander lifts and shifts direct and indirect fires as necessary to cover the movement of the assault element and to seal off the enemy withdrawal (see Figure 3-3).

- **Bypass**. The troop may bypass the enemy force and continue the reconnaissance to further develop the enemy situation throughout the depth of the zone while maintaining an element of surprise. The decision to bypass is based on the higher commander's intent for the operation. The troop should leave an element in contact with the enemy force; however, the more elements in contact with the bypassed enemy mean the less reconnaissance will be conducted.

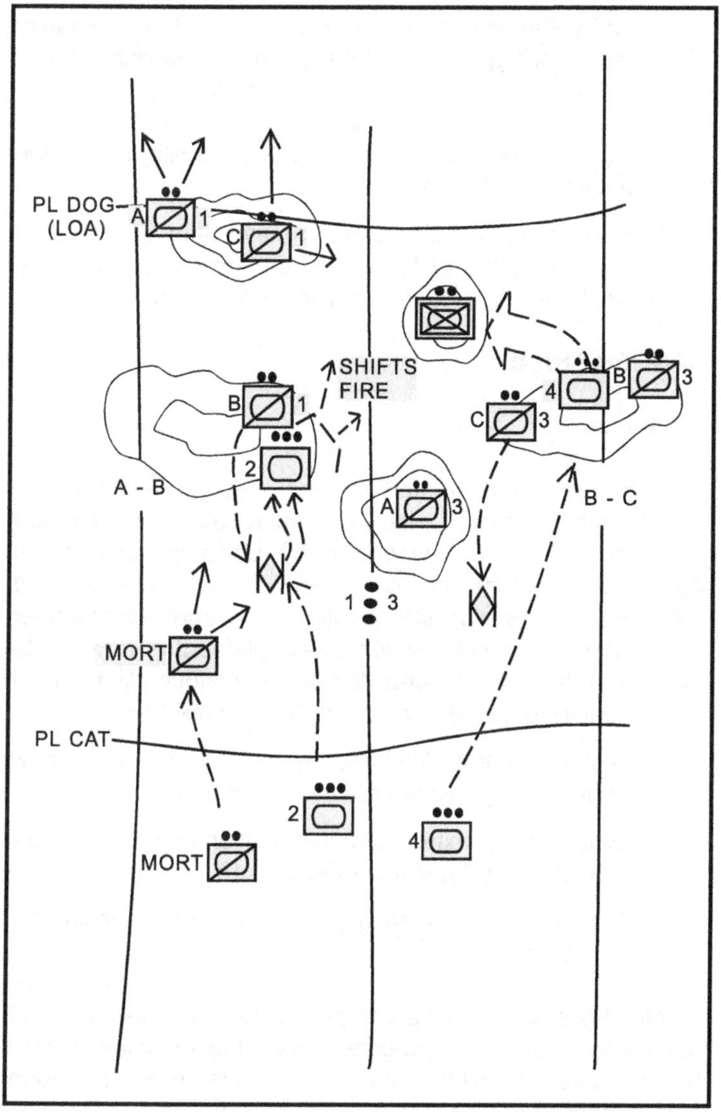

Figure 3-3. Troop positions for hasty attack.

- **Hasty Defense**. If the troop cannot conduct a hasty attack or bypass the enemy, it may establish a hasty defense to fix the enemy while waiting for further orders. If this is the troop's course of action, the squadron commander has the responsibility to develop the situation further.

The overriding considerations in selecting a course of action are the intent of the squadron commander and the troop's ability to complete the mission with minimum losses.

DANGER AREAS

Obstacles

Obstacles encountered during reconnaissance are treated as enemy contact, because all obstacles are assumed to be covered by enemy fire. Upon encountering an obstacle, scouts deploy to covered positions, report, and then begin to develop the situation. The troop commander may move his tank platoon(s)/AT platoon(s) forward to overwatch the obstacle while the scouts conduct dismounted and mounted reconnaissance to determine the following:

- Is the obstacle defended by the enemy? If so, how many enemy soldiers are there and where are they?
- What are the extent and composition of this obstacle? How deep, wide, steep, or long is it?
- Can the obstacle be bypassed, or will breaching be necessary?

Once the scouts have completed the reconnaissance of the obstacle and recommended a course of action (either bypass, hasty attack/breach, or hasty defense), the troop commander must choose the course of action that best meets the commander's intent and guidance for the mission. If the commander's decision is to bypass or breach the obstacle, the troop must mark the obstacle and any lane created through the obstacle (see Figure 3-4).

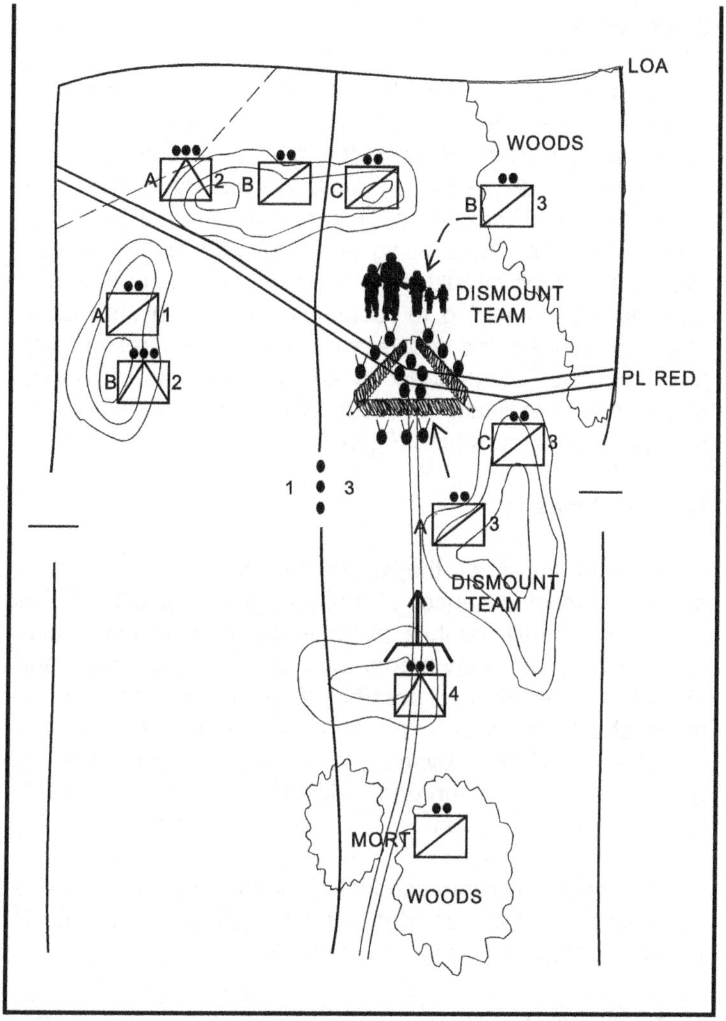

Figure 3-4. The troop develops the situation at the obstacle.

Open Areas

Open areas are dangerous for cavalry troops because they permit the enemy to observe or engage units as they move through them. When reconnoitering a route, zone, or area that includes an open area, do not try to force the reconnaissance through it. Treat it like a danger area. Reconnoiter the flanks using good covered and concealed routes. The troop should try to clear the far side of the open area before attempting to reconnoiter the open area itself. Open areas within the troop's zone or area of operations may force the troop commander to abandon the use of platoon boundaries to facilitate the reconnaissance to the flanks and rear of the open area (see Figure 3-5).

Built-up Areas

In most areas of the world the troop can expect to conduct reconnaissance of built-up areas (BUA). BUAs range from small hamlets to large towns, even portions of cities. BUAs are very dangerous for mounted units and should be avoided when possible. If the situation permits, the troop should conduct a reconnaissance of BUAs from a distance, and then bypass them. However, it may be necessary to move through a town as part of the troop's reconnaissance mission.

The troop's ability to conduct reconnaissance of a BUA is constrained by the lack of dismounted scouts. The troop is capable of finding the following in BUAs:

- Enemy vehicles.
- Enemy command and control facilities.
- Obstacles such as rubble, blown bridges, and craters.
- Logistics elements.
- Bypasses within the BUA.

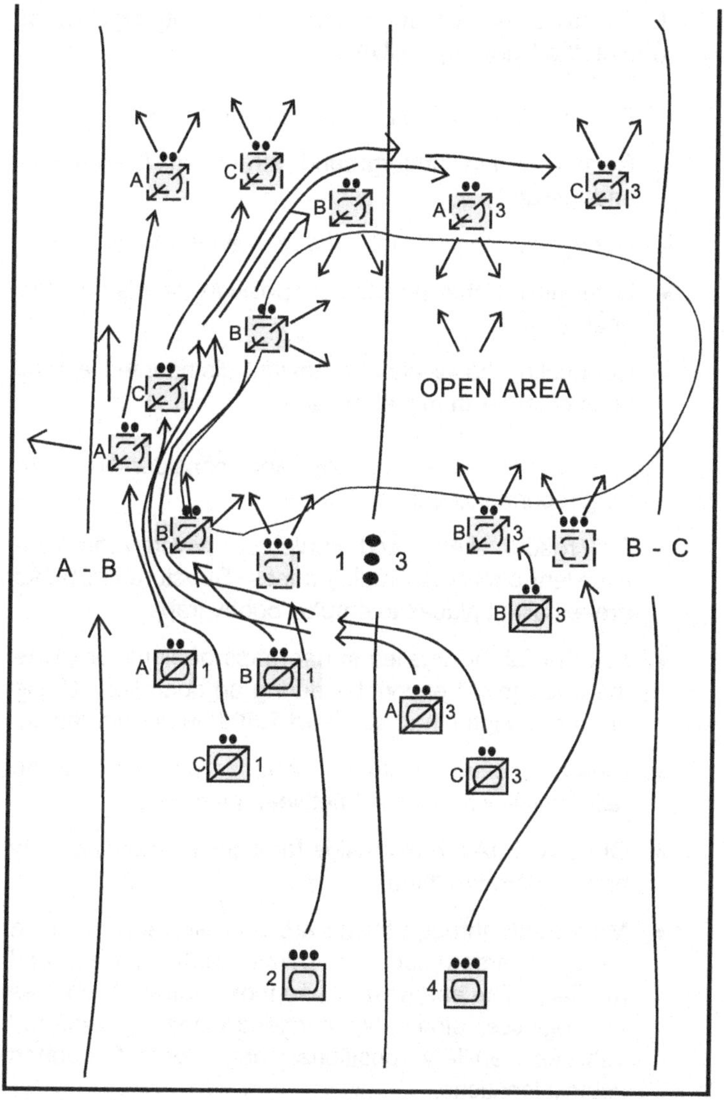

Figure 3-5. The troop reconnoiters to the flanks and rear of an open area.

If the troop is not augmented, it is not capable of performing the following in BUAs:

- Reconnoitering each building and street.
- Reconnoitering underground structures (basements and sewers).
- Determining the strength of dug-in enemy units.
- Determining the detailed disposition of dismounted enemy units.
- Conducting hasty attacks against dismounted enemy units of larger than platoon size.

The following are techniques and considerations for performing reconnaissance in BUAs:

- Dirt roads, alleys, and curves in paved roads are excellent places to employ mines. Structures in BUAs are excellent places to employ booby traps.
- Ask the S2 for detailed street maps of all major BUAs the troop might encounter during an operation. These are more useful than standard 1:50,000 military maps.
- Clearly define platoon zones. Do not divide responsibility for a street between platoons.
- Observe BUAs from outside for signs of enemy activity prior to entering them.
- Main roads through towns are normally accompanied by open areas such as parks, traffic circles, and medians. The enemy may use these same open areas for logistics elements, concentrations of armored vehicles, artillery positions, and combat support vehicle locations.
- Collateral damage constraints may limit the ability to employ some types of weapon systems.

LIGHT TROOP EXAMPLE

The squadron is conducting a zone reconnaissance. The troop zone contains a large town, 2 to 3 kilometers square, and the squadron main supply route, ROUTE GOLD, passes through the town. The troop will have to reconnoiter enough of the town to ensure there are no enemy who can disrupt movement along ROUTE GOLD.

On approaching the town, the troop has options on how to enter it. The techniques used will largely depend on the amount of time available for the mission. If time is limited, the scouts will halt short of the BUA, just off ROUTE GOLD, and observe the town from multiple vantage points before entering. If more time is available, the scouts will move well off the route and observe the BUA before entering. In any case, remember if enemy is present, he will probably cover the main road with fires. Once the platoon is inside the BUA, the main road may provide the best axis for the reconnaissance.

As the scout platoon observes the town, they look for any signs of enemy presence. The antitank platoons move into overwatch positions. The mortars establish a firing position in support of the scouts. Scouts reconnoiter to the flanks and rear of the BUA to find a bypass and establish flank and far side security (see Figure 3-6).

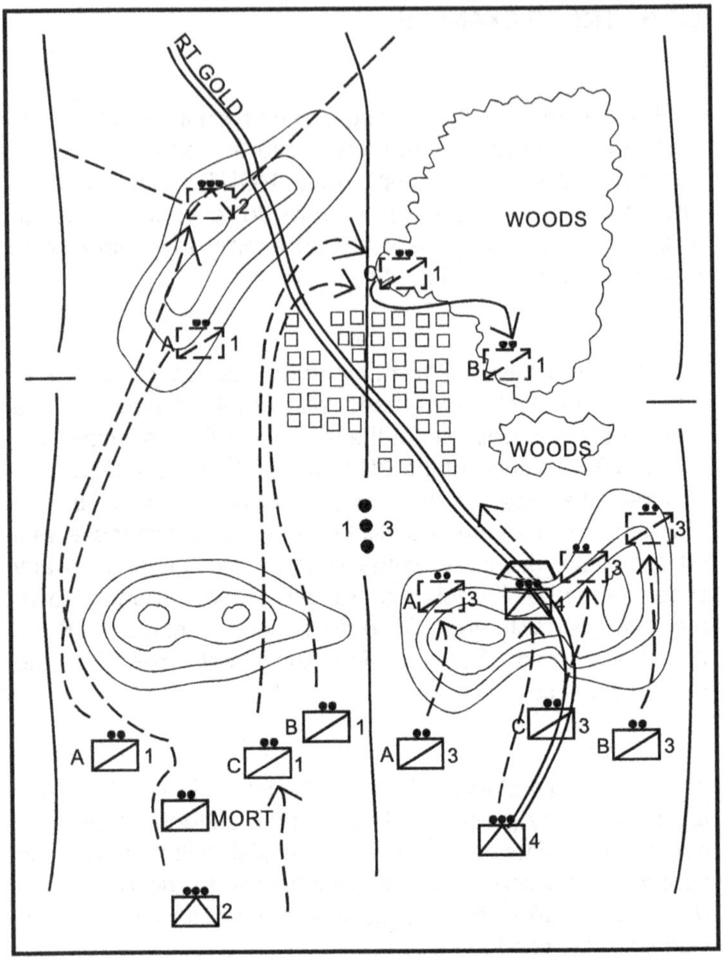

Figure 3-6. The troop securing the BUA with one scout platoon, ATs in overwatch, and one scout platoon preparing to reconnoiter the BUA itself.

Once the area surrounding the BUA is secured, the scouts move mounted to the last concealed position and dismount, searching for suitable entry points into the BUA. *If there is enemy contact*, the overwatching vehicles suppress with direct and indirect fires while the scouts move to a covered and concealed position. At this point the troop commander must decide to maintain contact and bypass, or continue to probe for entry points into the BUA. The determining factors for the commander's decision are the intent of the higher commander and the level of threat in the BUA. In most instances, cavalry troops maintain contact and bypass BUAs occupied by enemy forces. The enemy forces are handed off to follow-on forces for destruction.

If the troop does not encounter enemy contact on its initial reconnaissance of the outskirts of the BUA, the scout platoons continue their search for suitable entry points into the town. Once the scouts determine those entry points into the town are clear, they move into the BUA and reconnoiter nearby buildings (moving through side and back yards, alleys, and other off-street paths) and surrounding areas. The dismounted teams from the scout platoon do not enter or clear each building, but look for obvious signs of enemy presence. HMMWVs may be brought forward to overwatch the dismounted team's movement with caliber .50 machine guns or MK-19s (see Figure 3-7). The AT platoons should continue to overwatch from outside the BUA. After the scouts complete the reconnaissance of the BUA, the troop continues its reconnaissance in zone, ensuring that bypasses around the BUA are marked.

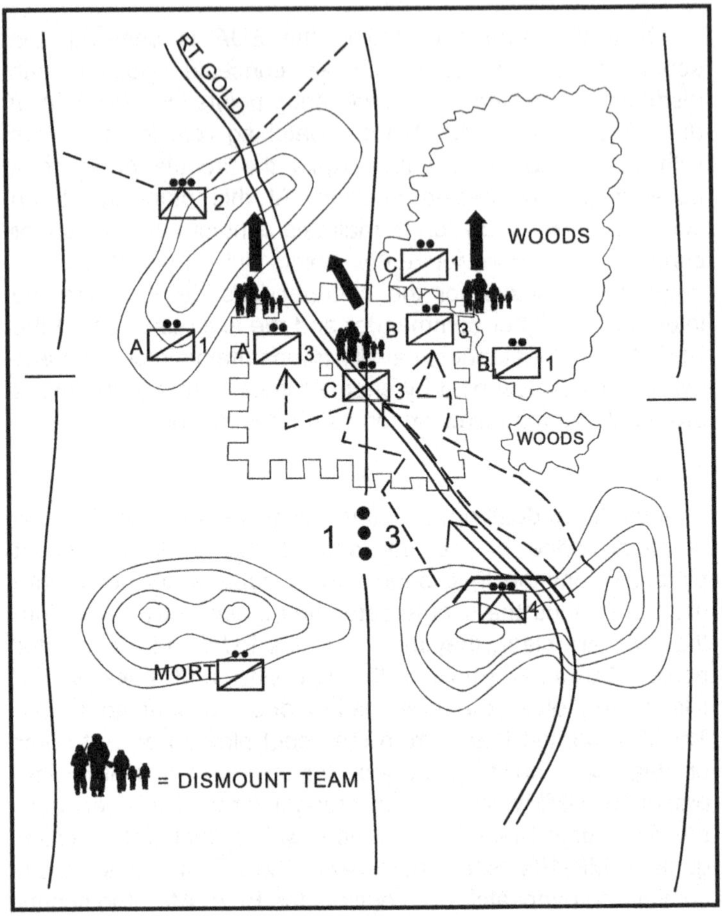

Figure 3-7. Scout section reconnoitering in a built-up area.

Section III. Route Reconnaissance

Route reconnaissance is a directed effort to gain detailed information about a specific route and the terrain on either side of the route that the enemy could use to influence movement along the route. Depending on the terrain surrounding the route, a route reconnaissance mission may be accomplished by a single scout platoon, or it may require the entire troop to cover the terrain and accomplish the required tasks.

A cavalry troop may be assigned the mission to conduct a zone reconnaissance (see Section IV of this chapter), but also have the task of reconnoitering a route that is within its zone. In this case, the troop mission would be zone reconnaissance. However, one of the subordinate units in the troop, scout platoon, or attached engineers may be given the route reconnaissance mission by the troop commander.

CRITICAL TASKS

During a route reconnaissance, the following critical tasks must be accomplished unless the SCO directs the troop to do otherwise:

- Reconnoiter and determine trafficability of the route.
- Reconnoiter all terrain the enemy can use to dominate movement along the route.
- Reconnoiter all BUAs along the route.
- Reconnoiter all lateral routes in the area of responsibility.
- Inspect and classify all bridges along the route.
- Locate fords or crossing sites near all bridges along the route.
- Inspect and classify all overpasses, underpasses, and culverts.

- Reconnoiter all defiles along the route; clear all defiles of enemy and obstacles within its capability, or locate a bypass.

- Locate mines, obstacles, and barriers, and within its capability, clear the route.

- Locate a bypass around BUAs, obstacles, and contaminated areas.

- Report route information.

- Find and report all enemy that can influence movement along the route.

TECHNIQUES

The SCO directs the troop to conduct a route reconnaissance as a mission or as a specific task in another mission. This section discusses route reconnaissance in the context of an assigned troop mission.

The troop commander considers several factors in formulating his concept.

- Start point, release point, and designation of the route.

- Mission to be performed at the start point and after reaching the release point.

- Time the mission is to start, and if required, be completed.

- Critical points along the route identified as checkpoints.

- Any constraints or restrictions.

- IPB information on the route, to include current enemy situation.

IPB provides critical information on the enemy and terrain. Enemy threats may be encountered in two basic forms.

- Ambushes along the route in close or restricted terrain or even tied to obstacles along the route.
- Attack by long-range direct or indirect fires from dominating terrain along the route.

Analysis of the terrain provides an indication of danger areas and the nature of the potential threat. From this information, the commander determines how much terrain on each flank of the route must be reconnoitered and the organization for combat. Any constraints or restrictions may also influence how much terrain is reconnoitered.

The troop commander adds control measures to provide an adequate framework for the mission. He places a boundary on both sides of the route far enough out to provide reconnaissance of the dominating terrain. A line of departure is placed perpendicular to the route short of the start point, allowing sufficient space to deploy into a tactical formation. An LOA is placed far enough beyond the release point to enclose dominating terrain that overwatches the release point. The commander uses additional phase lines to maintain a coordinated effort throughout the troop. Boundaries and phase lines are drawn along recognizable terrain features. The terrain features should be visible from both ground and air to facilitate air-ground integration. The commander uses other control measures as necessary for flexibility in the maneuver and execution of the mission (see Figure 3-8).

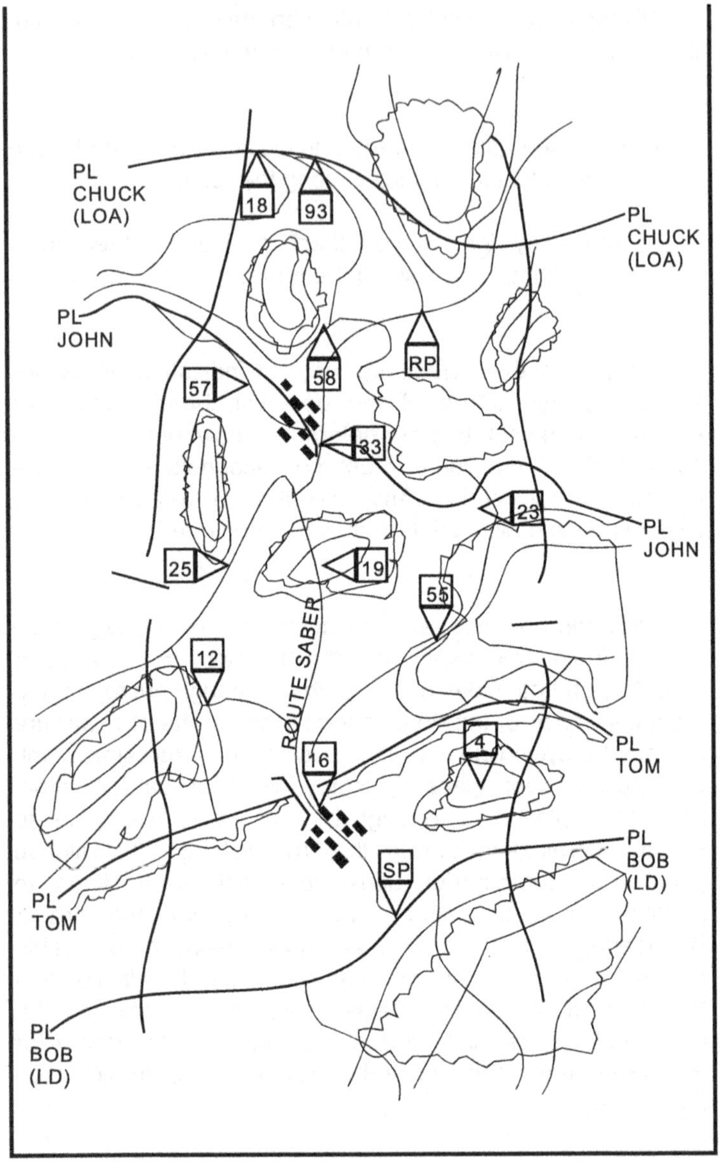

Figure 3-8. Troop graphic control measures for a route reconnaissance.

The cavalry troop normally performs a tactical road march to the line of departure and deploys to execute the reconnaissance of the route. Based on the amount of intelligence known about the enemy, the commander will determine how much security is required for the move forward to the line of departure. Also the commander should consider the effect his final disposition of forces will have on the troop's follow-on mission.

One technique for accomplishing all tasks as rapidly and safely as possible is as follows: Draw a 2.5-kilometer to 3-kilometer boundary on either side of the route. Near the start point, draw a line of departure from one boundary to the other. Next, draw an LOA about 3 kilometers forward of the release point from one boundary to the other. Add phase lines between the line of departure and LOA, as needed, to help control the progress of the troop. To control movement on the route, establish checkpoints. Use checkpoints to identify specific areas or locations to be reconnoitered. These checkpoints will serve as references to orient reconnaissance efforts.

One scout platoon moves out early across the line of departure to reconnoiter the terrain or BUAs on either side of the route. The other scout platoon reconnoiters the route itself, trailing 1 kilometer to 2 kilometers behind the lead scout platoon. This provides a good measure of security for the troop and the platoon working to reconnoiter and classify the route. The mortar section follows the lead scout platoon and remains positioned to range from 3,000 to 3,500 meters forward of the lead scout elements. The mortar section avoids moving on the route and stays on the flanks near the route. The tank/AT platoons follow and support behind the second scout platoon. They key their movement on the advance of the scout platoons. The tank/AT platoons should be kept back in depth to retain flexibility, because they must be ready to react anywhere in the troop zone. The troop trains follow about 2 kilometers behind the tank/AT platoons, and bound from one covered and concealed position to another. The troop CP displaces generally along the route using terrain that affords effective and continuous communications with troop elements and the squadron. The troop commander positions himself to observe the actions of the scout platoons.

EXAMPLE OF A TROOP ROUTE RECONNAISSANCE

The 1st platoon (scout) crosses the LD (PL BOB) using the two-section organization, and reconnoiters the terrain on either side of ROUTE SABER up to PL TOM. The platoon looks for the threat around the outskirts of the village, and explores routes into the village that intersect ROUTE SABER. The scouts also search the woods near checkpoint 4 for the enemy. The platoon scans the terrain north of the stream, and then moves up. One section locates a fording site west of the bridge at checkpoint 16, and the other section swims the stream on the east. The platoon makes a quick visual inspection of the bridge, then searches for the threat in the forested terrain near checkpoints 12 and 55. The mortar section crosses the LD next, about 2 kilometers behind the 1st platoon (scout), and establishes a firing position on the outskirts of the village. The 3d platoon (scout) crosses the SP when the 1st platoon (scout) crosses PL TOM.

The 3d platoon (scout) reconnoiters and classifies the route through the village, then moves up and classifies the bridge just north of town. The troop commander bounds closely behind the 3d platoon (scout) and observes their actions. The 2d and 4th platoons (tank) follow along the flank of ROUTE SABER about 1 kilometer behind the 3d platoon (scout), then move into the outskirts of the village. They observe the dominating terrain on the far side of the stream while the 3d platoon (scout) reconnoiters the bridge. The troop CP and trains remain in concealed positions south of the LD (see Figure 3-9).

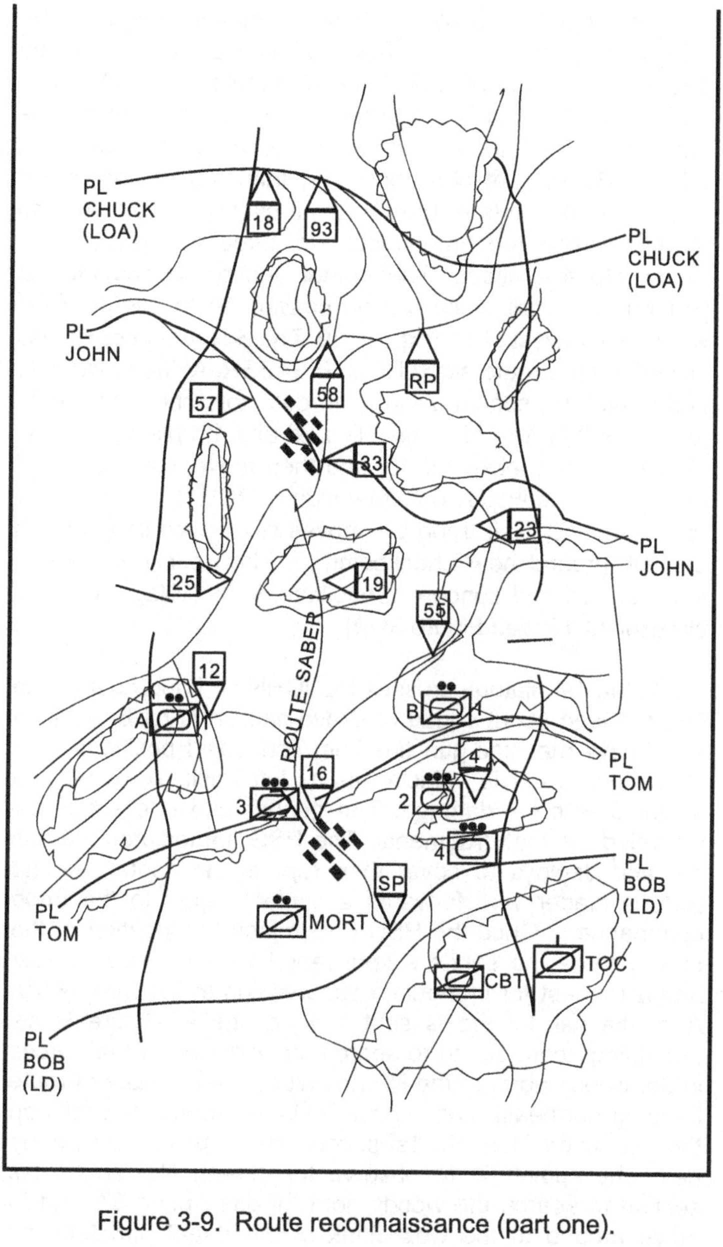

Figure 3-9. Route reconnaissance (part one).

The 1st platoon continues to reconnoiter in zone up to PL JOHN. The platoon searches the wooded areas for the threat, and explores the high-speed lateral routes to the east and west of ROUTE SABER. The platoon clears the shoulders and makes a visual inspection of the defile on ROUTE SABER at checkpoint 19, then moves forward to search the outskirts of the village at checkpoint 33. As with the first village, the platoon looks for threat forces along the roads into the village. The mortar section follows the 1st platoon and establishes a firing position on the edge of the woods southwest of checkpoint 19. The 3d platoon continues classifying the route along ROUTE SABER. At the defile near checkpoint 19, scouts using mine detectors check the route to ensure it is free of mines. The 2d and 4th platoons cross the bridge at checkpoint 16, and then move into concealed positions in the woods on either side of ROUTE SABER near checkpoint 19. The troop CP moves to a concealed position on high ground near checkpoint 12. The troop trains move into covered and concealed positions in the village south of checkpoint 16 (see Figure 3-10).

As the 1st platoon crosses PL JOHN and approaches the RP, the lead scout squad of Bravo section is hit by cannon and heavy machine-gun fire from concealed positions near checkpoint 93. The vehicle has its left track blown off and sustains external damage. The gunner and one scout are wounded by shell fragments. The PSG immediately returns fire and deploys to cover then reports the contact to the platoon leader who forwards a contact report to the troop commander. Once the PSG determines the location of the enemy force, he sends a spot report to the platoon leader and a request for immediate suppression to the troop FSO. After the call for fire is sent forward, the PSG directs his remaining scout squad to search for additional threat forces in the woods north of the RP and along the high-speed route heading northeast (see Figure 3-11). As mortar fire falls on the enemy position, the 1st platoon leader quickly moves up near checkpoint 58 to observe the threat. He directs his section to search the woods north of checkpoint 57, and to move around to the west flank of the threat and find out

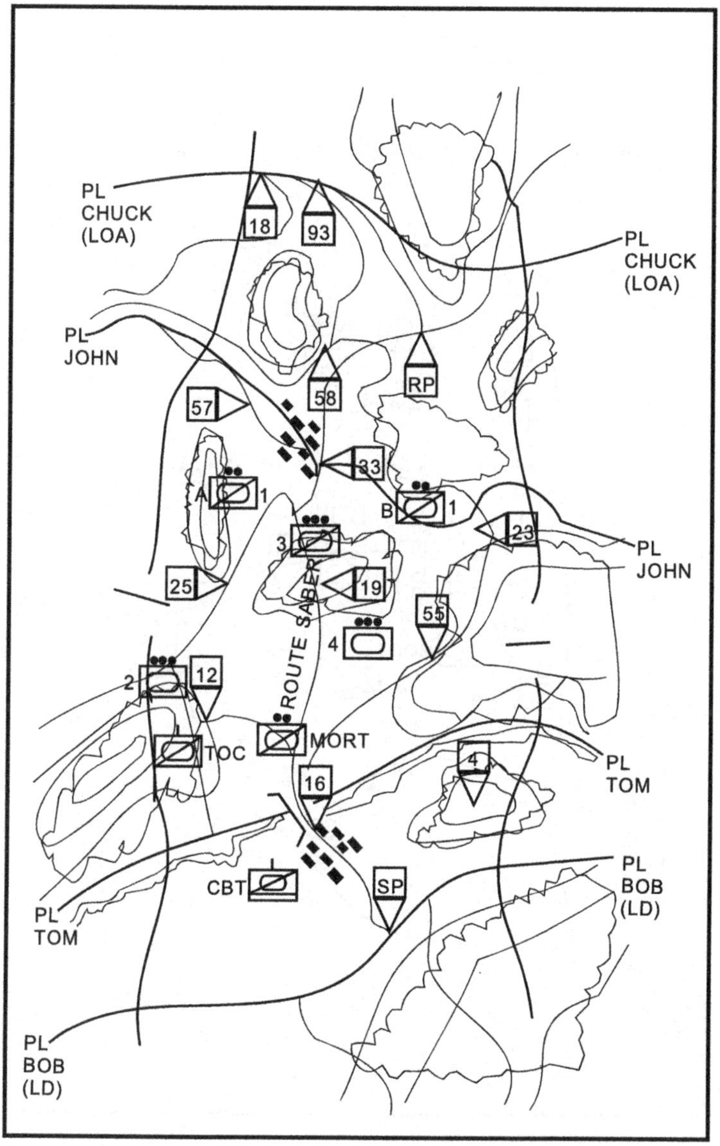

Figure 3-10. Route reconnaissance (part two).

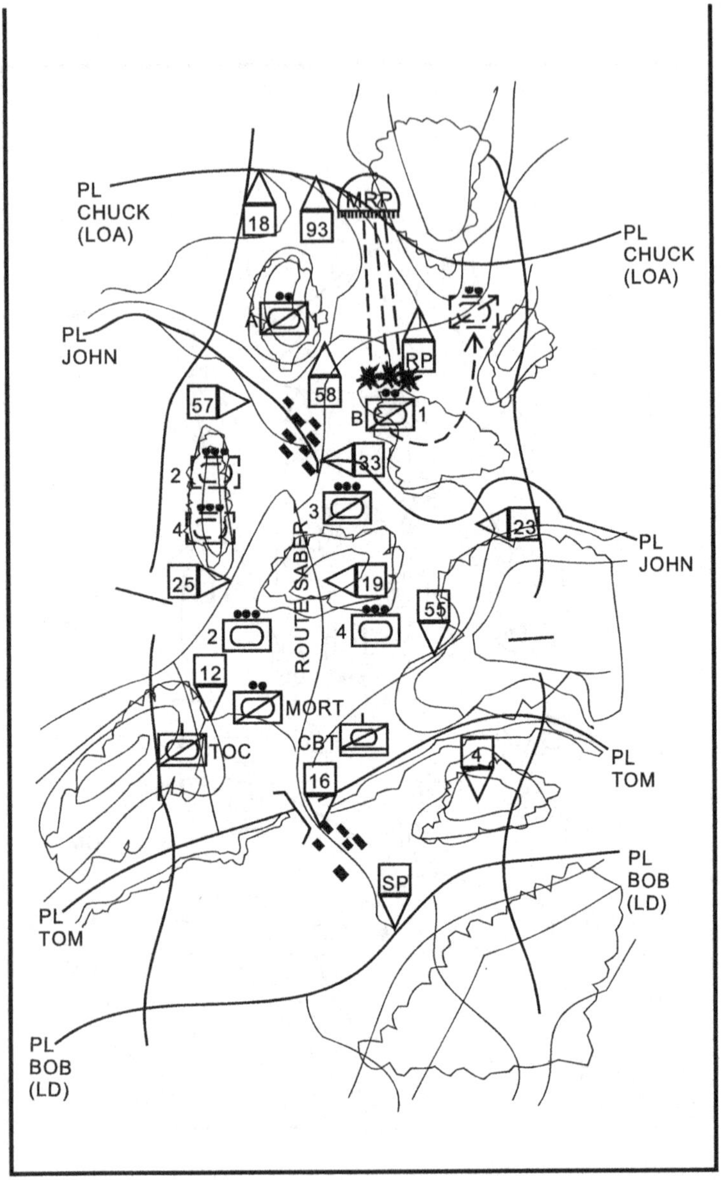

Figure 3-11. Route reconnaissance (part three).

what the platoon is up against and whether or not the threat force is supported by other elements in the village at checkpoint 18. He then reports to the troop commander. The troop commander immediately moves forward to join the 1st platoon leader and to see the situation for himself. En route, the commander tells the 2d and 4th platoons to move into concealed positions in the village at checkpoint 33, and to prepare to counterattack. The 3d platoon continues to reconnoiter the route through the village at checkpoint 33, and then moves into covered positions along ROUTE SABER just south of checkpoint 33. The troop CP holds in place and reports the situation to squadron. The 1SG quickly leads the M88 and the medic vehicle forward to treat and evacuate the injured scouts and their damaged vehicle.

Dismounted scouts from the 1st platoon determine that the threat force consists of three BMPs and one T-64 in prepared positions. Threat crews remain mounted and continue to slew turrets, scanning for more friendly elements. The threat reinforced platoon is oriented southeast toward the route. Their western flank is free of obstacles and unobserved. The troop commander decides to conduct a hasty attack (see Figure 3-12).

When the tank platoons complete their hasty attack, the troop commander gets permission to advance beyond the LOA and reconnoiter the enemy position. The 3d platoon returns to complete the reconnaissance of ROUTE SABER. As the platoon checks for mines along the route to the RP, they discover the route is mined just short of the RP. They find a bypass around the obstacle and forward a bypass report (Blue 10) to the troop XO. The platoon then organizes into two sections and moves out to occupy OPs overlooking the lateral routes at the east and west flanks of the troop near PL CHUCK. The troop CP moves into position on high ground near checkpoint 58. The 1SG directs activities of the troop trains, to include evacuation of injured soldiers and inoperable vehicles, security and evacuation of injured prisoners, requests for casualty replacements and resupply.

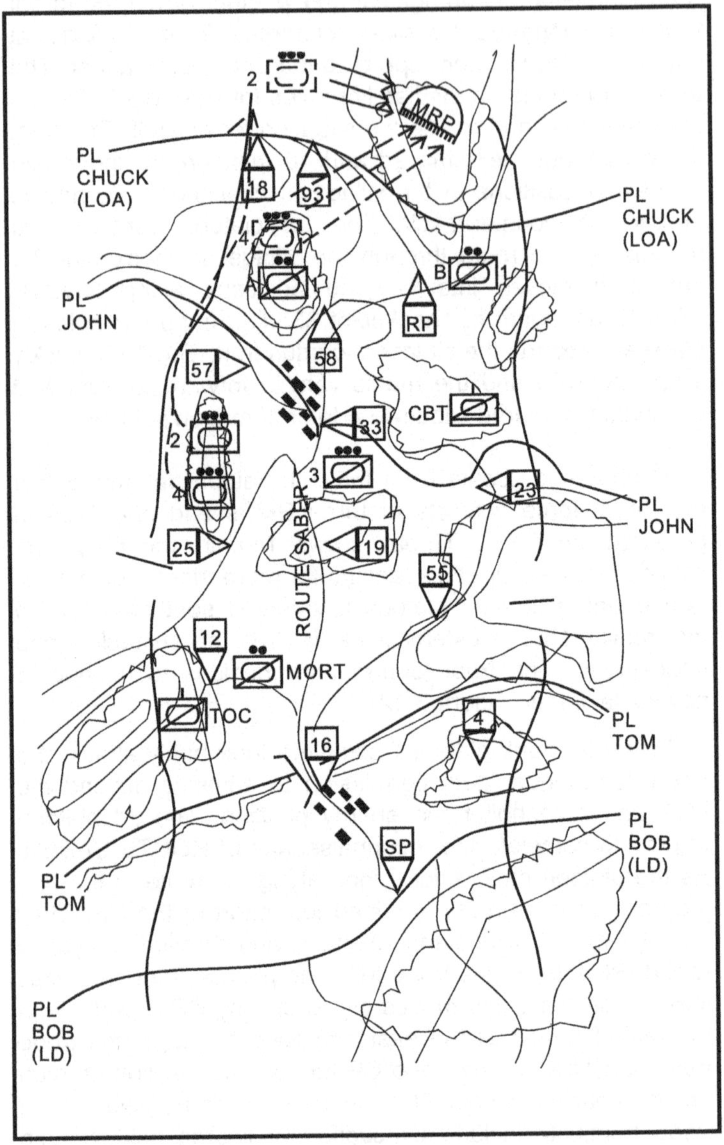

Figure 3-12. Route reconnaissance (part four).

Section IV. Zone Reconnaissance

Zone reconnaissance is the directed effort to obtain detailed information concerning all routes, obstacles, terrain, and enemy forces within a zone defined by boundaries. Obstacles include chemical and radiological contamination. A zone reconnaissance is assigned when the enemy situation is vague or when information concerning cross-country trafficability is desired. It is appropriate when previous knowledge of the terrain is limited or when combat operations have altered the terrain.

CRITICAL TASKS

Zone reconnaissance is a deliberate, time-consuming process. During a zone reconnaissance, the troop accomplishes the following critical tasks unless specifically directed otherwise by the higher commander:

- Reconnoiter all terrain within the zone.
- Inspect and classify all bridges within the zone.
- Locate fords or crossing sites near all bridges in the zone.
- Inspect and classify all overpasses, underpasses, and culverts.
- Locate and clear all mines, obstacles, and barriers in the zone within its capability.
- Locate a bypass around BUAs, obstacles, and contaminated areas.
- Find and report all enemy forces within the zone.
- Report reconnaissance information.

TECHNIQUES

The cavalry troop is usually assigned a zone reconnaissance as part of a squadron zone reconnaissance mission.

A zone reconnaissance mission is very time-consuming. Any commander who orders this mission must remember the number and complexity of the tasks to be accomplished. If he wants a faster tempo, he must prioritize reconnaissance tasks for the troop.

The cavalry troop can effectively reconnoiter a zone from 6 to 10 kilometers wide. If stretched any farther, the troop quickly exceeds its ability to accomplish the critical tasks.

Because the enemy situation is vague and knowledge of the terrain is limited to what is shown on maps or aerial photos, platoons will face unexpected situations around every corner; therefore, the course of action selected must also provide a good measure of protection for the troop as it executes the mission. When considering techniques of conducting a zone reconnaissance, the scheme of maneuver has to be flexible. The troop commander must convey his intent to subordinates so they can act quickly and without orders. He should employ forces, tank or AT platoons, in depth to give the troop flexibility and responsiveness.

When the troop receives a zone reconnaissance mission, the zone is usually identified by lateral boundaries. The line of departure and a reconnaissance objective or LOA are specified. The commander should divide the troop zone into two platoon zones for the scout platoons. Use caution when drawing the boundary. Make sure it is on easily identifiable terrain that is not a high speed avenue of approach. The number of critical tasks within the troop zone, and terrain restrictions should guide the commander in determining the location of the scout platoon boundaries. Doctrinal distances are not always the best solution if one platoon will be overtasked.

Add phase lines every 5 to 8 kilometers on easily identifiable terrain to control progress of the troop through the zone. Place contact points near the intersection of the boundary and all phase lines. Ensure the contact points provide at a minimum the potential for concealment from enemy observation to allow the exchange of information between flank units. Use checkpoints to focus the efforts of scout platoons and to assist in maneuvering the troop (see Figure 3-13). Use TIRS on maps as described in Chapter 2.

Use a troop vee or split-vee to conduct zone reconnaissance. As scout platoons reconnoiter the zone, the tank/AT platoons follow and support behind the scouts, keying their movement off the scouts' forward progress. The tank/AT platoons are kept far enough in depth to retain flexibility, and yet remain responsive to situations developed by the scout platoons. The distance from the scouts is determined by the terrain and enemy situation. Therefore, the distance the tanks/ATs move from the scouts may vary during the course of the operation. Control tank/AT platoon movement by one of the following methods:

- Give movement guidance to the platoon leaders. Have them key movement off the scouts.
- Move the platoons by CPs, battle positions, and hide positions.

Move the mortar section center of the troop zone to range forward of the scouts. The range forward of the scouts is determined by the scouts' ability to acquire the enemy. The mortars may move under the control of the FIST or eavesdrop on the troop net and key their movement off the progress of the scout platoons. Regardless of who controls their movement, the TOC and the FIST track the mortar position constantly to ensure they can support the scouts.

Note. If the mortar section is busy firing missions and computing fire data, they may not be able to stay abreast of the friendly situation. Therefore, the FIST or XO should be ready to control the mortars displacement.

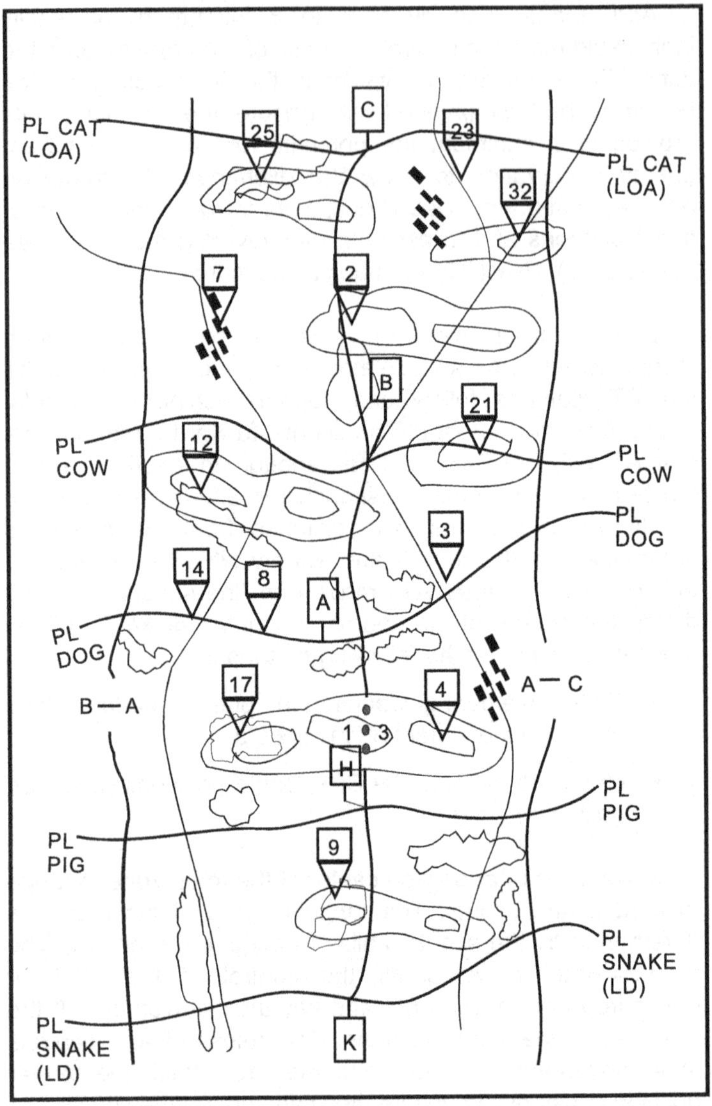

Figure 3-13. Zone reconnaissance graphic control measures.

The XO displaces the troop CP through the zone using terrain that affords effective and continuous communications with troop elements and the squadron. The 1SG moves the troop trains about 2 kilometers behind the tank platoons and bounds the trains from one covered and concealed position to another. The troop commander positions himself well forward to observe the actions of one scout platoon or the other. His position usually depends on where he expects initial enemy contact or problems to occur.

LIGHT TROOP EXAMPLE

The troop commander deploys the troop in a split-vee formation. The 1st and 3d platoons (scout) cross the LD, PL UTAH, abreast and reconnoiter the terrain in troop zone up to PL OHIO. The platoons determine both the open terrain and the trafficability of the route within their zone. Some hill slopes are greater than thirty percent and dense woods inhibit armor movement. Except for a marshy area north of checkpoint 9, the open terrain is dry and firm. Both routes will support heavy armor movement. The troop commander follows the 1st platoon. The mortar section eavesdrops on the troop command net to follow the progress of the scout platoons, and coordinates its moves with the troop FSO accordingly. As the scout platoons cross PL OHIO, the 2d and 4th platoons (antitank) cross the LD. They eavesdrop on the troop command net and key their movement on the progress of the scout platoons. The troop CP and the trains hold in place.

The scout platoons continue reconnaissance up to the river (PL BAMA). They take a close look at the dominant terrain near checkpoint 17 and at the village near checkpoint 4. As the platoons approach the river, scouts move into covered and concealed positions, dismount, and visually search the dominant terrain on the north side of the river. The 1st platoon inspects the bridge at checkpoint 14 and determines that it will support up to 40 tons. Scouts also verify that a good fording site exists near checkpoint 8. The 3d platoon confirms there is a fording site with a rock bottom just east of the blown bridge at checkpoint 3. It will support heavy armored traffic (see Figure 3-14).

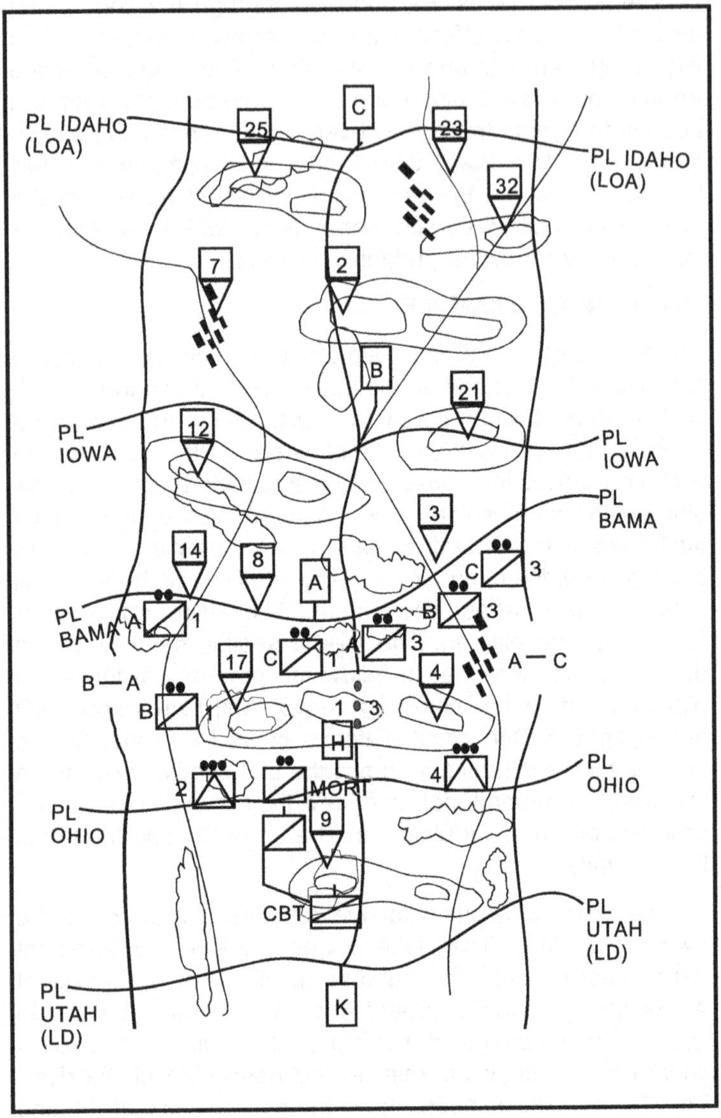

Figure 3-14. Zone reconnaissance (part one).

As the scout platoons continue reconnaissance towards PL IOWA, the 1st platoon observes two stationary BRDM-2s on the high ground near checkpoint 12. They look like a security outpost. The platoon engages with immediate suppression from the troop's 120-mm mortars. As the rounds impact, the outpost withdraws behind the ridgeline to the north. The 3d platoon determines the dominant terrain near checkpoint 21 is clear of threat forces. The mortar section crosses the river at checkpoint 8. The 2d platoon follows, crossing at checkpoint 8, while the 4th platoon uses the ford near checkpoint 3. The troop CP bounds forward to high ground near checkpoint 17. The troop trains cross the river and move into the woods on the northern bank.

The scout platoons continue reconnaissance while moving toward the troop LOA. As the 3d platoon crosses the ridge (PL IOWA), Bravo section receives heavy machine-gun fire from concealed positions south of checkpoint 2. The platoon immediately deploys, reports contact to the troop commander, and calls for indirect fire to suppress the threat (see Figure 3-15).

Under this protection, the platoon, using primarily a dismounted technique, reconnoiters to the flanks and rear to develop the situation. The troop commander acknowledges the report, and moves immediately to link up with the 3d platoon leader. The troop commander orders the 1st platoon to continue its reconnaissance to locate the enemy's flank and coordinate with the FIST to switch indirect fires to the squadron howitzer battery. The 3d platoon is organized in a three-section, three-vehicle configuration; the platoon leader is with the Charlie section. The platoon leader sends his Bravo section forward on the enemy's left and his Alpha section on the right to determine if the enemy is mutually supported by other threat forces from the flanks or rear. The Charlie section sergeant and a two-man dismount team move in closer and determine that the enemy consists of an MRP with three BTR 70s and one BMP-2. The Alpha scout section and scouts from 1st platoon identify the right flank of the threat and discover the flank is unprotected by obstacles and is exposed to direct fires.

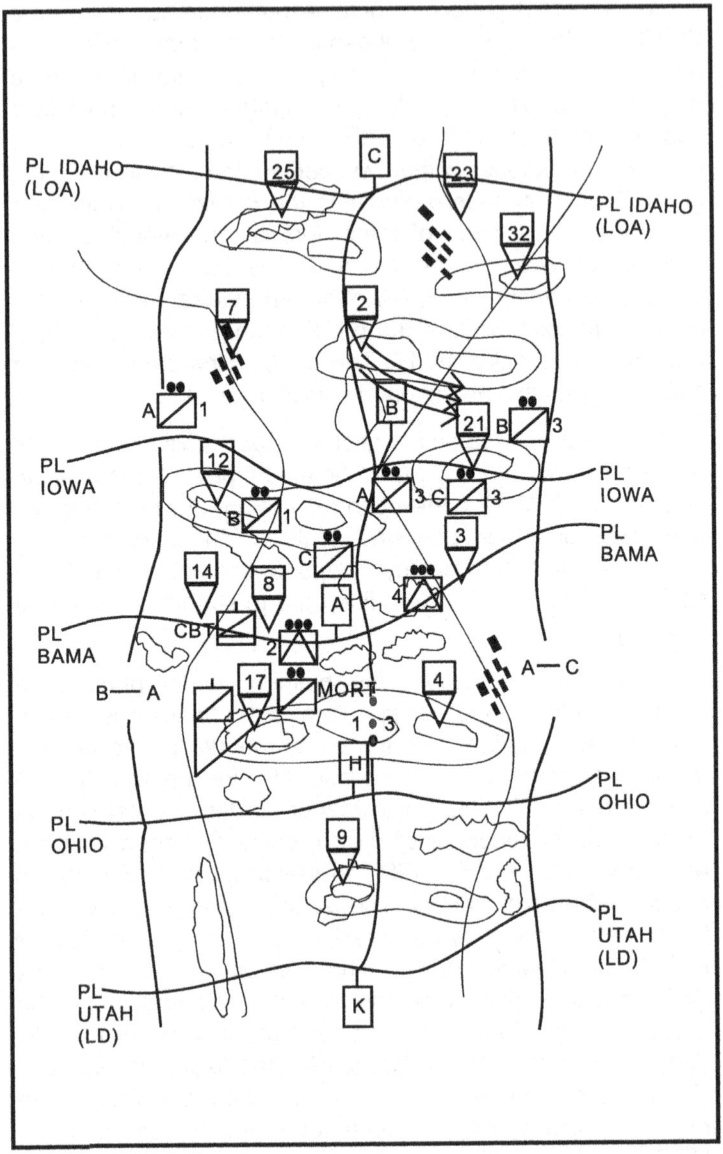

Figure 3-15. Zone reconnaissance (part two).

While en route to the 3d platoon's zone, the troop commander tells the 1st platoon to continue reconnaissance forward to PL IDAHO and determine if other enemy forces are providing mutual support to the enemy in the 3d platoon's zone. At PL IDAHO, he tells the platoon to establish a screen. The troop commander tells the 4th platoon to move over and join the 3d platoon Charlie section in overwatch of the enemy MRP, and prepare to overwatch with fires. The troop commander tells 2d platoon to prepare to conduct a hasty attack of the enemy MRP. The 1st platoon scout squad moves to checkpoint 12, links up with the 2d platoon, and guides them into position to attack the threat platoon.

The 2d platoon reports they are set in the assault position. The 4th platoon reports they are set in the overwatch position with the 3d platoon Charlie section. The 3d platoon leader reports that his dismount teams from his Alpha and Bravo sections are out of the target area and set. The 1st platoon leader reports he is set in a screen along PL IDAHO. The troop commander orders the FSO to fire for effect on the enemy position at his command. The commander issues orders and guidance for limits of fire and signals for the initiation, shifting, and lifting of fires. The troop commander issues the fire command for indirect fire over the command net and follows with a fire command for direct fire for the 4th platoon and the Charlie section of 3d platoon. The indirect fire from the squadron's howitzer battery and the troop mortars impact on the enemy position and is immediately followed by the fire from four TOWs, MK-19s, and M2 machine guns. The commander then issues the orders for 2d platoon to initiate its move to the attack-by-fire position followed by the order for the support element (4th platoon, Charlie section of 3d platoon) to shift fires. The troop FSO shifts fires to the rear of the enemy position. The 2d platoon moves forward and engages the enemy target from the flank to complete the destruction of the enemy force (see Figure 3-16).

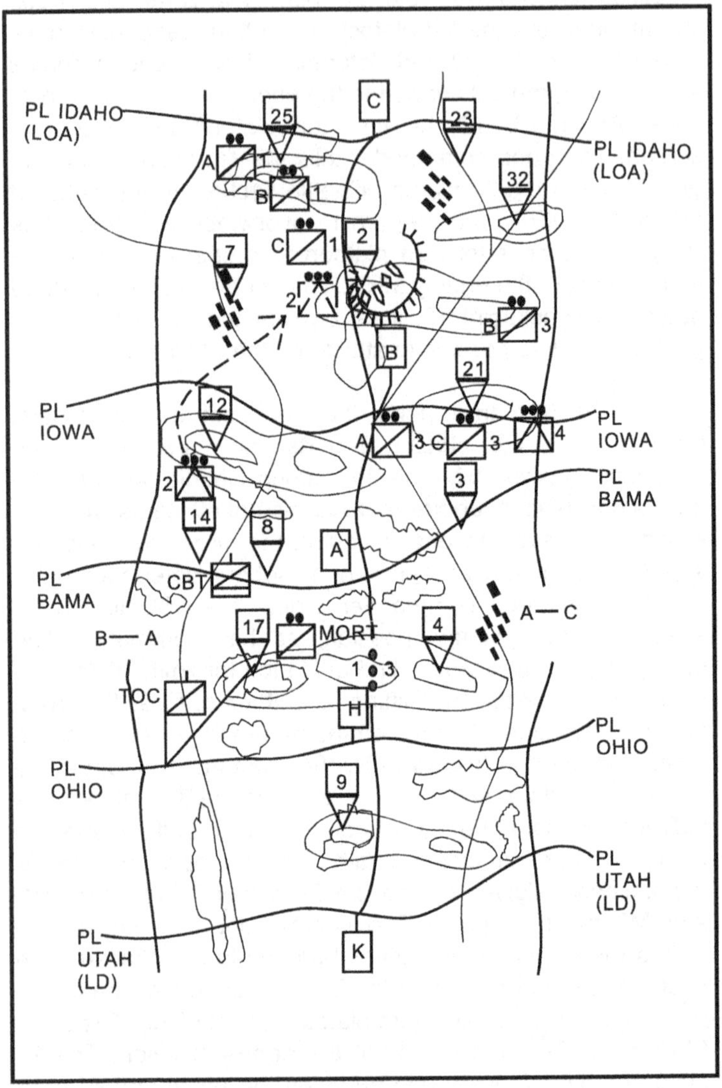

Figure 3-16. Zone reconnaissance (part three).

The antitank platoons consolidate in covered and concealed positions west of checkpoint 2. Scout dismount teams from the 3d platoon move in quickly, capture two wounded prisoners, and search the vehicles and personnel. The 3d platoon leader orders one scout squad to stay and secure the prisoners, and then completes his reconnaissance up to the LOA. The 1SG leads the medics to the location of the 3d platoon scout squad and the EPWs. The XO collects and transmits final reconnaissance reports to the squadron.

AIR-GROUND COORDINATION

Cavalry troops will often conduct a coordinated reconnaissance with an air cavalry troop. Because of the air cavalry troop's ability to maneuver faster and look deeper than ground cavalry, they are often placed to the flanks or forward of ground cavalry. The air cavalry troop should focus its reconnaissance efforts on areas that impede ground cavalry movement.

The air cavalry troop quickens the pace or tempo of the ground reconnaissance because it provides the ground unit with added security during movement and increased situational awareness once contact is made. The air cavalry troop can gain initial contact with the enemy or a critical piece of terrain. Once the air troop develops the situation with his assets, he can pass off the information/contact to the ground troop (see Figure 3-17).

- Ground troop and air troop coordinate zone reconnaissance operations.
- Air troop divides the zone into platoon zones.
- Air troop will use checkpoints to identify critical areas. Checkpoints should enhance ground troop movement throughout the zone.
- Platoon on the right reconnoiters forward of ground troop, identifies platoon ford site, and conducts route reconnaissance.
- Platoon on the left reconnoiters forward of ground troop, supports the other platoon in conducting route sweep, checks lateral routes in the zone, and looks for sites along the obstacle (river/road).
- ACT commander and ground troop commander identify critical areas where air-ground coordination is required or needed.
- ACT commander positions himself close to the ground troop commander (if needed).

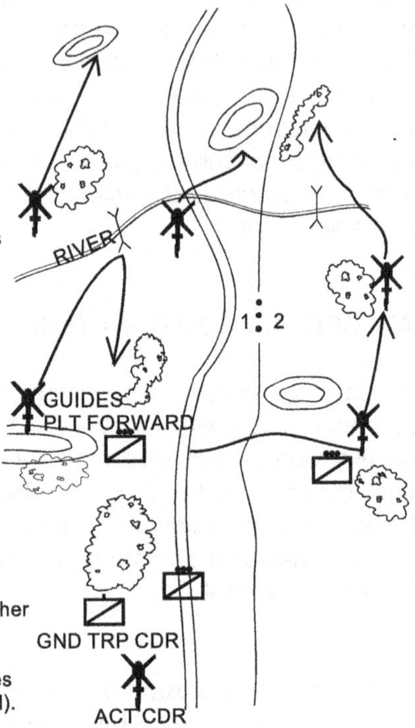

- If the ford site or bridge site is blocked, ACT elements reconnoiter the site forward of the location. They also find a bypass to the obstacle and lead ground troops to bypass the location. The ACT will start the screen forward of ground elements.

Figure 3-17. Air-Ground troop coordinated reconnaissance.

Section V. Area Reconnaissance

An area reconnaissance is a specialized form of zone reconnaissance. It is a mission conducted to gain detailed information about terrain features and enemy forces within a specified area or point that other forces intend to occupy, pass through, or avoid. A commander usually calls for area reconnaissance before he sends his forces into or near an area to avoid being surprised by actual terrain conditions or unexpected enemy forces.

CRITICAL TASKS

During an area reconnaissance, the following critical tasks must be accomplished unless the SCO directs the troop to do otherwise:

- Reconnoiter all terrain within the area.
- Inspect and classify all bridges within the area.
- Locate fords or crossing sites near all bridges within the area.
- Inspect and classify all overpasses, underpasses, and culverts.
- Locate and clear all mines, obstacles, and barriers in the area within its capability.
- Locate a bypass around BUAs, obstacles, and contaminated areas.
- Find and report all enemy within the area.
- Report reconnaissance information.

TECHNIQUES

An area reconnaissance is conducted like a zone reconnaissance. When the troop receives an area reconnaissance mission, the assigned area is identified as the terrain inside a solid, continuous boundary. Planning the movement to the area is the first step. Select the route(s), establish a march order on each route, and specify a start point, checkpoints, and a release point. Use a movement technique that keeps the troop moving quickly and securely. If possible, avoid contact with the enemy while en route. Report and bypass. Be careful when approaching the area to begin the reconnaissance. Use common sense. Avoid known enemy forces outside of the area where reconnaissance will be conducted; start somewhere else.

As in a zone reconnaissance, enclose the area within a troop zone. Draw a line of departure, an LOA, and lateral boundaries. Divide the troop zone into two platoon zones. Add phase lines along identifiable terrain to control movement through the area. Place contact points at the intersection of the platoon boundary and all phase lines. Place TIRS on the map. To identify specific areas or features, use checkpoints for reference (see Figure 3-18).

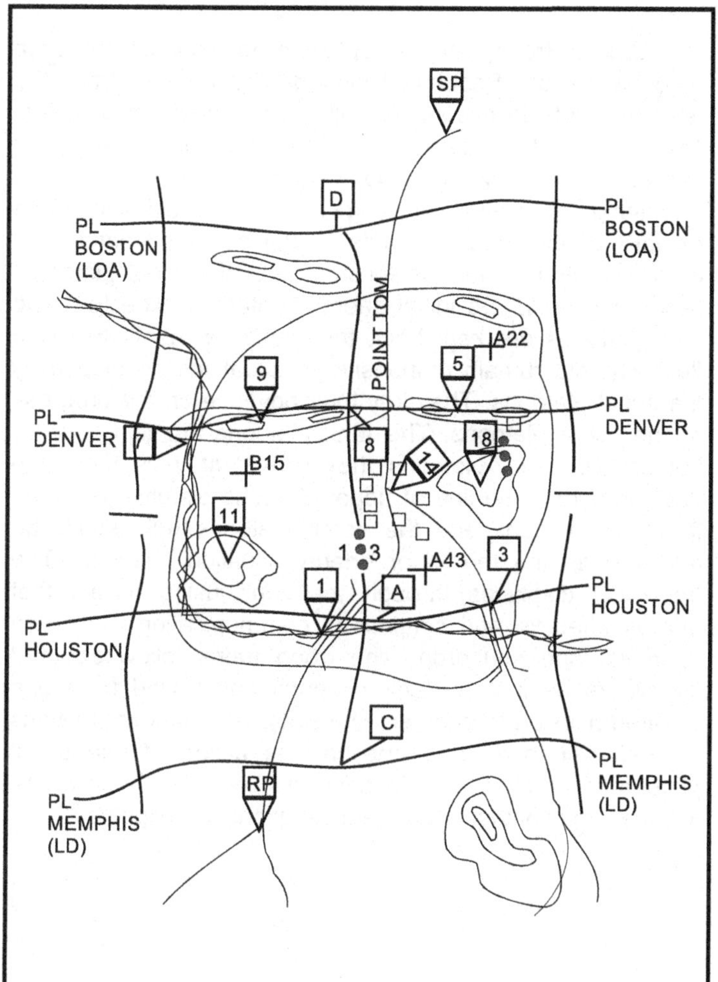

Figure 3-18. Area reconnaissance.

Use a troop vee or split-vee to conduct an area reconnaissance. Scout platoons deploy abreast from the release points to accomplish all the reconnaissance tasks. They move out across the line of departure first. The mortar section moves through the center of the troop zone, remaining in position to range from 3 to 3.5 kilometers forward of the lead elements in the scout platoons. The mortar section could also follow one of the scout platoons, depending on where initial enemy contact is expected. The tank platoons are kept back far enough in depth to retain flexibility, but remain responsive to situations developed by the scout platoons. They key their advance on the progress of the scout platoons. The tank platoons are told how far behind the scout platoons they should stay as the troop deploys across the line of departure. The distance between the tank platoons and the scout platoons will usually be adjusted as the terrain or enemy situation changes. The troop CP displaces through the zone using terrain that affords effective and continuous communication with troop elements and squadron. The troop trains follow about 2 kilometers behind the tank platoons and bound from one covered position to another. The troop commander positions himself well forward to observe the action of one scout platoon or the other. The location usually depends on where initial enemy contact or problem situations are expected.

Chapter 4

Security

Security is an essential part of all offensive and defensive operations. Cavalry provides security for the commander along an exposed front, flank, or rear of the main body where a threat may exist. Surveillance is continuous during security operations. Even during security missions that involve fighting the enemy, the scouts' primary task remains gathering information. Scouts do this by establishing OPs, conducting patrols, and performing reconnaissance.

Counterreconnaissance is an inherent task in all security operations. Counterreconnaissance is not a mission. It is the sum of actions taken at all echelons to counter enemy reconnaissance and surveillance efforts through the depth of the area of operations. Counterreconnaissance denies the enemy information about friendly units. It is both active and passive and includes combat action to destroy or repel enemy reconnaissance elements.

CONTENTS

		Page
Section I.	Purpose, Fundamentals, and Capabilities	4-2
Section II.	Screen	4-5
Section III.	Guard	4-31
Section IV.	Cover	4-32
Section V.	Route Security	4-33
Section VI.	Area Security	4-33
Section VII.	Convoy Security	4-34

Section I. Purpose, Fundamentals, and Capabilities

Security includes screening operations, guard operations, covering force operations, area security operations, convoy security operations, and route security operations. The squadron performs screen, guard, and route security missions. Covering force operations are normally the mission of a cavalry regiment. Separate brigades or task organized divisional brigades may perform cover operations as well.

Both the heavy and light cavalry troops perform two security missions (screen and convoy security) independently or as part of their parent squadron. Both also participate in guard, cover, and route security missions as part of their parent squadron or regiment. Either troop will normally perform reconnaissance, screen, defend, delay, attack, or a combination of these missions in support of their parent squadron or regiment (see the applicable section of Chapters 3, 5, and 6 for specifics).

PURPOSE

Security operations are designed to obtain information about the enemy and to provide reaction time, maneuver space, and protection to the main body. Security operations are characterized by conducting continuous reconnaissance to reduce terrain and enemy unknowns, gaining and maintaining contact with the enemy to ensure continuous information, and providing early and accurate reporting of information to the protected force.

FUNDAMENTALS

Five fundamentals are common to all cavalry security operations.

Orient on the Main Body. As a security force, the troop will be operating at a specified distance from a main body, between it and a known or suspected enemy force. If the main body moves, the troop also moves. The troop commander must know how the main body commander intends to maneuver his forces and where he wants the troop in relation to his movement. The troop commander maneuvers his troop to positions where he can provide the needed security.

Perform Continuous Reconnaissance. The troop's security and the security of the main body come in large measure from knowing everything about the terrain and the enemy within the troop's area of operations (AO). Reconnaissance and continuous patrolling go hand in hand with security operations. Cover all the ground in the AO. Determine what the terrain will allow the troop *and* the enemy to do.

Provide Early and Accurate Warning. Early and accurate warning of enemy approach is the cornerstone of security operations. The main body commander needs as much time as possible to shift and concentrate his forces to meet and defeat an unexpected enemy attack. Put observers in positions that afford long-range observation of expected enemy avenues of approach. Use aeroscouts and ground surveillance radar (GSR), if available, to enhance their ability to see. Place remote sensors in the ground to monitor avenues of approach that cannot be easily observed. If possible, send dismounted or mounted patrols forward of OPs to extend their ability to see, providing additional reaction time for the main body commander.

Provide Reaction Time and Maneuver Space. All security operations are designed to provide reaction time and maneuver space for the main body so it can deal effectively with an unexpected enemy attack. The troop provides early warning to the main body commander and sometimes has to fight hard to buy time and space so the main body's combat power can be concentrated on defeating the enemy.

Maintain Enemy Contact. Once visual or physical contact with the enemy is gained, do not allow him to break the contact. Maintain enemy contact and continue to report his activities until told to stop.

CAPABILITIES

Capabilities of the heavy troop include—

- Screen up to a ten-kilometer-wide sector.
- Maintain continuous surveillance of up to six battalion-size avenues of approach.*
- Can establish up to 12 short-duration OPs.

Capabilities of the light troop include—

- Screen up to a ten-kilometer-wide sector.
- Maintain continuous surveillance of up to six battalion-size avenues of approach.*
- Can establish up to 16 short-duration OPs.

*The maximum six long-duration OPs either cavalry troop can occupy is a function of personnel required to perform the following tasks at each OP:

- Man the actual OP.
- Maintain radio communications with the OP and with the platoon leader.
- Provide local security for the vehicles.
- Conduct dismounted patrols as required.
- Conduct resupply.
- Perform maintenance.
- Sleep/rest.

The ability to perform the above tasks simultaneously for periods in excess of 12 hours requires at least 9 to 10 personnel (collocation of two scout squads for M3-mounted scouts, or three scout squads for HMMWV-mounted scouts).

Section II. Screen

Screen is the most common security mission heavy and light cavalry troops conduct. Both troops conduct screen missions for their parent squadrons or other forces to—

- Provide early warning of enemy approach.
- Provide real-time information, reaction time, and maneuver space to the protected force.
- Destroy enemy reconnaissance elements within their capability (perform counterreconnaissance).
- Impede and harass the enemy.

The screen mission provides the least amount of protection of any security mission, and is appropriate when operations have created extended flanks, when gaps between forces exist and cannot be secured in force, or when required to provide early warning over gaps that are not considered critical enough to require security in greater strength. A commander normally assigns cavalry this mission when he needs time to respond to an unexpected enemy attack, and cannot afford to commit other forces to the task.

The screen mission is defensive in nature. As such, both heavy and light cavalry troops screen the front, flanks, and rear of a stationary force but only to the flanks or rear of a moving force. Screening operations are not performed forward of a moving force. Zone reconnaissance or movement to contact is the appropriate cavalry troop mission suited to the requirements of the offensive force.

CRITICAL TASKS

A screen mission has certain critical tasks that guide planning. To achieve the intent of a screen mission, the troop must accomplish the following critical tasks:

- Maintain continuous surveillance of all battalion-size avenues of approach into the troop sector under all visibility conditions.

- Destroy or repel all enemy reconnaissance patrols (if within the unit's capability and directed by the higher commander).

- Locate lead elements of enemy order of battle and determine the direction of movement of each.

- Maintain contact with, report the activity of, and impede and harass the enemy while displacing.

EXTENDED SCREENS

Both the heavy cavalry troop and the light cavalry troop normally screen a sector up to ten kilometers in width. However, METT-T may dictate an extended screen across frontages in excess of the norm. Either troop's ability to accomplish its critical tasks, or its ability to screen in depth, can diminish rapidly as frontages increase.

Examples of extended screens are described below.

- Heavy cavalry troop screens 20 kilometers of southern bank of unfordable river crossed by four bridges in sector.

- Light cavalry troop screens 25 kilometers of desert terrain, from dominant ridge.

DEPTH

Depth is also important in a screen. The term *"screen line"* is descriptive only of the forward trace along which security is provided. Depth allows an enemy contact to be passed from one element to another without requiring displacement. Depth is advantageous to—

- Destroy an enemy reconnaissance patrol without compromising critical OPs.
- Prevent an enemy from penetrating the screen line too easily.
- Prevent gaps from occurring when OPs displace or are lost.
- Maintain contact with moving enemy without compromising OPs.
- Prevent enemy templating of the screen line.

Depth is achieved primarily by positioning OPs, particularly where there are limited avenues of approach. Tank platoons, AT platoons, the mortar section, and attached elements positioned behind the screen line establish local security and provide surveillance. The degree to which depth can be attained is a function of many factors, which include—

- Higher commander's intent and concept as expressed in—
  - Graphical trace of the screen line (LOA).
  - Engagement criteria.
  - Destruction criteria.
  - Displacement/disengagement criteria.
- Width of the sector.
- Depth of the troop sector.

- Terrain and avenues of approach it will support.
- Attachments and detachments.

Screening is largely accomplished by establishing a series of OPs and conducting patrols to ensure adequate surveillance of the assigned sector. Screens are active operations. Stationary OPs are only one part of the mission. Employing patrols (mounted and dismounted), aerial reconnaissance, ground-based sensors, intelligence from space-based sensor systems, and OPs relocated on an extended screen ensure that continuous overlapping surveillance occurs. Inactivity in an immobile screen promotes complacency.

PLANNING CONSIDERATIONS/GUIDANCE TO SUBORDINATES

The enemy situation is often vague when planning a screen. The troop should develop plans that are flexible enough to react to any enemy course of action, particularly the worst case. Planning considerations for a screen should include a detailed description of how contact with the enemy reconnaissance will be gained then how and where it will be destroyed. Planning should also cover the method of displacement once the main body of the enemy force has been identified and how that force will be handed off to the main body in the main defensive belt. Because of the need for flexibility, screen operations will often begin to inherit the characteristics of defense or delay missions. To cover the displacement of scouts, some elements of the troop may be required to execute missions such as delay or defend.

Screen operations at troop level usually occur in four phases.

- Movement to and occupation of the screen line.
- Surveillance and counterreconnaissance.

- Gaining and maintaining contact with the enemy main body displacement of screen.
- Rearward passage of lines.

Higher command guidance should address each phase of the operation and cover at least the following:

- Location/orientation/width of the screen.
- Duration of the screen.
- Method of movement to and occupation of the screen line.
- Location and disposition of the friendly force being screened.
- Engagement/destruction criteria.
- Displacement/disengagement criteria.
- Follow-on missions.
- Positioning and orientation guidance for GSRs (if attached).

The following must be considered when developing and completing the plan and executing the screen mission:

Time Screen Must Be Established. The time the screen must be set and active will influence the troop's method of deploying to and occupying the screen line.

Movement to the Screen Line. If the screen mission is the result of a previous tactical maneuver such as zone reconnaissance, the troop will essentially be postured to begin screening from present positions. This situation occurs frequently, and may be the result of a FRAGO to halt at a specified phase line.

If the troop is not currently sited on the screen line, obviously, deployment to the screen line must occur before actually beginning the screen mission. Time determines the method of occupying the screen line. Thorough analysis of METT-T will determine which deployment technique or combination of techniques best meets mission requirements.

Trace and Orientation of Screen Line. The initial screen line is depicted as a phase line and often represents the forward line of own troops (FLOT). As such, the screen line may be a restrictive control measure for movement (limit of advance); coordination/permission would be necessary to move beyond the line to establish OPs or to perform reconnaissance. When occupied, OPs are sited on or behind the phase line. OPs should be given specific orientation and observation guidance.

Initial OP Locations. The squadron or troop commander may determine tentative initial OP locations to ensure effective surveillance of the sector and designated named area(s) of interest (NAI). At a minimum, the troop commander designates a primary orientation for the scouts during the conduct of the screen. Scouts, once set on the screen line, will report their location to the troop TOC and verify they are in compliance with the commander's orientation guidance. The scouts who occupy each OP always retain the responsibility to modify the location to achieve the commander's intent and guidance for orientation. The OPs are positioned along or behind the screen line.

OPs are generally categorized based on their expected duration of employment as either—

- Short-duration (less than 12 hours)

  or

- Long-duration (more than 12 hours).

OPs may be either mounted or dismounted. Mounted OPs maximize use of vehicular optics, weapon systems, and speed of displacement, but are more readily detected by the enemy. Dismounted OPs provide maximum stealth at the expense of speed of displacement, and vehicle-mounted optics and weapons.

A heavy cavalry troop can occupy up to 12 short-duration OPs (one per scout squad, six per scout platoon). For extended periods of time, the heavy troop can occupy six OPs (one per scout section, three per scout platoon).

A light cavalry troop can occupy up to 16 short-duration OPs (one per scout squad, eight per scout platoon). For extended periods of time, the light troop can occupy six OPs (one per scout section, three per scout platoon).

Width and Depth of the Screened Sector. The troop sector is defined by lateral boundaries extending out to a limit of advance (the initial screen line), forward of a rear boundary. The troop sector is established by the squadron or unit being screened. The troop rear boundary may be a squadron phase line and may serve as a battle handover line (BHL) to control passing of responsibility for the enemy to the protected force. The troop's ability to gain depth decreases as screened frontage increases.

Locations of Subsequent Screen Lines. The squadron or troop commander uses additional phase lines to control the operation. These phase lines may serve as subsequent screen lines. Displacement to the subsequent screen lines is event driven.

Scout Platoon Sectors. Assign clear responsibility of identified avenues of approach and designated NAIs. The nature of a screen normally requires both scout platoons to deploy abreast.

Tank/AT Platoon Sectors. Position the tank (heavy troop) or AT (light troop) platoons in the scout platoons' sectors. They may occupy hide or battle positions along avenues of approach. The tank/AT platoons remain responsive to the troop commander. They are the primary direct-fire killing asset.

Force to be Screened. The troop must orient on the force it is securing. If the main body is moving, the troop must move to maintain the screen's position relative to the main body.

Reinforcements. Any unique requirement posed by the mission may require assets not organic to the troop. GSR and engineers are common attachments at troop level.

- **GSR.** During screen operations, GSR is used to augment scout OPs and to add depth to the screen. GSRs should be attached to scout platoons, and the commander should provide the scout platoon leader with positioning and orientation guidance.
- **Engineers.** If engineers are attached to the troop, the troop commander should assign them with priority of mission and priority of effort in support of commander's guidance. During screen operations, engineers will normally dig survivability positions for scouts and tanks. Also engineers may emplace obstacles in support of the counterreconnaissance battle or assist the troop with displacement of the screen once contact has been established with the enemy body.

Special Requirements or Constraints. Specify all requirements for observing any NAI identified during the intelligence preparation of the battlefield (IPB). Task subordinate platoons as required. Specify the following in the troop OPORD:

- Engagement criteria.
  - What size force will scouts engage and destroy (if any)?
  - Where will this action occur?
  - What size force will the tank or AT platoons engage and destroy?
  - Where will this action occur?

- Disengagement criteria.
  - What event will cause scouts to displace from the initial screen line?
  - How will scouts maintain contact with the enemy while displacing?
  - What event will cause the tank or AT platoons to displace to subsequent or alternate positions?

Indirect Fire Planning. Fire planning integrates artillery and mortar fires. Position the troop mortars to fire up to two-thirds of their maximum range forward of the initial screen line. A wide sector may require the troop commander to position them to provide effective coverage of the most likely avenue of approach determined by IPB. The troop FSO plans artillery fires to adequately cover any gaps in mortar coverage.

Direct Fire Planning. Based on his analysis of the terrain, the troop commander determines where to engage the enemy (engagement areas). He also determines the location of battle positions that provide observation, fields of fire, and cover and concealment that support each engagement area.

Positioning of C2, CS, and CSS Assets. The troop commander positions himself to observe the most dangerous enemy avenue of approach. The troop TOC positions itself in depth to provide continuous control and reporting during initial movements. After the screen line has been reestablished in depth following displacement from the initial screen line, the TOC can reposition. Combat trains position behind masking terrain close enough for rapid response. They are best sited along routes providing good mobility laterally and in depth.

Patrol Requirements. Patrols may be required to cover gaps between OPs. The troop commander tasks the scout platoon leaders to perform specific patrols.

Coordination. The troop commander coordinates his concept with air cavalry troop (ACT) commanders who may be operating the same ground, flank troop commanders, tank company (heavy cavalry) and AT company (light cavalry) commanders, and other unit commanders as appropriate.

TECHNIQUES

Graphics

The following is a list of common graphic control measures used to control screen missions:

- Boundaries.
- Phase lines.
- Checkpoints.
- Contact points.
- NAIs.
- OPs.
- Mortar firing positions.
- Battle positions.
- Hide positions.
- TIRS.
- Unit symbols.

Movement to the Screen Line

In deploying to the screen line, the troop commander must deal with the competing requirements to establish the screen quickly to meet mission requirements and to provide the necessary level of security for the troop in doing so. The troop moves to the screen line using one of three basic methods—a tactical road march, zone reconnaissance, or movement to contact.

Tactical Road March. The troop conducts a tactical road march to a release point behind the screen line. From the release point, platoons deploy to occupy initial positions. This method of deploying to the screen line is the fastest, but least secure. It is appropriate when enemy contact is not expected and time is critical, or when an air cavalry troop is conducting zone reconnaissance forward of the ground troop.

Movement to Contact. The troop conducts a movement to contact from a line of departure to the initial screen line. This method is slower than a tactical road march, but more secure. It is appropriate when enemy contact is likely, time is limited, or when an air cavalry troop is conducting zone reconnaissance forward of the ground troop.

Zone Reconnaissance. The troop conducts a zone reconnaissance from a line of departure to the initial screen line. Given adequate time, this method is preferred as the troop can clear the zone of any enemy and enables platoons to become thoroughly familiar with the terrain. The troop can reconnoiter potential subsequent OP locations, battle and hide positions, and mortar firing positions, for example, as they move to the screen line. A zone reconnaissance is appropriate when time is available and information about the enemy or terrain is unknown.

Security Drill

A security drill is a series of rehearsed actions (battle drills) a scout platoon or cavalry troop takes to maintain contact with the main body of an advancing enemy force. It is used when collapsing the screen line to subsequent OP positions or when transitioning from a screen mission to a delay or defend mission.

At platoon level, OPs gain contact with the enemy main body, then report and prepare to displace to a subsequent position. When the enemy force reaches the OP's break point (point where the OP must displace or his position/movement will compromise him to the enemy), the OP passes off the responsibility to track the enemy to another OP in depth. The platoon displaces its OPs to subsequent positions in depth while maintaining contact with the enemy.

At troop level, the security drill combines the collapse of the initial screen line with the actions of organic tank or AT platoons. Scout platoons may perform platoon security drills initially, consolidating some or all of their combat power at a battle position to aid execution of a troop-level engagement.

At platoon and troop levels, conduct of security drills is tempered by the commander's overall concept, intent, and scheme of maneuver. Enemy actions (events) drive security drill execution (response) (see Figure 4-1).

STATIONARY SCREEN

Take a close look at the high-speed avenues of approach into the sector. Divide the sector into two platoon sectors. Make sure the platoon boundary is on easily identifiable terrain. Do not split avenues of approach with a platoon boundary or place the boundary on a road. Place NAIs, TIRS, or checkpoints beyond the screen line to focus surveillance from OPs. If needed, add additional phase lines to control withdrawal of the troop at 5- to 8-kilometer intervals. Place contact points at the intersection of the platoon boundary and all phase lines. Place TIRS on the map or overlay as described in Chapter 2.

Deploy the scout platoons abreast and establish a series of OPs along or behind the initial screen line, as terrain allows, but never forward of it without permission. Make it clear to the scout platoon leaders which avenues of approach (depicted as NAIs or checkpoints) they are to observe. If unobservable areas between OPs need to be routinely checked, have the scout platoon leaders prepare patrol plans for approval and subsequent execution.

Position the mortar section to fire 3 to 4 kilometers forward of the initial screen line, oriented on the expected enemy avenue of approach. Establish subsequent firing positions for the mortar section back through the sector. Position each tank (heavy) or AT (light) platoon in one of the scout platoon sectors, or have the tank/AT platoons consolidate. Position the tank/AT platoons in hide positions or battle positions in depth behind the scout platoons, oriented on the expected enemy avenue of approach. Establish subsequent positions for the tank/AT platoons back through the sector to support the scheme of maneuver.

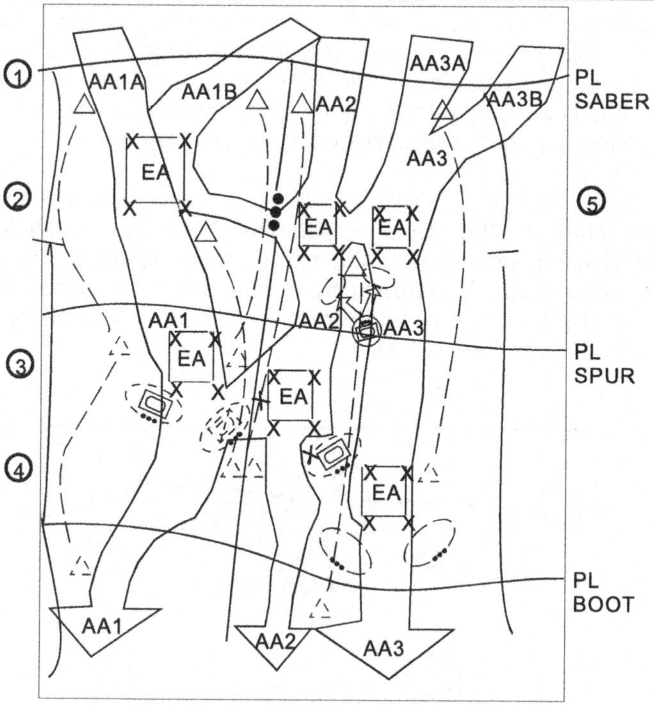

INDEX:

1. Initial screen line.
2. Scout platoon engagement area dependent on commander's intent.
3. Subsequent OPs.
4. Troop consolidates fire power of tanks and scouts to execute engagement while maintaining contact.
5. Scouts maintain contact and execute engagement with tank platoon.

Figure 4-1. Scout platoon/cavalry troop security drill.

Have the XO position the command post on terrain that affords good FM radio communications with troop elements and squadron headquarters. If possible, the command post should be positioned behind the subsequent screen positions. This allows the TOC to remain in position during the initial collapse of the screen line. Establish tentative subsequent command post sites back through the troop sector. The XO retains the authority to adjust the actual TOC location to maintain effective communications.

The 1SG first positions the troop trains within 5 to 8 kilometers of the initial screen line, then he establishes subsequent locations for the trains to bound back through the troop sector. The troop commander positions himself well forward where he can best observe and control the actions of the troop (see Figure 4-2).

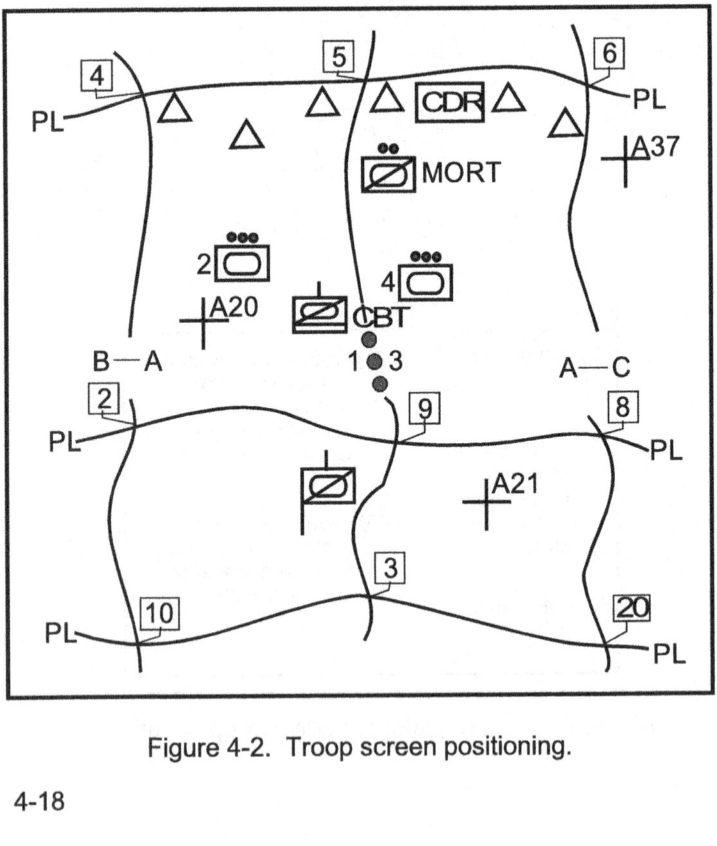

Figure 4-2. Troop screen positioning.

MOVING FLANK SCREEN

The troop may conduct a moving flank screen by itself or as part of the squadron. It may be tasked to screen the exposed flank of the squadron while the squadron conducts a movement to contact, a hasty attack, or a zone reconnaissance (see Figure 4-3). The troop may also participate in a squadron mission to screen or guard another combined arms force (see Figure 4-4).

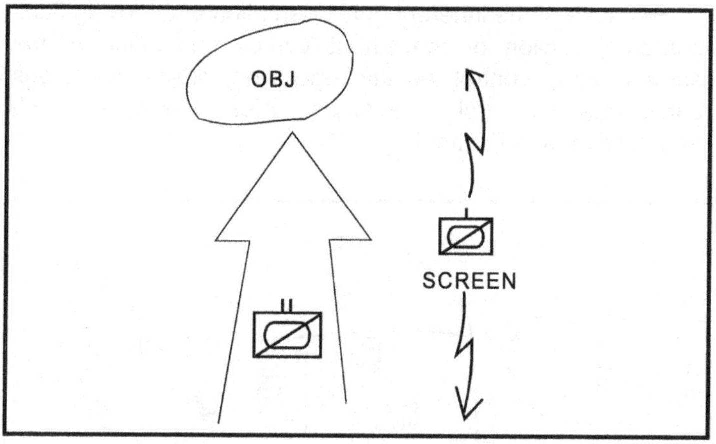

Figure 4-3. Squadron movement to contact; troop flank screen.

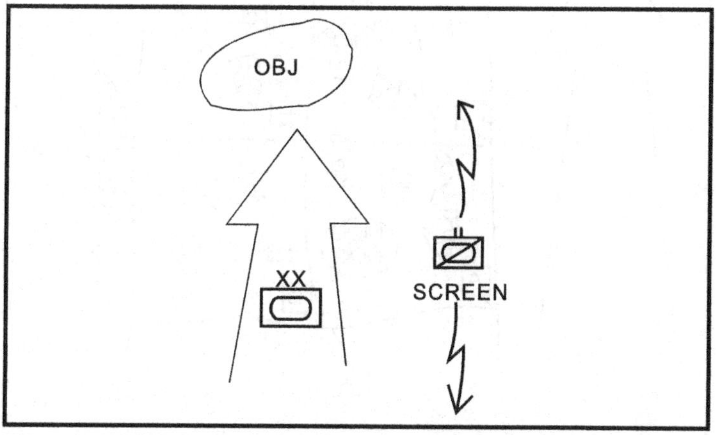

Figure 4-4. Squadron flank screen.

A moving flank screen uses the same techniques as when screening a stationary unit. Position the scout platoons to maintain continuous surveillance on the avenues of approach, the mortar section to cover likely avenues of approach with indirect fire, and the tank/AT platoons in depth to destroy or repel enemy reconnaissance units.

Applying Graphic Control Measures

Because of the inherent dual orientation of a moving flank screen (direction of movement versus orientation of the screen line), control of the operation poses numerous challenges. Control measures must facilitate both orientations (see Figure 4-5).

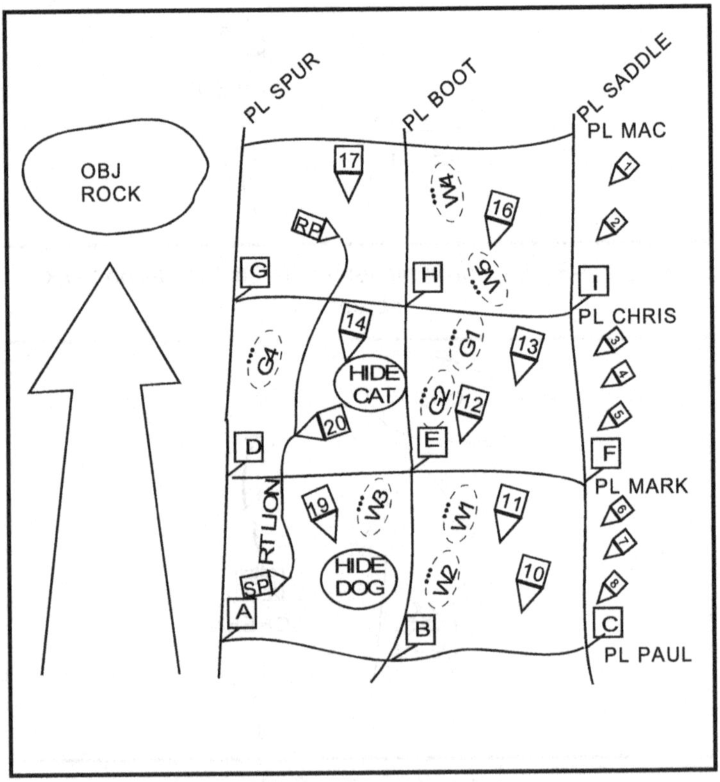

Figure 4-5. Troop moving flank screen graphics.

Follow the procedures below when applying graphic control measures to a moving flank screen.

- Use phase lines to control the scout platoon's movement (placed perpendicular to the screen line). Plan to use these phase lines as on-order boundaries for subordinate platoons if enemy contact is gained. Place phase lines no more than 5 kilometers apart (corresponding to the width of a scout platoon's screen frontage). Do not divide avenues of approach with them.

- Use additional phase lines rearward of (parallel to) the initial screen line to control retrograde movement (toward the protected force). Plan to use these phase lines as subsequent (on-order) screen lines.

- Use objectives, checkpoints, or axes of advance to control the movement of the tank/AT platoons for movement. Position and plan to use these objectives or checkpoints as battle positions for the tank/AT platoons if the scouts make enemy contact at the screen line. Plan subsequent (on-order) battle positions between the screen line and the protected force.

- Use mortar firing positions or checkpoints to control movement of the mortar section. Position these firing positions rearward of the screen line where they allow the mortars to fire two-thirds maximum range forward or cover likely avenues of approach. Plan subsequent (on-order) mortar firing positions between the screen line and the protected force.

While this number of graphic control measures and required planning may seem excessive, they provide maximum flexibility in terms of mission execution. The troop commander can issue simple FRAGOs to adjust the plan to the enemy situation.

Repositioning the Screen

The troop must reposition to stay oriented on the force it is securing. Movement along the screen line is determined by the speed of the protected force. Movement is conducted by one of three techniques—continuous marching, bounding by platoons (alternately or successively), and bounding by OPs (alternately or successively).

Continuous Marching. This technique is appropriate when the protected force is moving quickly and contact is not likely. It is the least secure movement technique.

Deploy both scout platoons abreast with the two tank or AT platoons, the mortar section, and the remainder of the troop in depth (between the screen line and the protected force). The trace of the screen line is essentially the route of advance for the two scout platoons in column. Have the tank/AT platoons and the remainder of the troop move along a designated route or axis of advance (see Figure 4-6).

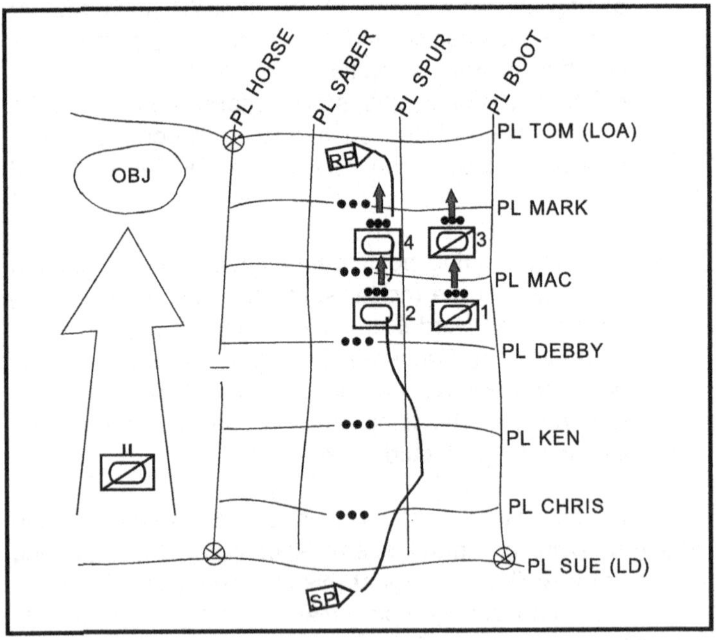

Figure 4-6. Troop moves by continuous marching.

Bounding by Platoons. This technique is appropriate when the protected force requires greater protection than afforded by continuous marching, is not moving quickly, or knows enemy contact is possible. Bounding platoons alternately may leave temporary gaps in the screen line as they move. Bounding platoons successively is more secure but slower than bounding platoons alternately.

Deploy both scout platoons abreast with the two tank/AT platoons, the mortar section, and the remainder of the troop in depth (between the screen line and the protected force). Alternately bound one scout platoon around (to the rear of) the other to assume new positions along the screen line (see Figure 4-7), or successively bound the scout platoons along the screen line (see Figure 4-8). Have the tank/AT platoons occupy designated positions sequentially or alternately.

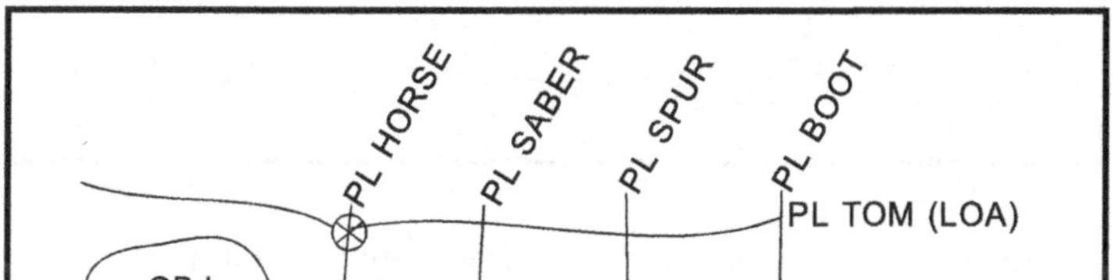

Figure 4-7. Troop moves by alternately bounding platoons.

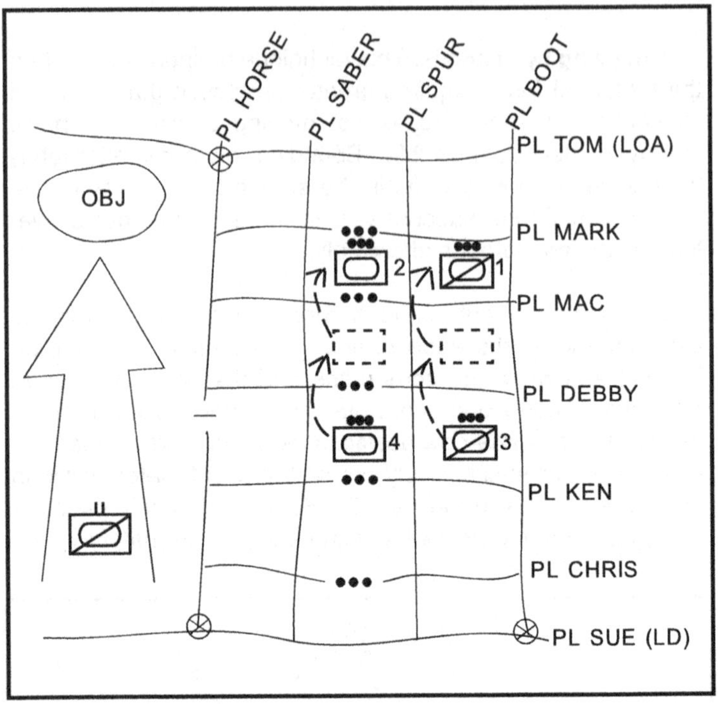

Figure 4-8. Troop moves by successively bounding platoons.

Bounding by OPs. This technique is appropriate when the main body is moving slowly, contact is possible, and maximum security is required. Bounding OPs alternately will disrupt the integrity of the scout platoons as OPs bound to their next position. Bounding OPs successively is easier for the scout platoons to control.

Deploy both scout platoons abreast with the two tank/AT platoons, the mortar section, and the remainder of the troop in depth (between the screen line and the protected force). Alternately bound the rearmost OP around (to the rear of) the other OPs (both platoons) to assume a new position along the screen line (see Figure 4-9), or successively bound the OPs along the screen line (see Figure 4-10). The number of OPs on the screen line may be reduced, as two or more may be bounding at any given time. The rate of advance of the protected force will determine this. Have the tank/AT platoons occupy designated positions sequentially or alternately.

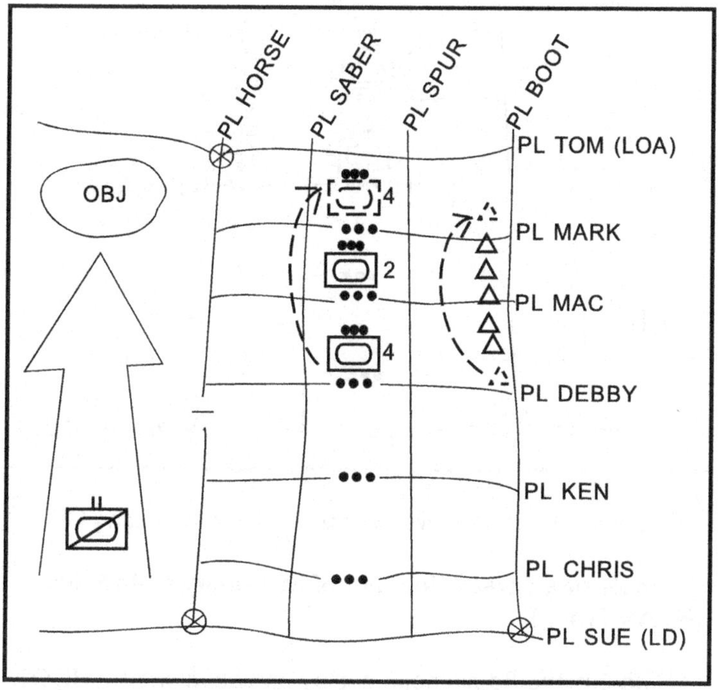

Figure 4-9. Troop moves by alternately bounding OPs.

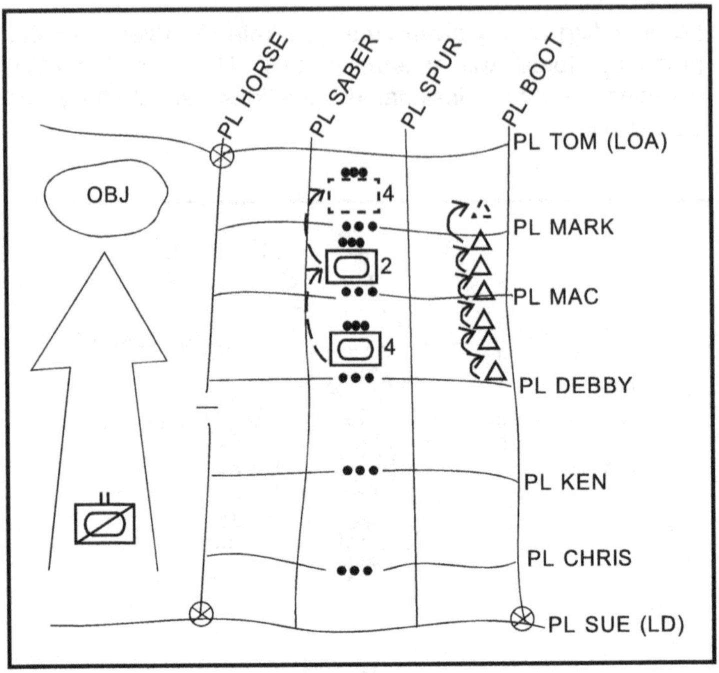

Figure 4-10. Troop moves by successively bounding OPs.

Example of a Screen Forward of Stationary Main Body (Heavy Troop)

The 1st and 3d platoons deploy abreast along PL BOOT, the initial screen line, and establish OPs to maintain continuous surveillance of the high-speed avenues of approach into the troop sector. Route 220 in the 1st platoon sector is the most likely route of enemy approach. Scout platoons execute their approved patrol plans between OPs, and periodically check the areas that cannot be observed. The mortar section lays its guns in a centrally located position about 3 kilometers behind the initial screen line and orients on Route 220. The 2d platoon occupies a hide position behind the 1st platoon. The 4th platoon occupies a hide position north of checkpoint F. Both positions provide good lateral movement and several maneuver options. The troop CP positions south of TIRS A21. The 1SG positions the troop trains in the woods just north of contact point 8, which is near a good lateral road network. The troop commander and FIST collocate well forward with the 1st platoon leader, overlooking Route 220 (see Figure 4-11).

The 1st platoon leader reports two BMPs and one T-64 bounding across the open terrain, moving southwest astride Route 220 about 2,000 meters east of PL BOOT. He maintains contact with the patrol as it crosses PL BOOT and enters the troop sector. The troop commander orders 2d platoon to attack the patrol by fire. The 2d platoon moves into hull-down positions and ambushes the patrol. All three vehicles are hit and damaged. The 1st platoon leader sends his PSG to capture the surviving BMP crew members and to search the vehicles for vehicle unit symbols, maps, sketches, and any other items of intelligence value. The PSG links up with the 1SG, who has moved forward. He hands over the prisoners and material, and returns to his position on the screen line. The 1SG evacuates the prisoners and captured material to the squadron combat trains. The 3d platoon reports no contact in its sector. The XO reports the situation to the squadron.

The troop commander tells the troop to be alert for other divisional reconnaissance patrols in the sector, and to expect regimental reconnaissance within the next hour. The XO eavesdrops on the squadron OI net and monitors reports from Troop B that indicate it has encountered several divisional reconnaissance patrols. The XO passes this information to the troop commander.

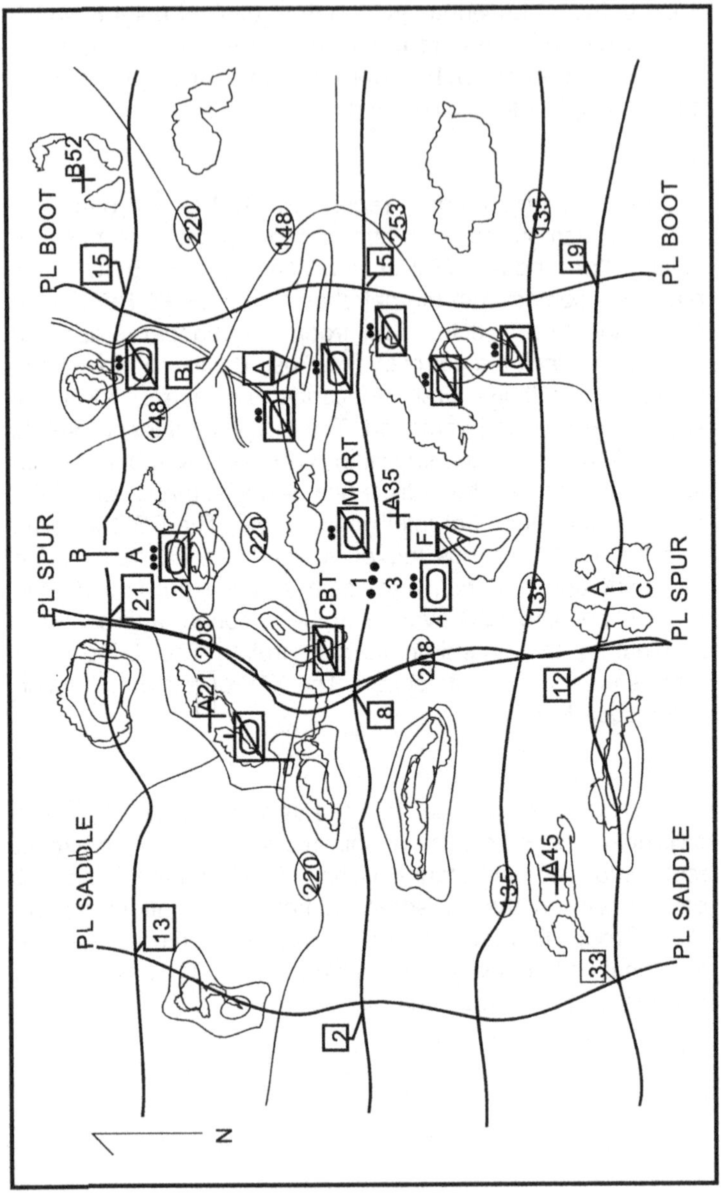

Figure 4-11. Troop screen.

Within the next 45 minutes, both scout platoons report reconnaissance patrols of two BMPs each in their sectors. The troop commander maneuvers a tank platoon into position to ambush each patrol. Three of the four BMPs are quickly destroyed. One BMP in the 1st platoon's sector is damaged, but is able to move into a covered and concealed position in the wooded area west of checkpoint A. Artillery begins to fall on the dominant terrain east of PL SPUR. The XO reports the situation to squadron.

The troop commander orders the 1st platoon to regain contact with the damaged BMP. The 1st platoon leader sends Bravo section to locate the damaged BMP. Bravo section moves into the wooded area and quickly finds and destroys the BMP.

The troop commander orders the scout platoons to move one of their scout sections back in sector in anticipation of the arrival of enemy reconnaissance elements. Both scout platoons reposition a scout section back towards PL SPUR while the troop commander shifts his tank platoons, mortar section, and trains.

The 1st platoon later reports a reinforced MRP moving southwest along Route 220. The 1st platoon continues to observe the enemy reconnaissance elements and reports their progress to the troop. The FSO engages the enemy with mortar high explosive (HE) and smoke rounds to disrupt their movement and keep them buttoned up. The troop commander tells the 4th platoon to move north and be prepared to counterattack the enemy by fire. He quickly outlines its route using TIRS. The enemy reconnaissance elements rapidly move in an attack formation toward the terrain previously occupied by the 2d platoon. It appears that the enemy elements are attempting to secure the dominant terrain in the 1st platoon sector and fix the tank platoon so the lead company of the enemy advance guard can maneuver around it.

The FSO continues to suppress the enemy with smoke and HE while it attempts to maneuver toward its objective. The troop commander directs 2d platoon into position to establish a base of fire and fix enemy elements while 4th platoon counterattacks by fire from the south. The FSO coordinates mortar and artillery fires to suppress and isolate enemy elements as the tank platoons execute their counterattack.

While the tank platoons complete the destruction of the enemy, heavy artillery begins to impact along and east of PL SPUR. The 1st platoon scouts report a tank platoon followed by six BTR-60s moving west along Route 220. The troop commander reports this information directly to the squadron commander. He orders the troops to fall back to PL SPUR, and to begin coordination for battle handover and passage of lines. The troop commander orders the 3d platoon to begin to bound back to PL SPUR. He tells the 1st platoon to maintain contact with the enemy lead company as it falls back to PL SPUR. The troop commander maneuvers the tank platoons west of PL SPUR.

The 1st platoon leader bounds his scout sections back to PL SPUR as he maintains contact with the lead company. The FSO engages the MRC with mortar and artillery fires to slow and disrupt its advance. The XO keeps squadron informed of the situation in the troop sector.

The troop XO moves back to the contact point to begin coordination for battle handover and passage of lines. The troop continues to maintain contact with the lead MRC as it maneuvers back in sector, while the FSO engages the MRC with mortar and artillery fires.

Section III. Guard

Cavalry squadrons conduct guard missions to provide another unit early warning of enemy approach, and to prevent enemy ground maneuver forces from coming within direct-fire range of the protected unit as it performs other missions or tasks. A guard is conducted to the front, flanks, or rear of a stationary or moving force. Many of the considerations for conducting a moving flank screen apply to conducting a moving flank guard.

A cavalry troop mission unique to moving flank guard operations is assigned to the lead troop of the squadron. This three-fold mission consists of the following tasks:

- Maintain contact with the protected force.
- Reconnoiter the zone between the protected force and the squadron route of advance.
- Reconnoiter the squadron route of advance.

The lead troop accomplishes these tasks by performing zone reconnaissance. The speed of the protected force determines how thoroughly the reconnaissance is performed. Assistance is required if the zone is too wide for the lead troop. An air cavalry troop may maintain contact with the protected force, or a following troop may perform route reconnaissance along the squadron's route of advance.

Note. For a detailed description of techniques for performing a squadron guard mission, see Chapter 4, FM 17-95.

The cavalry troop does not conduct guard operations independently. It normally performs reconnaissance, screen, defend, delay, and attack missions as part of a squadron guard operation. Review the appropriate chapter in this manual for specific mission details. See Chapter 3, Sections III and IV, Route and Zone Reconnaissance; Chapter 5, Section III, Hasty Attack; and Chapter 6, Sections II, III, and IV, Defend From a Troop Battle Position and Defend/Delay in Troop Sector.

Section IV. Cover

A covering force operates apart from the main body to develop the situation early and deceives, disorganizes, and destroys enemy forces. It accomplishes all the tasks of screening and guard forces. Unlike screening or guard forces, a covering force is a tactically self-contained force (that is, it is organized with sufficient combat support and combat service support assets to operate independently of the main body). Covering force operations can be either offensive or defensive in nature and can be conducted to the front, flanks, or rear of a stationary or moving force.

Cover operations are performed by the regiment (ACR or LACR) for the corps commander to provide the time and space he needs to make decisions and to achieve his operational objective.

Defensive cover operations, whether forward, flank, or rear, prevent the enemy from attacking the corps main body at the time and place and with the combat strength he chooses. Defensive cover operations are generally intended to destroy the enemy's initiative and set him up for defeat. Both heavy and light cavalry troops participate in defensive cover operations as part of their parent squadrons. Depending on the squadron commander's scheme of maneuver, they either screen, defend, delay, or attack.

Offensive cover operations are conducted by the regiment to seize and retain the initiative for the corps commander, and to allow him to attack decisively with the main body of the corps at the time and place he chooses. In offensive cover operations, the troop usually conducts movement to contact or zone reconnaissance missions. See the appropriate chapters in this manual for mission details.

Section V. Route Security

Cavalry squadrons and regiments conduct route security missions to prevent enemy ground maneuver forces from coming within direct-fire range of the protected route. A route security force operates on and to the flanks of a designated route. Route security operations are defensive in nature and, unlike guard operations, are terrain oriented. A route security force conducts reconnaissance, screens, attacks, defends, and occupies key locations along the route to prevent an enemy force from impeding, harassing, containing, seizing, or destroying traffic along the route.

Route security operations are not conducted independently by a cavalry troop. Both the heavy and light cavalry troops participate in route security operations as part of their parent squadrons and may conduct reconnaissance, screen, defend, delay, attack, or convoy security missions throughout the area of operations.

Section VI. Area Security

An area security force neutralizes or defeats enemy operations in a specified area. It operates in an area delineated by the headquarters assigning the area security mission. It screens, reconnoiters, attacks, defends, and delays as necessary to accomplish its mission. Area security operations may be offensive or defensive in nature and focus on the enemy, the force being protected, or a combination of the two.

Area security operations are conducted to deny the enemy the ability to influence friendly actions in a specific area or to deny the enemy use of an area for his own purposes. This may entail occupying and securing an area before the enemy can, or taking actions to destroy enemy forces already present.

The area to be secured may range from specific points (bridges, defiles) to areas such as terrain features (ridgelines, hills) to large population centers and adjacent areas. The factors of METT-T and unit capability will determine specific unit missions.

Area security missions are conducted by cavalry troops, squadrons, and regiments who employ the techniques of screen, guard, offense, and defense, depending on the nature and purpose of the mission.

Section VII. Convoy Security

Convoy security operations are performed as a minimum by a cavalry troop or a company team. Both heavy and light cavalry troops are suited to the requirements of protecting a convoy due to their organic reconnaissance capability and combat power. Both the cavalry troop and company team should be reinforced with engineers. METT-T considerations, such as restrictive terrain and limited time, may dictate a coordinated effort with air cavalry assets.

Convoy security operations are conducted when insufficient friendly forces are available to continuously secure lines of communication in an area of operations. They may also be conducted in conjunction with route security operations. A convoy security force operates to the front, flanks, and rear of a convoy element moving along a designated route. Convoy security operations are offensive in nature and orient on the force being protected.

A convoy security mission has certain critical tasks that guide planning and execution. To protect a convoy, the security force must accomplish the following critical tasks:

- Reconnoiter the route the convoy will travel.
- Clear the route of obstacles or positions from which the enemy could influence movement along the route.

- Provide early warning and prevent the enemy from impeding, harassing, containing, seizing, or destroying the convoy.

The convoy security force is organized into three or four elements to accomplish the following (see Figure 4-12):

- **Reconnaissance element.** The reconnaissance element performs tasks associated with zone and route reconnaissance forward of the convoy.
- **Screen element.** The screen element provides early warning and security to the convoy's flanks and rear.
- **Escort element.** The escort element provides close-in protection to the convoy. May also provide a reaction force to assist in repelling or destroying enemy contact.
- **Reaction force.** Provides firepower and support to the elements above in order to assist in developing the situation or conducting a hasty attack. May also perform duties of the escort element.

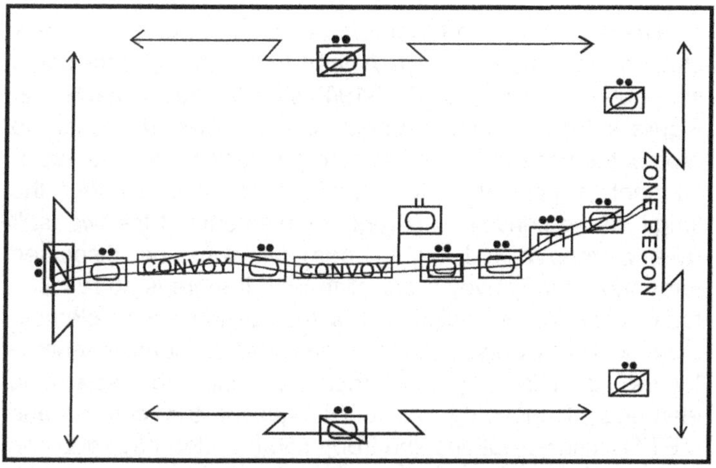

Figure 4-12. Cavalry troop conducts convoy security.

TECHNIQUES

The troop commander organizes and coordinates the efforts of his unit to fulfill the critical tasks associated with the convoy security mission.

Tasks of the reconnaissance element can usually be fulfilled by a single scout platoon. The troop commander ensures the reconnaissance element focuses on trafficability of the route and enemy forces that may influence movement along the route. METT-T may dictate the use of engineers to assist in reconnoitering and clearing the route. Convoy speed is determined by the pace of reconnaissance (METT-T). As a guide, the reconnaissance element should operate a minimum of 3 to 5 kilometers ahead of the main body of the convoy.

Tasks of the screen element can usually be fulfilled by a single scout platoon also. The troop commander ensures the screen element establishes OPs to provide early warning on critical portions of the route or key avenues of approach to the route. OPs have a limited ability to destroy enemy forces; therefore, their primary purpose is to acquire the enemy and direct reaction forces or indirect fire to destroy it.

Tasks of the escort element are best performed by tank platoons in a heavy troop. A light troop should consider a scramble of platoons (MK-19/M2/TOW—see Chapter 2, Figure 2-13) to gain the suppressive firepower necessary to protect the convoy. The troop commander ensures the escort element is positioned to provide security throughout the length of the convoy. This requires elements of the two tank (heavy troop)/mixed (light troop) platoons be dispersed throughout the convoy order of march. If there is no reaction force available or designated, a task of the escort element may be to provide reaction forces that respond to enemy forces identified by the reconnaissance or screening elements. Depending on the length of the convoy and METT-T considerations, the troop commander may keep one tank/mixed platoon consolidated or specifically designated as the reaction force.

Chapter 5

Offense

Offense is the decisive form of war. Brigades, battalion/task forces, and company teams are the principal offensive force for the corps or division. Cavalry units normally perform reconnaissance and security missions in support of corps and division offensive operations. Cavalry troops may perform certain offensive missions as part of a squadron, regiment, or other combined arms force. These offensive missions are normally performed during reconnaissance or security operations.

If required, cavalry troops may perform offensive operations within an economy-of-force role for a higher headquarters.

CONTENTS

		Page
Section I.	Purpose and Fundamentals	5-1
Section II.	Movement to Contact	5-2
Section III.	Hasty Attack	5-11
Section IV.	Deliberate Attack	5-22
Section V.	Raid	5-24

Section I. Purpose and Fundamentals

PURPOSE

The main purpose of the offense is to defeat, destroy, or neutralize the enemy force. Offensive operations are also undertaken to secure decisive terrain, to deprive the enemy of resources, to gain information, to deceive and divert the enemy, to hold the enemy in position, to disrupt his attack, and to set up conditions for future successful operations.

FUNDAMENTALS

Successful offensive operations have four fundamentals in common.

- **Surprise.** Strike the enemy at the time and place or in a manner that he least expects.
- **Concentration.** Mass available forces, strive for overwhelming superiority in men, weapons, and firepower. With concentration, however, vulnerability becomes a factor. A force that is dispersed is much more survivable. The commander must maintain a high sense of situational awareness to anticipate the conditions of battle that will allow him to mass at the critical point, kill the enemy, and quickly disperse to survive.
- **Tempo.** Tempo is the rate of speed of military action. Controlling or altering the rate is essential for maintaining the initiative. Tempo can be fast or slow dependent on the capabilities of the troop relative to that of the enemy. Commanders must adjust tempo to ensure synchronization.
- **Audacity.** Boldness in the plan's execution is key to success in offensive operations. Commanders should understand when and where they are taking risks, but must not become tentative when executing their plan.

Section II. Movement to Contact

A movement to contact is a mission undertaken to gain or regain contact with the enemy. Unlike a zone reconnaissance, which is focused on reporting detailed information on the terrain and enemy within a given zone, movement to contact is focused on finding the enemy. The critical tasks are geared for achieving fast movement and rapid location of enemy forces.

The cavalry troop normally conducts a movement-to-contact mission as the lead element of a squadron or combined arms force hasty attack. The troop may also conduct the mission as part of a squadron movement to contact when the squadron is leading the advance of another combined arms force, such as a brigade or division.

A movement to contact is characterized by rapid aggressive action. The troop commander must rapidly develop the situation and may be permitted to bypass enemy forces to maintain momentum.

CRITICAL TASKS

During a movement to contact, certain critical tasks will be accomplished. Unless the squadron commander gives guidance otherwise, the troop will—

- Reconnoiter and determine the trafficability of all high-speed routes within the zone.
- Inspect and classify all bridges, culverts, overpasses, and underpasses along high-speed routes. Identify all bypasses and fords that cannot support rapid heavy-armor movement.
- Clear all high-speed routes of mines, obstacles, and barriers within its capability, or locate a bypass.
- Find and report all enemy forces within the zone and determine their size, composition, and activity.

TECHNIQUES

The cavalry troop can perform movement to contact in a zone up to 10 kilometers wide. The size of the zone depends on the terrain and enemy situation. Scout squads, and in some cases sections, must be mutually supporting during movement based on the enemy situation. Mutual support between squads and sections is gained by reducing the size of the zone based on the terrain. The more restrictive the terrain the less mutual support becomes available.

When considering techniques for conducting a movement to contact, remember the basics.

- Always retain the ability to maneuver.
- Make contact with the smallest force possible.
- Employ forces in depth; stay flexible.
- Develop the situation rapidly once contact is made with the enemy force. Operate at a tempo that forces him to react to you, not you to him.

When the troop receives a movement-to-contact mission, the zone of action is normally identified by lateral boundaries. A line of departure and an objective or limit of advance are specified to orient the troop's movement. The objective is usually a terrain feature placed deep enough to ensure contact with the enemy. Divide the troop zone into two platoon zones. Think about where to draw the boundary. Make sure it does not split enemy avenues of approach and that it is on easily identifiable terrain. Do not use a road for a boundary. To control progress through the zone, add phase lines every 5 to 8 kilometers on easily identifiable terrain. Place contact points near the intersection of the platoon boundary and all phase lines. If specific areas need to be checked, identify them using checkpoints for reference. Place TIRS on the map.

Use a troop vee or split vee formation when conducting a movement to contact. Deploy scout platoons abreast and have them accomplish all the related reconnaissance tasks. Move them out first across the line of departure. The mission of the scouts is to find the enemy, develop the situation, and recommend to the troop commander the best course of action for the employment of the tank/AT platoons.

The mortar section, under control of the troop fire support officer or mortar section sergeant, follows about 2 kilometers behind the scouts, centered in the zone, to provide continuous coverage forward of the scouts. The troop command post, under control of the XO, displaces in the zone, positioning on terrain that affords effective and continuous communication with troop elements and the squadron. The troop trains, under control of the 1SG, follow about 2 kilometers behind the tank/AT platoons, and bound from one covered and concealed position to another. The troop commander positions himself well forward to observe one scout platoon or the other. His position is usually located where he expects initial enemy contact or difficult situations.

Heavy Troop

Due to the survivability and firepower of scout platforms (M3) in the heavy troop, the tank platoons may be kept in considerable depth (1,000 to 3,000 meters) behind the scouts until enemy contact is made. Once contact is made and the scouts have developed the situation/recommended a course of action, the troop commander may then employ his tanks, using firepower and maneuverability to fix the enemy or destroy it by hasty attack.

Light Troop

In the light cavalry troop, the AT platoon(s) should be kept close (500 to 1,500 meters) to the scout platoons. Upon contact, the HMMWV scouts will require more time to develop the situation than Bradley-mounted scouts. The AT platoon(s) can assist the scouts in developing the situation by overwatching the scouts. Also the TOW takes time to employ, so reaction time is slower than with the tank. During movement to contact, the light troop commander must be able to execute his actions on contact very quickly to maintain the momentum of the operation. A technique the light troop might use is outlined below.

- Scouts make contact with an enemy force.
- The troop commander immediately moves one AT platoon forward to overwatch the scouts as they develop the situation.

- If the situation dictates, the AT platoon suppresses with direct fires to allow the scouts to maneuver.

- Once the situation is developed, the troop positions to either fix the enemy for the follow-on force or destroy it by a hasty attack.

EXAMPLE OF MOVEMENT TO CONTACT (HEAVY TROOP)

The troop is ordered to reestablish contact with elements of a withdrawing enemy mechanized infantry platoon. As the troop deploys in a split-vee formation, the 1st and 3d platoons cross the line of departure abreast, north and south respectively, and begin to reconnoiter in troop zone up to PL BOOT. The scout platoons determine the trafficability of high-speed routes within their zones, and search for the enemy in areas that dominate the routes. The bridge across the creek at checkpoint B will support only 20 tons, so the 1st platoon finds a solid-based fording site 100 meters south of the bridge. Thus far, the high-speed routes are not damaged or blocked. The 3d platoon reports finding tracks of what appears to be three BMPs heading northwest from Hill 244. The 1st platoon reports tracks of an enemy company-size force heading west along Route 220. Track patterns indicate BMPs and T-64s/T-72s. The mortar section eavesdrops on the troop command net, and moves with the scout platoons, staying back about 1.5 kilometers, centered in the troop zone. The 2d and 4th platoons are paired with their sister scout platoons and key their movement on the scout platoon's progress, staying back about 1 kilometer and using terrain that affords good cover and concealment. The troop CP and trains hold in place (see Figure 5-1).

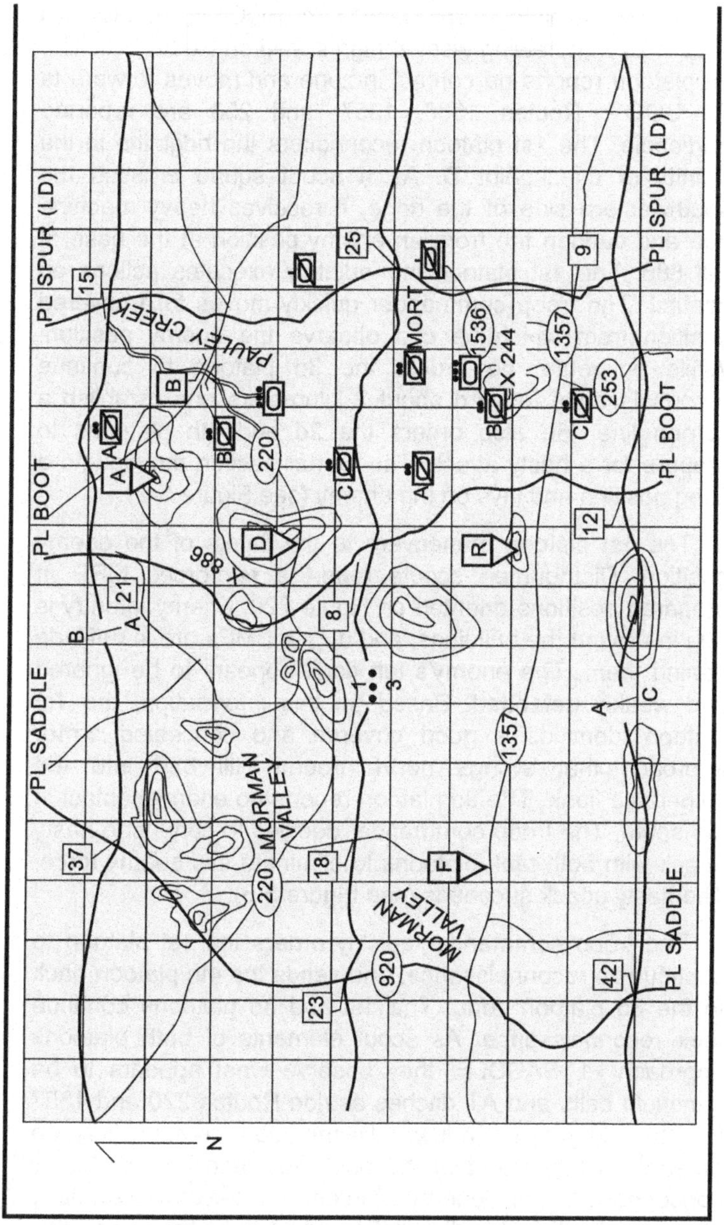

Figure 5-1. Movement to contact.

As the 1st platoon approaches PL BOOT, it reports two BRDM-2s withdrawing west at high speed on Route 220. The 3d platoon reports no contact in zone and moves forward to PL BOOT. Routes 1536, 1357, and 253 are reported trafficable. The 1st platoon reconnoiters the ridgeline in the vicinity of checkpoint D. As a scout squad crosses the southwestern side of the ridge, it receives heavy machine gun and cannon fire from an enemy position at the base of Hill 886. The 1st platoon immediately executes actions on contact. The troop commander quickly moves to a covered position from which he can observe the enemy position. While en route, he orders the 3d platoon to continue reconnaissance forward about 2 kilometers and establish a screen line. He also orders the 2d and 4th platoons to prepare for a hasty attack. The mortar section moves into a firing position and lays on the enemy (see Figure 5-2).

The 1st platoon maneuvers to the flanks of the enemy position. Dismounted scouts report a reinforced MRP in prepared positions oriented on Route 220. Enemy infantry is dug in around the buildings, and dug-in BMPs are in defilade behind them. The enemy's left flank appears to be ignored and weakly defended. Based on this information, the 1st platoon identifies a good covered and concealed armor approach that swings north around Hill 886 into the vulnerable flank. The 3d platoon reports no enemy contact in the south. The troop commander decides to conduct a hasty attack with both tank platoons to eliminate this enemy force. The hasty attack succeeds (see Figure 5-3).

The troop commander quickly orders the 1st platoon to continue its reconnaissance, and sends the 4th platoon back to the 3d platoon zone. The 1st and 3d platoons continue their reconnaissance. As scout elements of both platoons approach PL SADDLE, they observe what appears to be minefield belts and AT ditches astride Routes 220 and 1357 in the Morman Valley. Dismounted reconnaissance determines the extent of the obstacles, and reveals heavy concentrations of enemy forces in several company strongpoints along the dominating ridgeline by Highway 920.

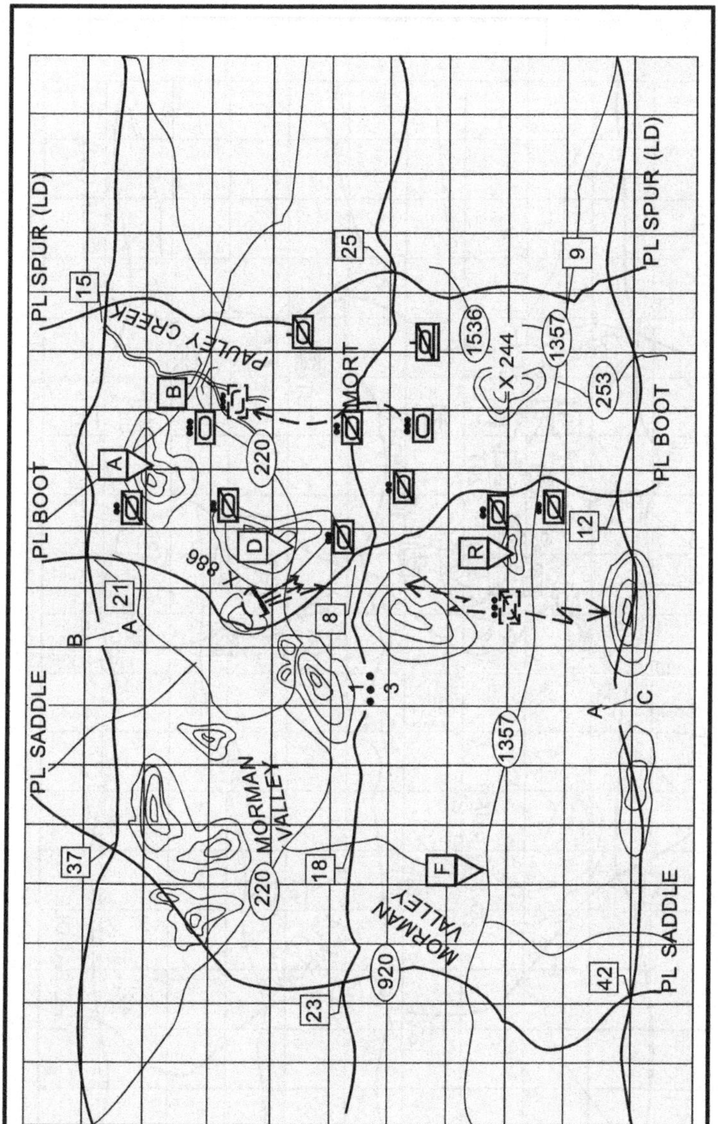

Figure 5-2. Actions on contact.

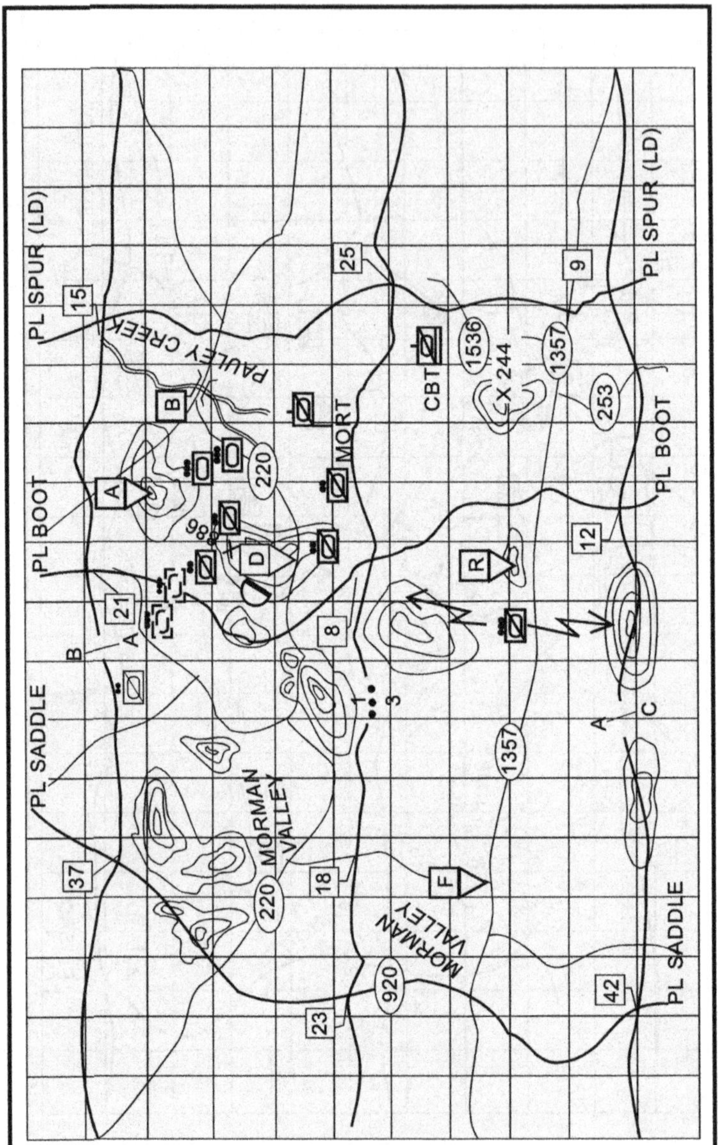

Figure 5-3. Hasty attack.

The troop commander takes a look. En route, he orders both scout platoons to determine if any gaps or boundaries exist between the enemy positions. After seeing the situation, the troop commander calls his squadron commander and recommends the troop establish a hasty defense and prepare to assist in battle handover and forward passage of following attack forces (see Figure 5-4).

Section III. Hasty Attack

A hasty attack is conducted with a minimum of preparation to defeat an enemy force that is not prepared or deployed to fight. It is a course of action routinely employed in cavalry operations to seize or retain the initiative, or to sustain the tempo of operations. A hasty attack can be executed while the troop is engaged with a zone reconnaissance mission or movement to contact. (See Chapter 3, Section II, and Section II of this chapter for other examples of hasty attack.)

CRITICAL TASKS

To execute a hasty attack, the following critical tasks must be accomplished:

- Reconnoiter and determine the size, composition, and orientation of the enemy force.
- Determine if the objective enemy force is supported by other units nearby.
- Find a high-speed covered and concealed approach into the enemy's flank(s).
- Establish a maneuver element (usually one or both tank platoons) to move to a position of advantage over the enemy and attack him by fire.

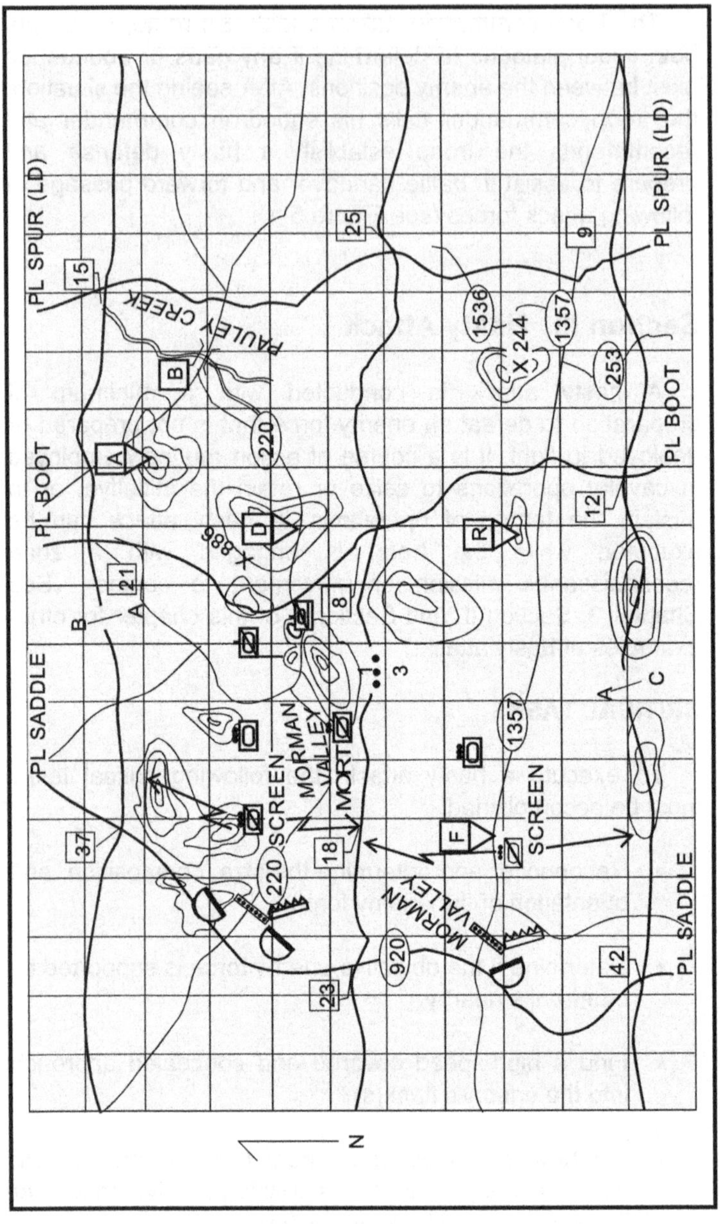

Figure 5-4. Hasty defense.

- Establish a base-of-fire element (usually one or both scout platoons) to defeat or suppress all observed enemy AT weapons with long-range direct and indirect fires before the maneuver force deploys into its attack.

- Isolate the objective enemy force from other mutually supporting units with indirect fires (usually with smoke and HE ammunition).

- Attack the enemy by fire or by fire and movement, and defeat him.

- Once the attack is completed, immediately establish hasty defensive positions and OPs on high-speed avenues of approach into the troop position.

TECHNIQUES

Each critical task has a time at which it will be accomplished in relation to all other critical tasks. A good hasty attack depends on the commander's sense of timing and on his ability to employ his forces to accomplish the tasks in the proper sequence. The commander has to synchronize—to concentrate and apply different forms of combat power against the enemy at the right times and places. The decision to conduct a hasty attack is usually made after an enemy force's reconnaissance and dispositions show that winning requires a quick strike, with little preparation. Tactics for conducting a hasty attack have three common features.

- Known or suspected enemy AT weapons are suppressed and destroyed with direct and/or indirect fires before the maneuver force is committed.

- The enemy is forced to fight in two directions.

- The enemy is suppressed and unable to react.

SETTING UP THE TROOP FOR A HASTY ATTACK

While conducting other missions, scouts will often make contact with an enemy force. In developing the situation, a scout platoon may recommend hasty attack as a course of action to the troop commander, who decides to execute the recommended course of action. The troop commander issues FRAGOs that will position forces to execute an attack simply and effectively.

The scout platoon in contact continues to reconnoiter the enemy's position and to accomplish its reconnaissance tasks. One section of the scout platoon remains in contact with the enemy. The other scout platoon continues its reconnaissance up to a limit of advance established by the troop commander. Both scout platoons continue to develop the situation further by looking for the presence of other enemy units, to the flanks or rear, supporting the enemy contact.

The FIST moves to a good position to see the battlefield and to control the indirect fires. The FSO places the mortar section on terrain where it establishes a firing position and prepares to suppress the enemy position.

The size and strength of the enemy may require the use of one or both tank/AT platoons. The commander may choose to do one of the following options:

- Move one tank/AT platoon into an overwatch position and attack by fire with one.
- Keep one in reserve behind the scout platoon not in contact and attack by fire with one.
- Attack by fire with both platoons.

The scouts determine a good attack position and attack-by-fire position for the tank/AT platoon(s). A scout squad from the scout platoon in contact moves to a link-up point with one or both tank/AT platoon(s) to guide them into the attack position.

The troop commander moves to collocate with the scout platoon in overwatch. The 1SG moves medics and recovery vehicles close to the battlefield. The XO assists the commander in control of the troop and keeps the higher commander informed.

INDIRECT FIRES IN SUPPORT OF A HASTY ATTACK

Indirect fires should complement the troop's scheme of maneuver. In the light troop, dependent on availability of supporting indirect-fire systems, indirect fires may prove to be the best weapon of destruction. The troop can acquire and engage the enemy with indirect fire from positions offering good protection from enemy direct/indirect fire. Because the light troop is in covered and concealed positions, the survivability of the troop is enhanced. Indirect fires must also be controlled to prevent fratricide. The troop commander must determine the following:

- Who will control the indirect fires during the hasty attack?
- Who will initiate indirect fires onto the objective?
- Who will lift and shift the indirect fires to subsequent targets?
- What will the signal be for lifting and shifting indirect fires?
- Are there any restrictive fire measures or restrictive fire areas?

There are many different answers to these questions. METT-T will determine which answer works best. Under most conditions the commander or the FSO will be in the best position to control the engagement of indirect-fire systems.

The troop commander should use available indirect fires from mortars and supporting artillery to—

- Suppress the enemy while scouts are maneuvering to develop the situation.
- Screen enemy observation of scouts or assault element during the conduct of the hasty attack.
- Isolate the enemy contact by firing HE and smoke between the enemy force and any possible supporting positions.
- Shift indirect fires off the objective to block enemy withdrawal routes.

DIRECT FIRES DURING HASTY ATTACK

When executing the hasty attack the troop must combine indirect fires with direct fires. The direct-fire systems available to the heavy troop commander are highly destructive. The high rate of fire from the M3 and M1 allows the commander to quickly place effective fires into the objective area while stationary or moving.

The systems available to the light troop commander can be just as destructive when used in combinations and against a comparably equipped enemy. The light troop commander should strive to engage the enemy with combinations of weapons. Using the bounding technique and move-set drills by platoons, the troop commander can place effective suppressive and destructive fires into the objective area. (The high rates of fire from the M2 caliber .50 and MK-19 should be used to suppress the enemy and cover the TOW systems as they engage.) See Figure 5-5.

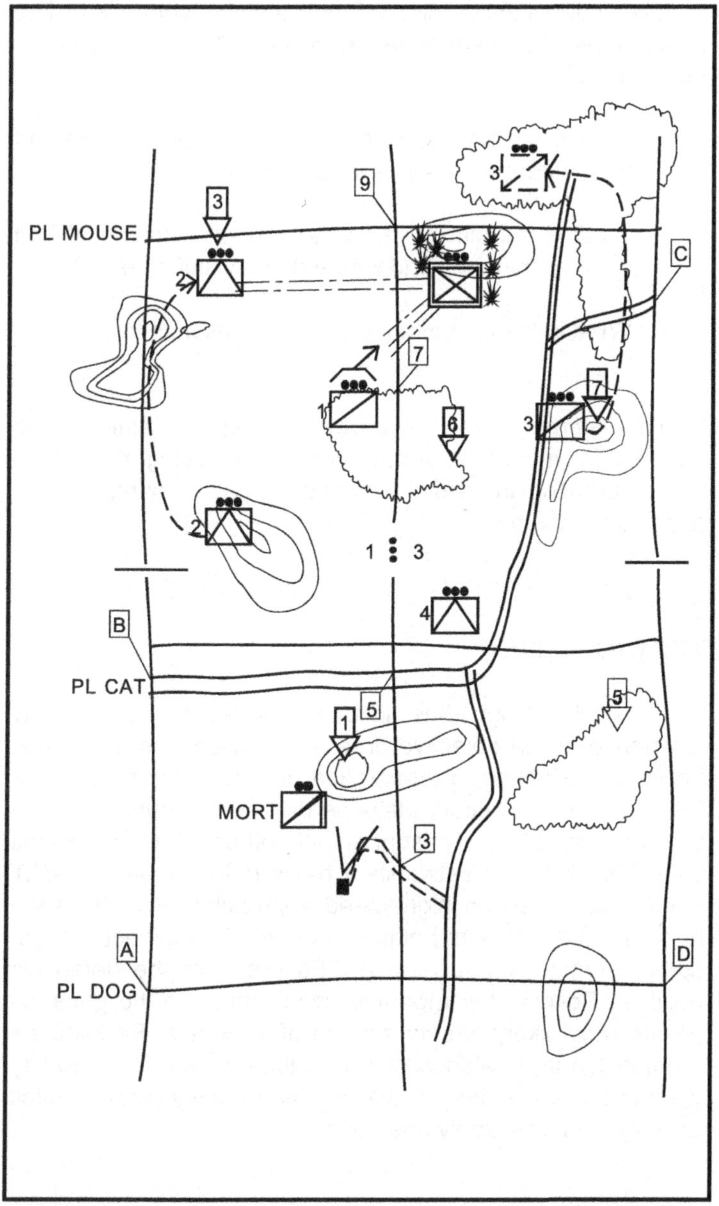

Figure 5-5. Light cavalry troop using fire and maneuver.

Like indirect fires, direct fires must be controlled. The commander must determine the following before executing a hasty attack:

- Who will initiate direct fires into the objective area and from where will they be initiated?
- What is the aim point for the support element? When do they shift fires and where do they shift fires to?
- What is the limit of advance for the assault force?

To determine the answers to these questions the commander must look at the enemy he is facing, the terrain he is operating in, and the forces he has to bring to bear against the enemy.

THE OBJECTIVE

Once the troop has initiated its attack and enemy resistance in the objective area has ceased, the troop may begin consolidating in preparation of continuing its mission. Have one or both scout platoons dismount scouts to sweep the area quickly for prisoners and other items of tactical value. Immediately establish a hasty defense with tank/AT platoons oriented on high-speed approaches into the troop position, and with scout platoons oriented on expected high-speed enemy approaches at OPs out from the defensive position. Position the mortar section and have the guns lay on the most likely enemy avenue of approach. Redistribute ammunition and balance crews as time allows. Replace key leaders quickly. If time is available and enemy counterattack unlikely, begin resupply operations.

EXAMPLE OF A HASTY ATTACK

A troop is conducting a zone reconnaissance moving toward its limit of advance (LOA). As the 3d platoon crosses the ridge (PL SABER) in its zone, Charlie scout section is hit by enemy tank cannon and heavy machine-gun fire from concealed positions near checkpoint 2. The platoon immediately takes cover and reports contact to the troop commander. The troop commander acknowledges and immediately moves to link up with the platoon leader. The platoon sergeant calls for indirect fire to suppress the enemy force. Under this protection, the platoon leader orders dismounted reconnaissance through the woods on the enemy's right to find his flank. He then sends his Bravo scout section forward on the enemy's left to find his other flank and to determine if the enemy is mutually supported by other enemy forces near checkpoint 2. The platoon leader and two scouts dismount, move in closer, and determine the enemy force consists of a reinforced MRP, two tanks, and three BMPs, in a hasty defensive position. Scouts report what appear to be minefields forward and to the flanks of the enemy position. The Alpha section identifies the right flank of the enemy position and discovers the flank is not protected by obstacles and not covered by direct-fire weapons (see Figure 5-6).

En route to the 3d platoon's zone, the troop commander tells the 1st platoon to continue reconnaissance forward to the LOA and find out if other enemy forces are providing mutual support to the enemy platoon in the 3d platoon's zone. He also tells the platoon leader to establish a screen along the LOA in his platoon zone. He then tells the 4th platoon to join the 2d platoon near checkpoint 12, using the ridgeline to cover its move, and prepare to conduct a hasty attack. Using TIRS, he outlines a route that runs from checkpoint 12 around the western edge of the village to checkpoint 7, then to a position from which it can engage the enemy on his most vulnerable flank (see Figure 5-7).

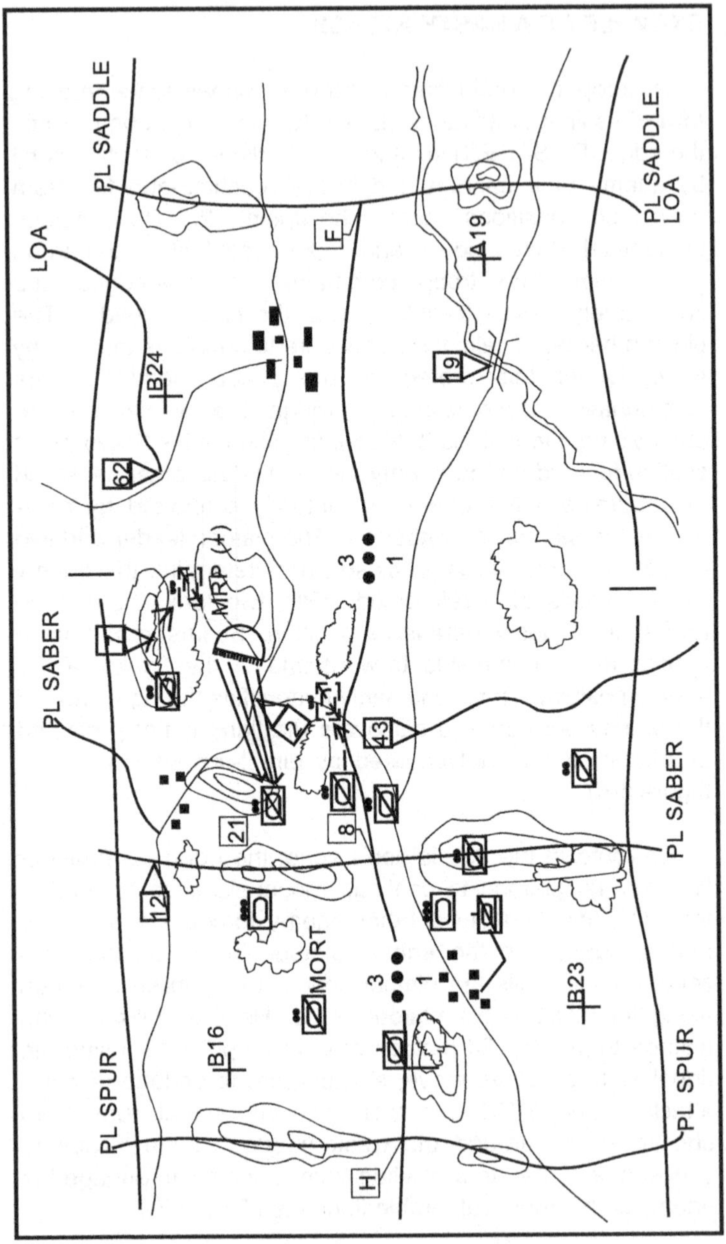

Figure 5-6. Develop the situation.

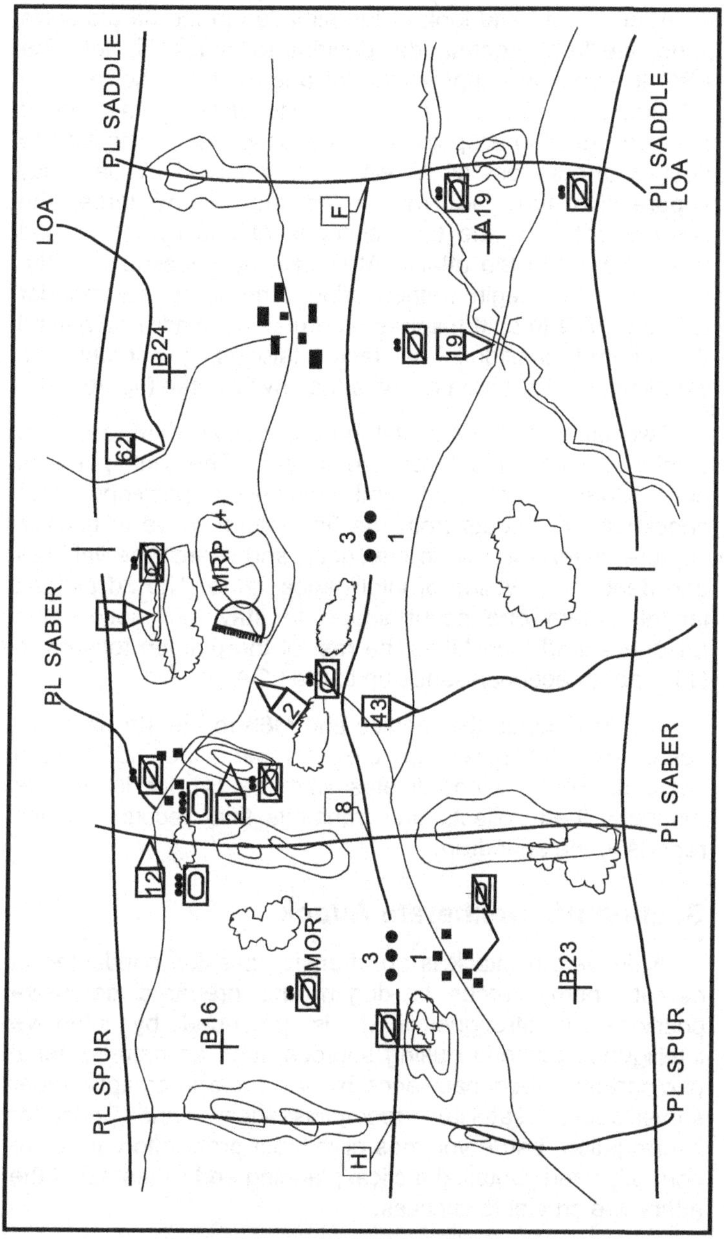

Figure 5-7. Set for hasty attack.

After a first-hand look at the situation in the 3d platoon's zone, the troop commander decides to attack. He tells the FSO to continue suppressing the enemy. Next, he tells the 3d platoon leader to establish a base of fire from covered positions near checkpoint 21. The 3d platoon is tasked to destroy or suppress all AT weapons observed. While heavy suppressive fires are raining on the enemy force, the commander joins the tank platoons at checkpoint 12 and leads them into the attack. After passing checkpoint 7, the tank platoons begin maneuvering. The troop commander tells the FSO to shift his fires on routes of withdrawal behind the enemy position. The tank platoons go around the woodline, deploy abreast, and attack by fire (see Figure 5-8).

Two tanks in the 4th platoon are hit and damaged; one soldier is killed and three are injured. The tank platoons consolidate in covered and concealed positions near checkpoint 7. Scouts from the 3d platoon move in quickly, capture several wounded prisoners, and search the vehicles and dead for anything of intelligence value. The 3d platoon leader orders one scout squad to stay and secure the prisoners, and then takes the rest of the platoon forward to finish zone reconnaissance up to the LOA.

The 1SG leads the medics and M88 to the 4th platoon's casualties and picks up the prisoners the 3d platoon secured. The XO coordinates with the S4 for immediate resupply, then collects and transmits final reconnaissance reports to the squadron.

Section IV. Deliberate Attack

A deliberate attack is an offensive mission conducted to defeat enemy forces in dug-in and prepared defensive positions or strongpoints. It is preceded by intensive intelligence gathering using sources such as patrols, aerial photographs, reconnaissance by fire, radio intercepts, radar surveillance, satellite reconnaissance, and prisoner interrogation. Large volumes of indirect preparatory fires are normally used. Detailed tactical planning and rehearsal of the attack are crucial to success.

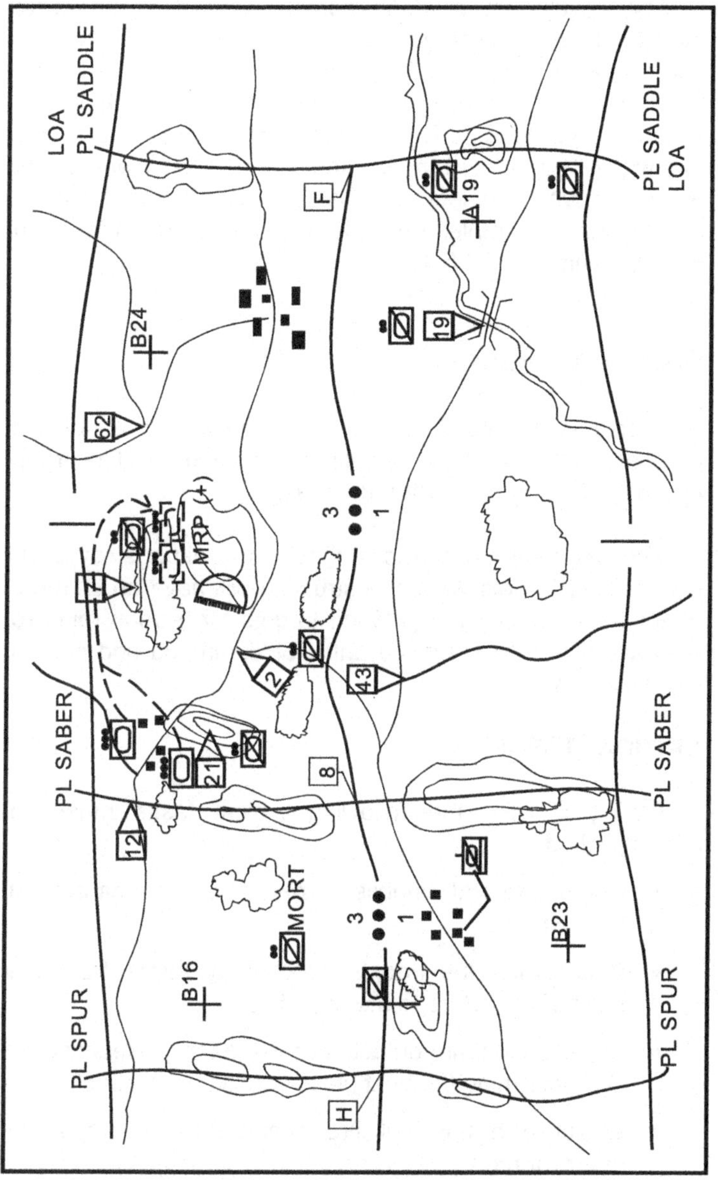

Figure 5-8. Tank platoons move into position for the hasty attack.

The cavalry troop alone cannot conduct a deliberate attack against prepared defensive positions. It usually participates with its parent squadron or other combined arms force as one of the three major elements in a breaching and assault operation. The troop can serve as the *overwatch/support element,* the *breaching element* (if reinforced with engineers), or the *assault element*. Doctrine for attacking a complex obstacle or a strongpoint is found in FM 71-1 and FM 17-95-10.

Section V. Raid

A raid is an attack into enemy territory for a specific purpose, with no intent to gain or hold terrain. The troop returns to friendly lines after the attack.

The cavalry troop normally conducts a raid against clear, well-defined targets whose destruction will have an adverse effect on an enemy force. Good targets for a troop raid are enemy artillery, rear service units, or command and control headquarters.

CRITICAL TASKS

During a raid, the following critical tasks must be accomplished:

- Coordinate and establish a restricted fire area around the raid area.
- Reconnoiter the troop's direction of attack from the point of departure to the objective.
- Establish visual contact with the enemy. Reconnoiter and determine the best place to initiate the attack.
- Establish a base of fire element to suppress the enemy force.

- Establish a maneuver element (usually one or both tank platoons) to move to a position of advantage over the enemy and attack them by fire.
- Suppress the enemy force with long-range direct fires while the maneuver element moves up to begin its attack.
- Establish surveillance on enemy avenues of approach to the objective, where follow-on enemy maneuver units would most likely appear.
- Attack the enemy by fire or by fire and movement to defeat them. Finish quickly.
- Establish security to protect the withdrawal of the remainder of the troop.
- Return rapidly to friendly lines and enter at the designated passage point.

TECHNIQUES

A raid achieves its purpose through speed, violent execution, and surprise. The troop's maneuver should produce that effect. There will probably not be a chance to reconnoiter before the attack; speed and surprise generate the security needed. Consequently, the troop might not always take the mortar section, combat trains, or command post. Vehicles disabled on a raid may be left behind in the interests of speed and security for the entire troop. When the troop receives a FRAGO to conduct a raid, the commander should receive the following graphic control measures from the S3 or squadron commander: a point of departure along the line of departure or line of contact, a direction of attack, an objective, a route of withdrawal, and a passage point to enter friendly lines. Analyze the terrain and the enemy situation, then issue the troop FRAGO.

The FSO should coordinate with the squadron FSO to establish a restricted fire area around the raid area. The 1SG should coordinate for immediate resupply of Class III and Class V when the troop returns. Have one scout platoon take the lead and reconnoiter the troop direction of attack from the point of departure on the line of departure to the objective. The lead platoon should establish visual contact with the raid objective to confirm the troop's scheme of maneuver and establish surveillance of the objective. The other scout platoon may act as the base-of-fire element, and follow about 1 kilometer behind the lead platoon. Have the lead platoon guide it into an overwatch position to deliver long-range suppressive fires on the enemy. Consolidate the tank platoons, move forward to join them, then follow about 1 kilometer behind the second scout platoon.

Once the base-of-fire element is set, have the platoon suppress the enemy with direct and indirect fires. Guide the maneuver element into a position to attack by fire. Finish quickly and move out.

Tell the scout platoon performing reconnaissance and surveillance tasks to screen and protect the withdrawal of the remainder of the troop back to friendly lines. Have the other scout platoon lead the way back on the designated route of withdrawal, and have the tank platoons follow. Stay back and supervise the conduct of the rear screen until all troop elements have entered friendly lines. Consolidate and resupply the troop with fuel and ammunition as soon as possible. Get ready to fight again.

EXAMPLE OF A RAID (HEAVY TROOP)

The squadron is defending in sector astride a likely Threat regimental zone of attack. The three armored cavalry troops are deployed abreast with the tank company in reserve. The squadron frontage is about 10 kilometers wide. Threat forces attack from the march. Squadron elements in the northern half of the sector are softened with long-range preparatory fires delivered by rocket and artillery forces. As artillery fires increase in intensity, the fire support element and the advance guard battalion of the regiment deploy from the march into prebattle formation and then attack. The advance guard battalion is quickly defeated by the concentrated fires of the northern cavalry troop and a surprise counterattack by the tank company. The squadron rapidly repositions to meet the assault of the regimental main body. Within one hour, the regimental main body approaches the line of contact with two MRBs abreast. Their axis of advance indicates most of their combat power will be concentrated in the northern half of the squadron's sector. Artillery suppressive fires increase as the Threat battalions deploy into attack formation. Aeroscouts report two 122-mm artillery battalions of the regimental artillery group moving into firing positions about 4 kilometers behind the attacking combat maneuver units. In this situation, with good defensible terrain, the squadron commander decides to concentrate the combat power of the troops, the tank company, and his available artillery in the north to defeat the advancing maneuver forces. He orders the southern cavalry troop to conduct a raid behind enemy lines to destroy the regimental artillery group simultaneously.

The troop commander quickly analyzes the terrain, and then issues a FRAGO to the troop. He tells the 1st platoon, which is currently on the screen line, to take the lead and reconnoiter the troop's direction of attack from the point of departure on the LD to the objective area. The platoon is tasked to establish visual contact with the Threat artillery units and to locate the best covered and concealed approach into their flanks or rear. The troop commander tells the 3d platoon to follow about 1 kilometer behind the 1st platoon and establish a base of fire to suppress and fix the artillery units. He orders the 2d and 4th platoons to consolidate rapidly near the point of departure. He tells the mortar section to remain in the troop sector. Priority of fires goes to the 1st platoon. As the 1st platoon approaches the point of departure, the troop FSO coordinates with the squadron FSO to establish a restricted fire area around the raid area. Once it is established and confirmed by the FSO, the troop commander orders the 1st platoon forward (see Figure 5-9).

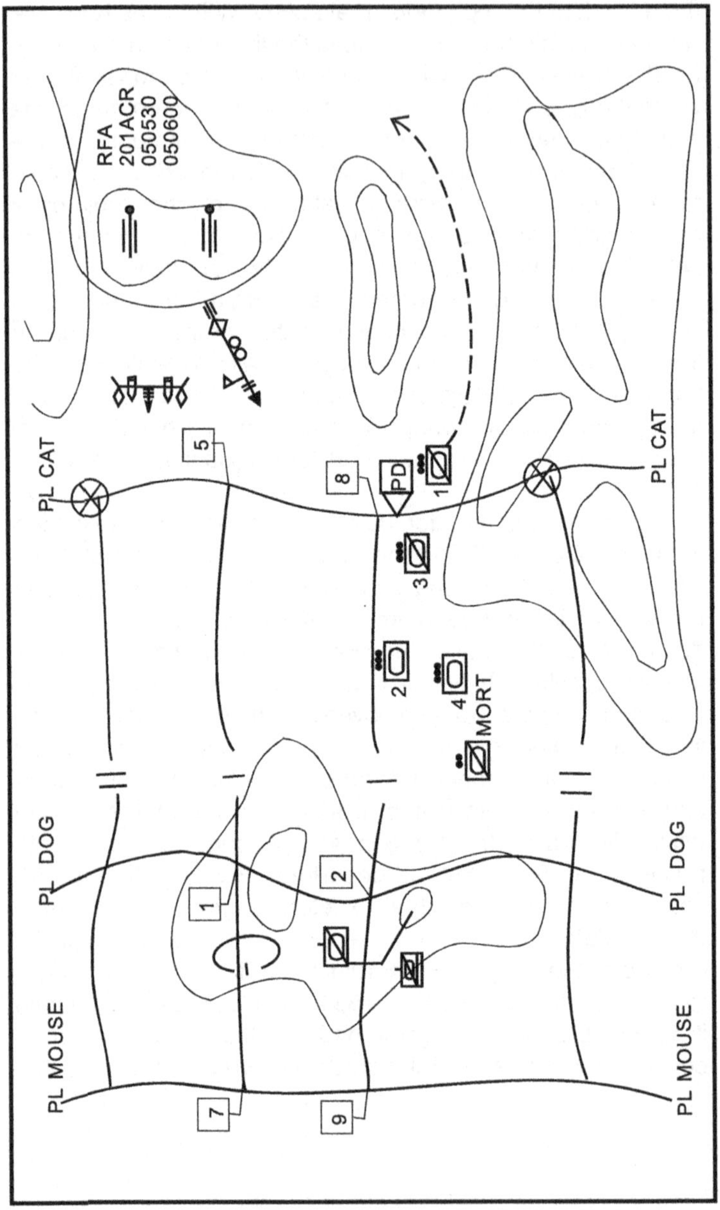

Figure 5-9. Heavy troop raid (part one).

The 1st platoon makes visual contact in about 10 minutes, and continues to reconnoiter for a good axis of attack. The 3d platoon moves into an overwatch position about 1,800 meters away and lays down a curtain of suppressive fire with chain guns, TOW missiles, and mortar fire. While suppressive fires are delivered, the tank platoons maneuver into the left rear of the artillery units, then deploy abreast and attack the Threat artillery battalions by fire (see Figure 5-10).

After silencing the battalions, the troop commander immediately orders the 3d platoon to lead the troop back to friendly lines along the designated route of withdrawal. The 2d and 4th platoons follow while the 1st platoon screens the withdrawal of the troop. The troop commander joins the 1st platoon and supervises the conduct of the rear screen until all troop elements have reentered friendly lines (see Figure 5-11). The troop returns to defensive positions, and the 1SG begins resupply actions.

EXAMPLE OF A RAID (LIGHT TROOP)

The light troop should consider the raid target when planning its operation. A combination of weapon systems may best suit the destruction of the raid objective rather than operating in pure platoons. A mix of weapon systems may also assist in security of the objective during the raid. The troop scramble should be considered as a possible task organization when planning a raid.

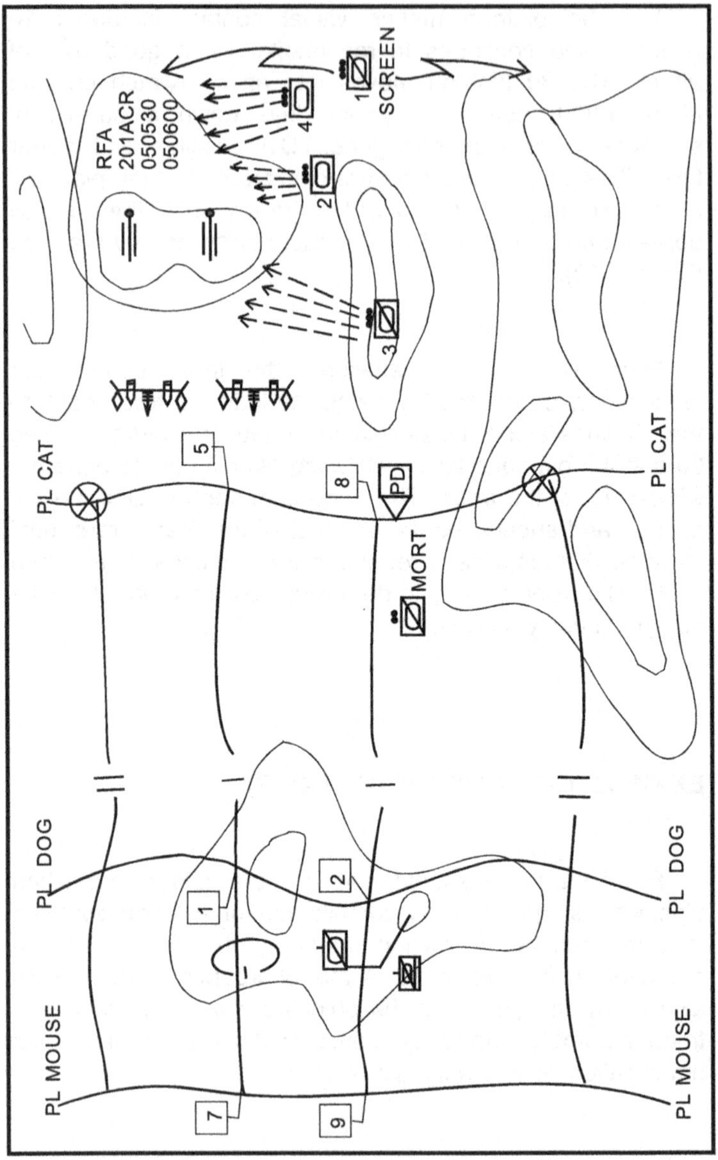

Figure 5-10. Heavy troop raid (part two).

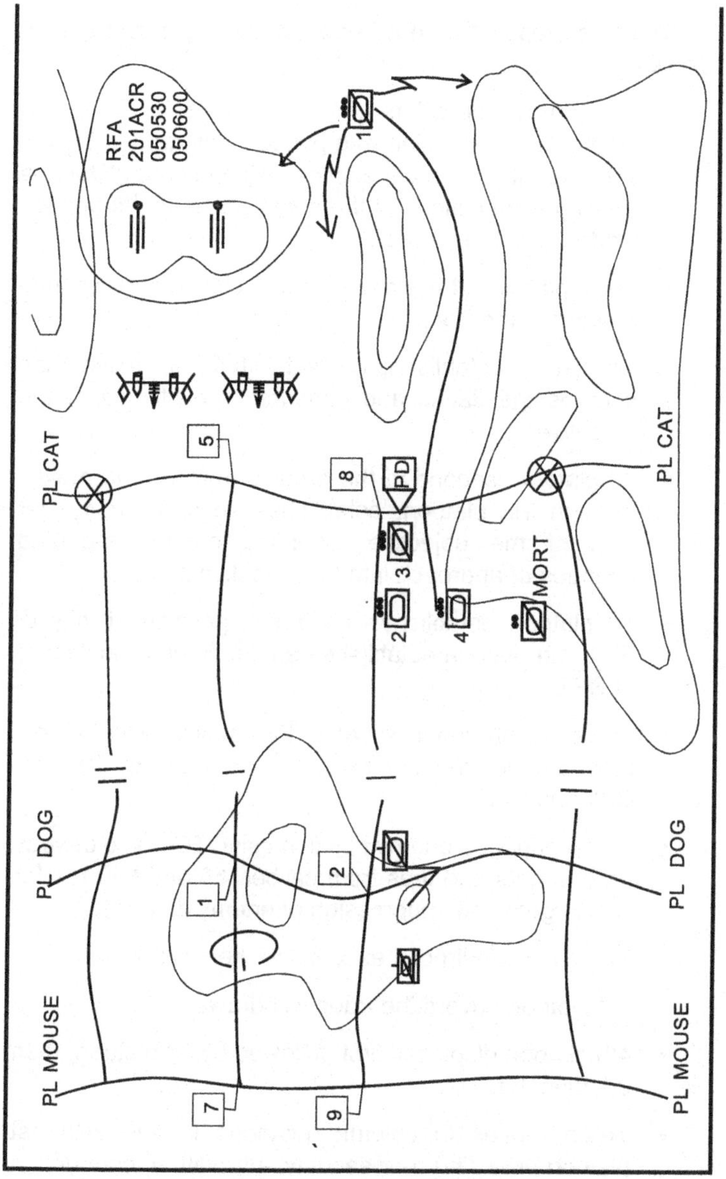

Figure 5-11. Heavy troop raid (part three).

Example steps of a light troop conducting a raid are as follows:

- Troop moves in column, platoons in a vee. Troop is a scramble organization with platoons integrating the AT vehicles into the scout sections. Each scout platoon is running two sections of three vehicles plus the platoon leader. See Figure 5-12.
- Lead platoon (1st platoon) conducts reconnaissance forward of the troop.
- 1st platoon establishes contact with the objective, then secures the flanks and conducts surveillance of the objective.
- 2d platoon, second in the order of movement, moves through 1st platoon, establishes security on the far side of the objective, orienting on the mounted avenues of approach into the objective area.
- 3d platoon establishes an attack position vicinity of BP1. 4th platoon establishes an attack position vicinity of BP2.
- Troop commander initiates the attack with indirect fires. 3d and 4th platoons move and attack by fire from BP1 and BP2.
- Troop engages enemy position using TOWs to destroy hard targets and fires from caliber .50 and MK-19s for soft targets and suppression of enemy defenses.
- Troop uses indirect fires to cover their withdrawal.
- 1st platoon covers the troop withdrawal.
- 4th platoon displaces first, followed by 3d platoon, then 2d, then 1st.
- Troop moves in column, platoons in vee with 1st platoon providing rear security. 4th platoon conducts a rapid zone reconnaissance forward of the troop as it moves toward friendly lines. See Figure 5-13.

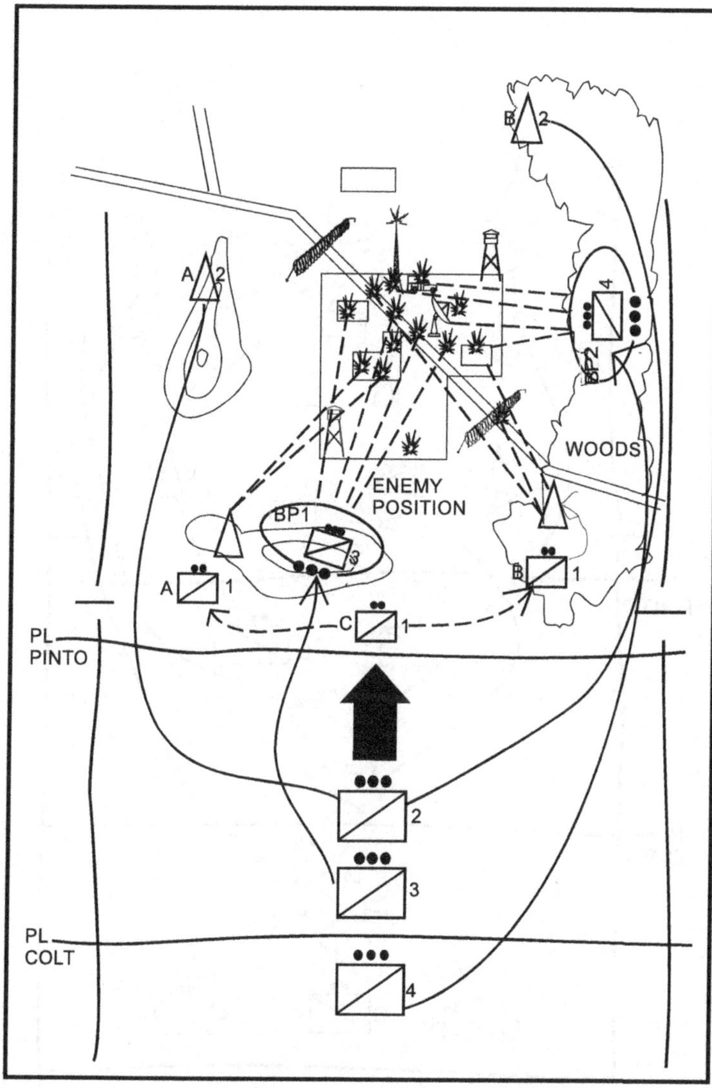

Figure 5-12. Light cavalry troop conducts a raid using a troop scramble.

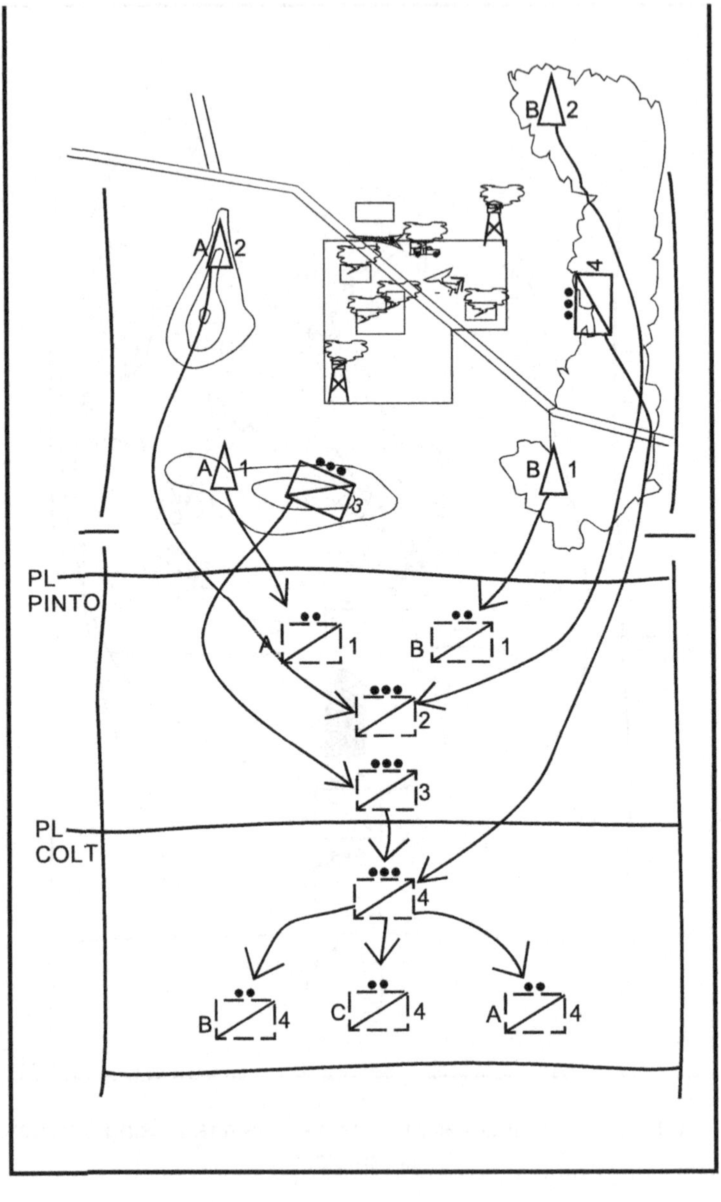

Figure 5-13. Light cavalry troop completes the raid and returns to friendly lines.

Chapter 6

Defense

Defense is not the decisive form of war. While defense can deny success to the enemy, it seldom assures victory. The defense is, however, the stronger form of war due to the inherent advantages of the defender. Army operations recognize the strength of the defense, but emphasize the necessity to transition to the offense quickly.

Brigades, battalion task forces, and company teams are the principal defensive forces for the corps or division. Cavalry units normally perform security missions for the defense or reconnaissance missions to support attacks. Cavalry units frequently perform defensive operations as a part of security missions, or when required to defend in an economy-of-force role for a higher headquarters.

Contents

		Page
Section I.	Purpose, Fundamentals and Schemes of Maneuver	6-2
Section II.	Defend From a Troop Battle Position	6-6
Section III.	Defend in Troop Sector	6-10
Section IV.	Delay in Troop Sector	6-22

Section I. Purpose, Fundamentals, and Schemes of Maneuver

PURPOSE

The immediate purpose of defensive operations is to defeat an enemy attack and create opportunities to go on the offense. For both the heavy and light cavalry troops, defensive operations may also achieve one or more of the following:

- Gain time.
- Concentrate forces elsewhere.
- Attrit enemy forces as a prelude to offensive operations.
- Control key or decisive terrain.
- Retain tactical, strategic, or political objectives.

In some cases, the troop may be required to defend because it cannot muster enough combat power to attack.

FUNDAMENTALS

Eight fundamentals are common to cavalry troop defensive operations.

Depth. Position platoons in depth, and place obstacles in depth. Depth allows the troop to—

- Gain enemy contact early.
- Perform counterreconnaissance tasks.
- Ascertain enemy direction of attack/intentions.
- Develop the situation, providing reaction time and maneuver space to concentrate combat power when and where it is needed.

Dispersion. Deploy subordinate elements as far apart as possible without losing their ability to concentrate (mass) firepower against the enemy. The more dispersed the troop, the harder it is for the enemy to mass fires against it as a whole. However, do not allow the enemy to concentrate its forces or fires against isolated elements.

Security. The troop may employ passive or active measures, or a combination of techniques. All must be considered in the defensive plan. Examples of passive and active security measures follow:

Passive	Active
Disperse vehicles and platoons.	Screen/Establish OPs.
Use camouflage/ cover and concealment.	Perform mounted/ dismounted patrols.
Impose radio listening silence.	Establish GSR posts.
Use hide positions.	Establish M8 chemical alarm net.
Enforce noise and light discipline.	
Minimize movement.	
Do not position in likely target areas.	

Maximize Terrain Advantages. Study the terrain. Reconnoiter it from both the troop commander's and the enemy's view, if possible, to determine—

- Avenues of approach.
- Reconnaissance avenues of approach.
- Restricted/severely restricted areas.
- Defiles (canalizing terrain).

- Engagement areas.
- Battle positions.
- Subsequent and alternate battle positions.
- Hide positions to support battle positions.
- OP positions forward of defensive positions.
- Subsequent and alternate OP positions.
- Positions where obstacles can be tied in with natural obstacles to turn, disrupt, or block the enemy.
- Positions that facilitate counterattacking by fire or by fire and maneuver into the flanks and throughout the depth of the enemy.
- Routes to and from each position.

Stop Enemy Rate of Advance. Offense is based on two principles—speed and mass. Develop a defensive plan that blunts the momentum of the enemy attack through the use of obstacles and fires (direct and indirect). If most of his combat power is killed, the enemy will be forced to dig in and establish a hasty defense or withdraw from combat.

Mass Combat Power at the Right Place and Time. In order to defeat a massed attack, the troop must mass fires against the enemy where and when he is least able to escape its effects. The ability of the troop to mass combat power when and where it is needed is a function of—

- Early warning/reaction time.
- Responsive/rapid maneuver by subordinate elements.

Force the Enemy to Fight in Two Directions. When engaging the attacking force, maneuver platoons into positions that force the enemy to turn and fight in two or more directions. This will force him to split his fires, preventing him from concentrating fires, and to expose his vulnerable flanks.

Counterattack. Draw the enemy into structured engagement areas and attack him *en masse* with overwhelming firepower to destroy him quickly and decisively. Wrest the initiative from him. Maneuver forces to exploit the situation. Attack by fire and by fire and maneuver into his vulnerable flanks and throughout the depth of his formation.

DEFENSIVE SCHEMES OF MANEUVER

There are three basic schemes of maneuver the commander can use in designating a course of action for a defensive mission. These schemes of maneuver center on the use of battle positions and sectors for subordinate platoons, or a combination of the two.

Defend from a Troop Battle Position. Both heavy and light cavalry troops may defend from a battle position. This method is be used when the enemy situation is clear, when there is only one avenue of approach, or when the troop commander must coordinate subordinate unit fires. In this scheme of maneuver the troop commander retains most of the authority for fighting the battle. The troop commander must understand his squadron commander's intent and concept to prevent holding the troop in place and risking its destruction.

Defend in Troop Sector. Both the heavy and light cavalry troops may defend in troop sector. This method may be used when the enemy situation is vague, when there is more than one avenue of approach, or when subordinate platoons require more freedom or action. In this scheme of maneuver, the troop commander delegates much of the responsibility for fighting the battle to his subordinate platoon leaders. He maintains control through the use of effective control measures and a clearly understood intent and concept.

Defend by Combination of Battle Positions and Sectors. Both heavy and light cavalry troops may defend using a combination of battle positions and sectors. This is the most common method of defense for a cavalry troop as it provides the troop commander maximum flexibility to maneuver and mass fires throughout the depth of the sector when changes in METT-T dictate.

Section II. Defend From a Troop Battle Position

Either the heavy or light cavalry troop may defend from a troop battle position. This mission is normally assigned when the squadron commander elects to concentrate the direct fires of the troop or squadron within an engagement area. The troop cannot maneuver outside the position without the squadron commander's permission. Within the battle position, the troop commander positions his platoons to concentrate all direct fires where the squadron has specified. The troop fights to retain the position unless ordered by the squadron commander to counterattack or withdraw. The troop may still retain the task to screen/maintain contact with the enemy forward of the battle position, depending on the squadron commander's intent. If so, the tank/AT platoons will initially be positioned within the troop battle position, and the scout platoons may assume adjacent positions within the battle position following execution of a security drill (see Security Drill paragraph in Chapter 4, page 4-15).

CRITICAL TASKS

To defend a troop battle position, the following critical tasks must be accomplished:

- Decide where the enemy will be killed and designate the engagement area.
- Establish OPs oriented forward and to the flanks of the BP to gain contact with the enemy force and provide early warning.
- Establish primary and alternate platoon BPs to concentrate direct fires within the engagement area as directed by the squadron commander.

- Designate supplementary platoon BPs to cover other routes of enemy approach.
- Establish sectors of fire for each platoon.
- Reconnoiter and establish platoon routes from hide positions to platoon BP, and for withdrawal to subsequent platoon/troop BPs.

TECHNIQUES

The squadron commander assigns troop battle positions when he can mass the fires of two or more troops in a squadron engagement area. The size of a troop battle position can vary, but it should provide enough depth and maneuver space for platoons to maneuver into alternate/supplementary positions and execute local counterattacks.

The troop commander conducts a thorough terrain study before the troop occupies the position, keeping in mind where the squadron commander wants the troop to concentrate its fires.

- Designate primary and alternate positions for each platoon.
- Position platoons to achieve flanking fires along the avenue of approach. Consider the effective range of each platoon's weapon systems.
- Position platoons to mass direct fires within the engagement area and to provide mutual support.
- Position platoons to cover any dead space in the engagement area. If not possible, plan indirect fires to cover the dead space.
- Use TRPs to control fires and orient weapon systems for each platoon.
- Occupy the BPs from the rear. Establish OPs to support the squadron plan. Allow platoon leaders time to reconnoiter and position vehicles to place effective direct fires within their sectors of fire. If better

defensive terrain lies outside the BP, call the S3 or commander for approval to adjust the boundaries.

- Once platoons are set, inspect them to make sure each platoon is properly oriented and has good fields of fire. When satisfied, have the platoons reconnoiter routes to subsequent positions and select firing positions for each vehicle. If time and engineer support are available, dig prepared positions for vehicles. Have the platoon leaders prepare platoon fire plans for approval.
- After preparing the BP, have all elements except OPs move to hide positions to reduce the risk of enemy observation and to decrease their vulnerability to enemy fires.
- Maintain security. Position the command post where it has FM communications with subordinate elements and the squadron. Position the troop trains behind good cover that is out of direct-fire range and allows quick access to each platoon position. The troop commander is positioned where he can observe the engagement area and control the troop. The FSO should be nearby to ensure coordinated fire support.
- Identify trigger points/lines.
  - Upon the enemy reaching what terrain feature does the troop initiate indirect fires? Direct fires?
  - Upon the enemy reaching what terrain feature does the troop displace to prevent decisive engagement?
- If the troop has to disengage and displace under fire to a subsequent position, bound the troop back by platoon(s), consistent with the squadron scheme of maneuver. If the troop disengagement and displacement are covered by another element, the troop may move as a whole to a subsequent position.

See Figure 6-1 for an illustration of the techniques described above.

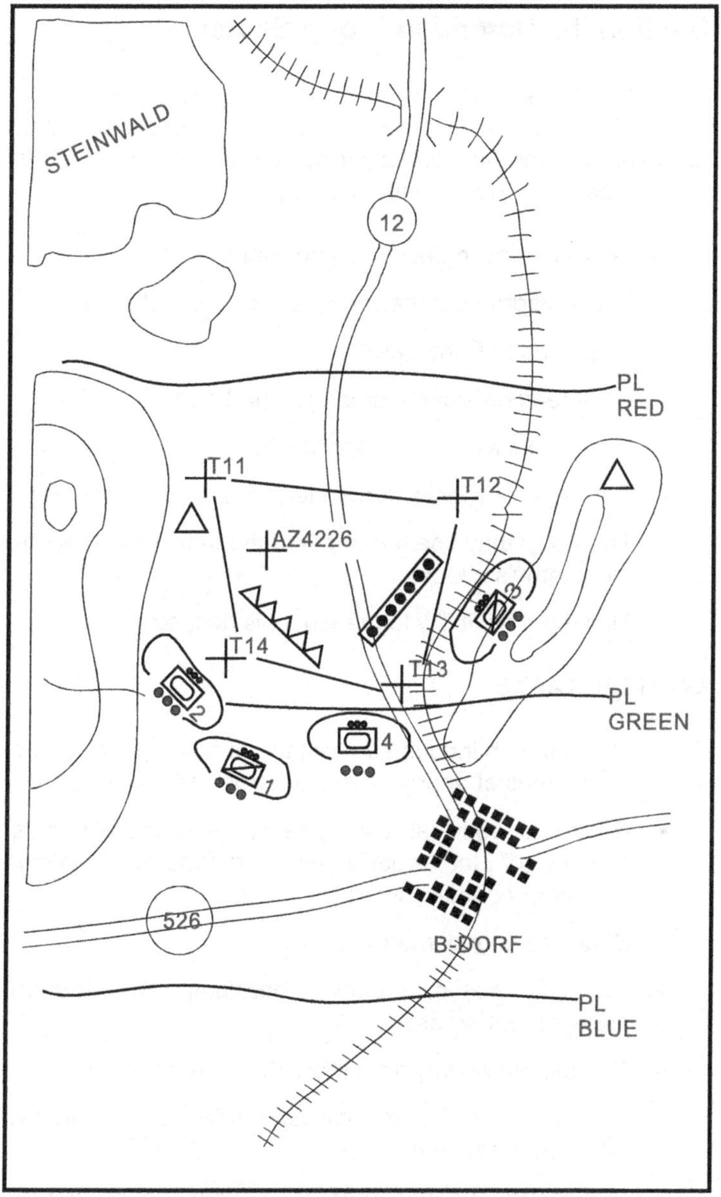

Figure 6-1. Troop engagement area.

Section III. Defend in Troop Sector

Both heavy and light cavalry troops may defend in sector. METT-T considerations determine optimal troop sector width; however, the troop is normally allocated a sector oriented on a single battalion-size avenue of approach.

Either troop may defend in sector when—

- The squadron cannot concentrate its fires due to—
  - extended frontages.
  - defending along a cross compartment.
  - multiple avenues of approach.
- Retention of specific terrain features is not necessary.
- The troop may use the depth of the sector to dissipate the enemy's attack.
- Maximum flexibility to maneuver is desired.

CRITICAL TASKS

- Maintain continuous surveillance of high-speed routes or avenues of approach into the troop sector (screen).
- Destroy or repel all enemy reconnaissance elements forward of the troop's initial defensive positions (counterreconnaissance).
- Structure engagement areas.
- Position platoon battle positions to support engagement areas.
- Engage the enemy from more than one direction.
- Determine criteria for initiating fires, counterattack, and disengagement.
- Prevent the enemy from penetrating the troop rear boundary or designated NPL (no penetration line).

TECHNIQUES

When given the order to defend in sector, the squadron will usually provide the following graphic control measures: *troop boundaries, an initial screen line, a rear boundary, phase lines, contact points between troops, TIRS, and TRPs that support any squadron engagement areas.*

Study the terrain in sector. Identify the terrain near the initial screen line from which OPs can maintain continuous long-range surveillance of enemy avenues of approach.

Determine where platoons can be positioned astride or on the enemy avenues of approach. Look for positions that provide good observation and fields of fire into the avenue of approach, and good cover and concealment for hide and defilade positions. Take a look at proposed platoon battle positions, and determine where troop fires can be massed on the avenue of approach. Use this portion of the avenue of approach to structure a troop engagement area(s).

The engagement area is where the troop will destroy an advancing enemy force. Establish a series of TRPs and use them to assign sectors of fire to each platoon. This allows the commander to control the fires of the troop and to achieve overlapping platoon fires.

With the assistance of supporting engineers, plan obstacles within the sector to support the defensive plan. Reinforce existing obstacles within engagement areas, and plan more obstacles to slow, canalize, or turn the enemy. Obstacles can buy the troop time to engage the enemy, and increase the effectiveness of indirect fires in the engagement area by compressing threat formations, slowing them down, and detaining them in the engagement area. Obstacles can give the commander time to maneuver platoons to counterattack or to move to subsequent positions. Plan obstacles in depth so the enemy gets bottled up in the engagement area and is confronted with a series of breaching operations. Make sure the troop can observe and place fires on all obstacles in the sector. Place the obstacles to achieve different effects. Obstacles emplaced on the reverse side of a hill or depression will cause the enemy to pile into them before he sees them. Strategically emplaced obstacles will cause the enemy to turn when he sees them, exposing his flanks to direct fires.

Give the FSO planning guidance so he can develop the troop indirect-fire support plan for the mission. Plan fires to support the scouts on the screen line forward of the troop. The scouts need indirect-fire support to engage enemy reconnaissance forces, to disrupt enemy lead echelon formations, and to attack follow-on forces. Plan indirect fires to engage enemy forces in the engagement area, when they are slowed by obstacles. These fires suppress, disrupt, and confuse the enemy and allow platoons to set up the direct-fire engagement. Plan indirect fires behind the engagement area to isolate the enemy. In addition, plan fires forward of the troop's positions to help disengage from the enemy in case the troop cannot stop him from initial positions. Identify rally points in sector behind the battle positions. Crews or troop elements that become separated or disorganized during battle move to these identified rally points to reassemble or reorganize.

Position platoons to maximize their weapons' effectiveness and crew/vehicle survivability based on the given terrain and the capabilities of the enemy. Platoons positioned at the base and along one or both flanks of the engagement area will force the enemy to fight in two or more directions.

Determine how to employ the scout platoons. Although the primary role of scout platoons is to conduct reconnaissance and screening in support of the troop, the troop commander may need to use their firepower to support troop defensive missions. Depending on sector width and number of avenues of approach, one scout platoon may be employed in a screen mission forward of the troop during a mission to defend in sector. It may fall back to a battle position after identifying the attacking enemy force (security drill), or it may stay forward of the troop, continuing to screen to identify follow-on forces. The other scout platoon may fight the attacking force from a battle position. The placement of the scout platoon's BP depends on the role the commander wants the platoon to play in the troop fight.

After making final adjustments to initial battle positions with the platoon leaders, plan alternate positions and subsequent positions in depth. Give platoon leaders time to reconnoiter covered and concealed withdrawal routes to their alternate subsequent positions.

Position the troop mortars where they can support the scouts on the screen line. Ensure they can engage targets from 3 to 3.5 kilometers beyond the screen line, or as far as the scouts can observe. Plan other mortar positions so that they can support the troop fight as the threat enters the engagement area. Also, plan positions through the sector.

The first sergeant positions the troop trains behind the initial troop battle positions, where they are responsive to troop needs but not vulnerable to direct fires. The XO positions the command post behind the initial platoon BPs on terrain that affords good FM radio communications with the troop elements and squadron headquarters. If possible, the XO positions the command post behind the subsequent troop positions; this reduces its vulnerability to fires, and allows it to remain stationary and maintain good FM communications while the troop displaces to other positions. The XO and first sergeant plan subsequent positions through the sector.

Position the FIST where it can maintain good digital FM communications with the supporting artillery unit. If possible, keep the FIST where it can use the laser designator to designate high-priority targets in the engagement area for Copperhead or other laser-guided munitions.

The troop commander must be in position to see the battlefield. To control the troop fires, he must understand which areas can and cannot be engaged by platoon battle positions.

The first critical task for a defend mission is to destroy or repel the threat reconnaissance. The scout platoon(s) on the screen line is responsible for identifying enemy reconnaissance forces, engaging them with indirect fire, and defeating them if possible. The other platoons may need to assist the scout platoon(s) in defeating the reconnaissance forces after the scouts identify them.

After destroying or repelling enemy reconnaissance forces, the troop is prepared to take on the lead echelon of the enemy force. Remember the scheme of maneuver. Let the enemy enter the engagement area and then mass the direct fires of the troop to strike a decisive blow. Depending on how the obstacles are set up, the troop commander may want to strike the enemy just before he reaches the obstacles; then, as the enemy deploys in reaction to troop fires, he hits the minefields and tank ditches. The commander may wish to wait until the enemy gets into the obstacles, and strike when he is confused and his formations are compressed.

Continue the fight by maneuvering platoons into alternate or supplementary positions while counterattacking by fire to complete the destruction of the lead echelon forces. If the troop is unable to defeat the enemy in the initial engagement area, it must be prepared to displace to subsequent positions. Do not allow the troop to become decisively engaged. Use the disengagement criteria from the scheme of maneuver to ensure adequate time to bound the troop back to subsequent positions by platoon(s). Keep one or two platoons in contact with the enemy, engaging him with direct and/or indirect fires. The other platoons move back in sector to subsequent positions. Once they are set, they engage the enemy with indirect fires or long-range TOW missile shots or tank main gun fire so the remainder of the troop can break contact and move to its subsequent positions.

Once the platoons are in their subsequent positions, make any needed adjustments to their positions or orientations based on what the enemy is attempting to do. Finish off the enemy from these positions, and then move forward and reoccupy initial or alternate positions, if possible.

EXAMPLE - DEFEND IN TROOP SECTOR (HEAVY TROOP)

The troop is given a mission to defend in sector and hold the threat forward of PL SABER. The troop commander completes his plan and issues the order. The 1st platoon moves forward and establishes a screen along PL SPUR. The 3d platoon prepares BP 20 at the base of the initial engagement area and occupies a hide position to the rear. BP 20 is about 2,500 meters away from the base of EA BEAR. The platoon's fires are concentrated in the direction of TRP 25. The 2d and 4th platoons occupy BP 30 and BP 40, respectively, along the eastern flank of EA BEAR. The 2d platoon orients on TRP 21, and the 4th platoon orients on TRP 27. The western flank is protected by a steep ridgeline. The troop command post is on high ground about 2 to 3 kilometers behind the initial defensive positions. The troop trains are just south of BP 30. The troop commander positions between BP 20 and BP 30, where he has a good view of all platoon positions and the engagement areas (see Figure 6-2).

A threat reconnaissance patrol of two BMPs approaches the screen line moving along the flanks of the high-speed route into the troop sector. As the patrol crosses PL SPUR, a 1st platoon scout element unmasks and quickly ambushes the patrol with cannon fire. One scout squad immediately searches the threat vehicles for intelligence information and captures two wounded prisoners. The troop commander tells the 1st platoon leader to expect a CRP within an hour. The rest of the troop stays in hide positions away from terrain most likely targeted by threat artillery units (see Figure 6-3).

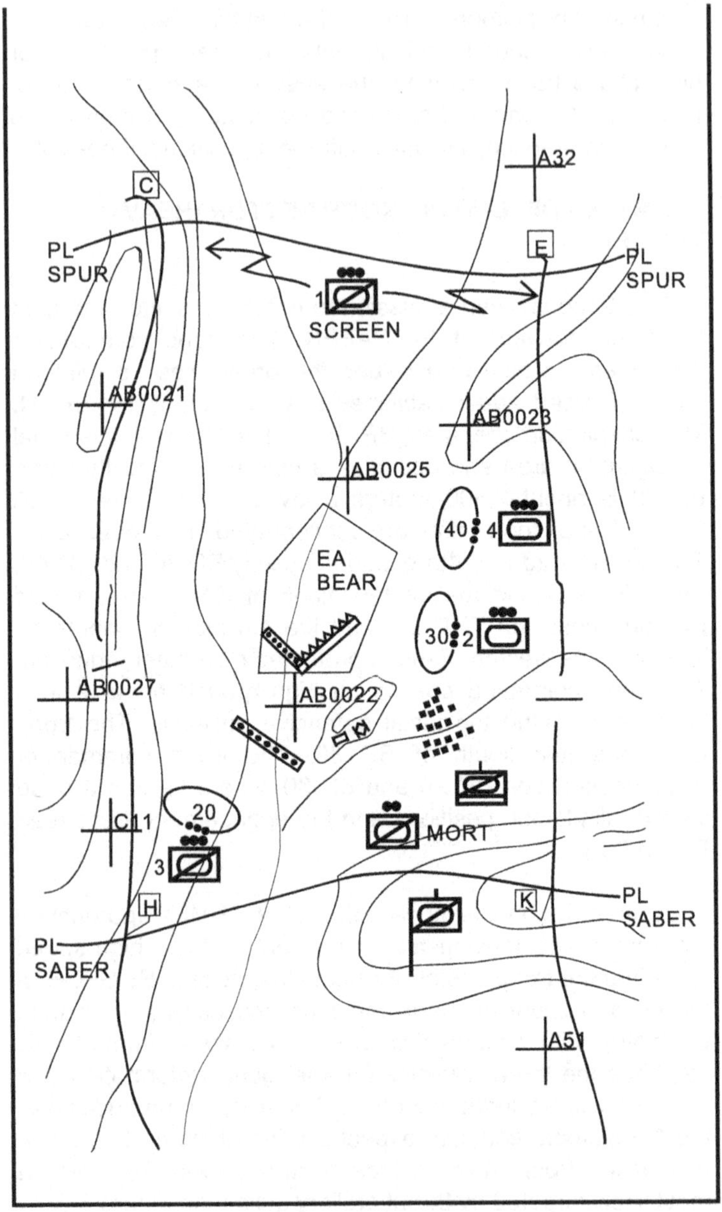

Figure 6-2. Defend in sector (part one).

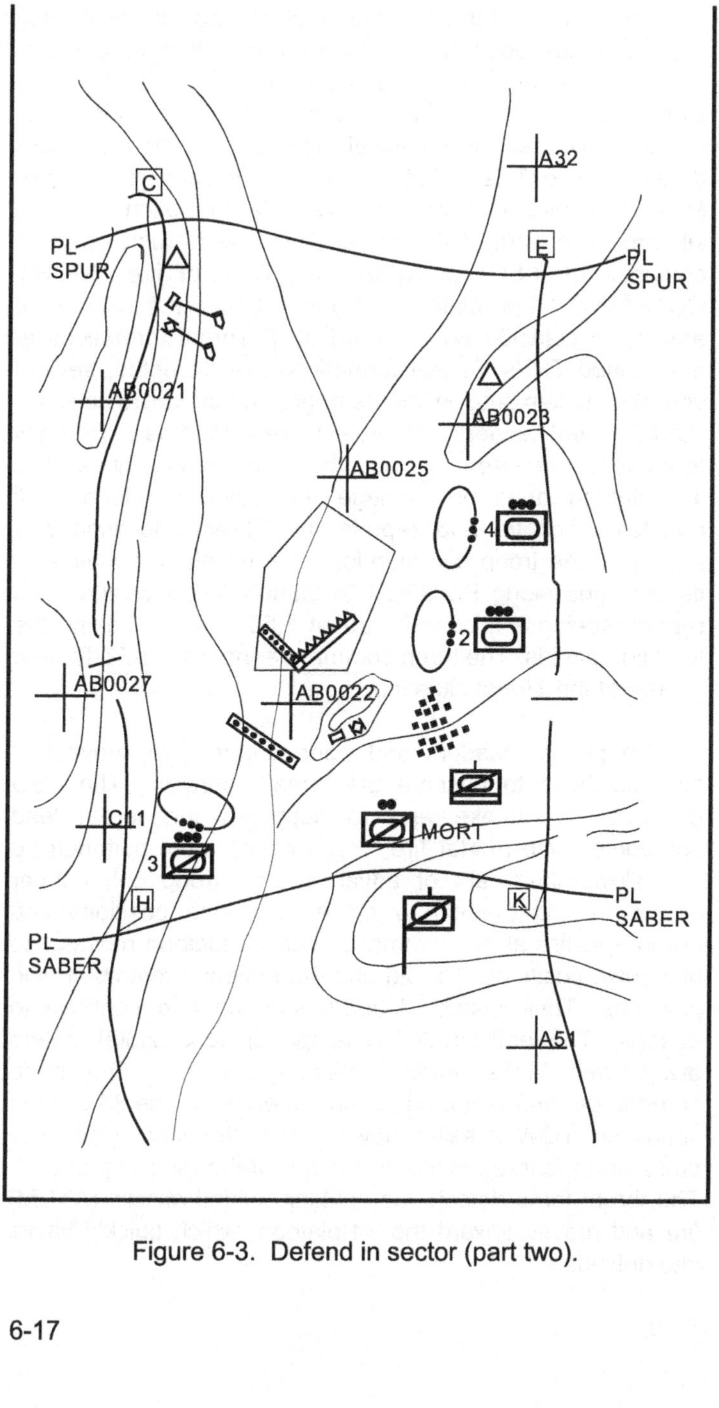

Figure 6-3. Defend in sector (part two).

About 45 minutes later, heavy concentrations of artillery and rocket fire begin falling along areas of high ground that dominate the avenue of approach near EA BEAR. Troop elements button up. When the suppression lifts, they immediately test for chemical agents. No chemicals are detected. The TCs unbutton. The 1st platoon reports two MRCs advancing abreast in prebattle formation about 2 kilometers forward of the screen line. The platoon sergeant contacts the FSO. Using the technique of fire "AT MY COMMAND," the platoon sergeant times the impact of artillery to coincide with the arrival of Threat formations at preplanned TRPs. Threat formations are disrupted, several vehicles sustain suspension damage, and the advance slows down. Threat leaders scramble to restore order, company formations are reformed, and the advance continues. The 1st platoon does not engage the advancing force, but maintains contact and reports the Threat's location and activity to the troop commander. The 1st platoon maintains its positions along PL SPUR to identify follow-on forces. It reports seeing a third MRC about 1,500 meters behind the lead companies. The troop commander now has a fairly clear picture of the Threat situation.

The platoon leaders and troop commander move into hide positions to observe the threat approach. The FSO continues to smoke and to suppress the threat lead companies with mortar fire, which keeps them buttoned up and slows their rate of advance. The troop commander orders the 3d platoon to move into firing positions and prepare to fire at his command. The 3d platoon moves into hull-down positions. The 2d and 4th platoons remain in hide positions. Their platoon leaders stay up and continue to observe. The lead threat MRCs appear about 2,800 meters away, with tanks leading platoon columns. The troop commander orders the 3d platoon to engage. The 3d platoon sends six TOW missiles down range, targeting mine-roller tanks and the threat platoon leader's BMPs (see Figure 6-4). The threat force detects the antitank guided missile (ATGM) fire and moves toward the 3d platoon, which quickly backs into defilade.

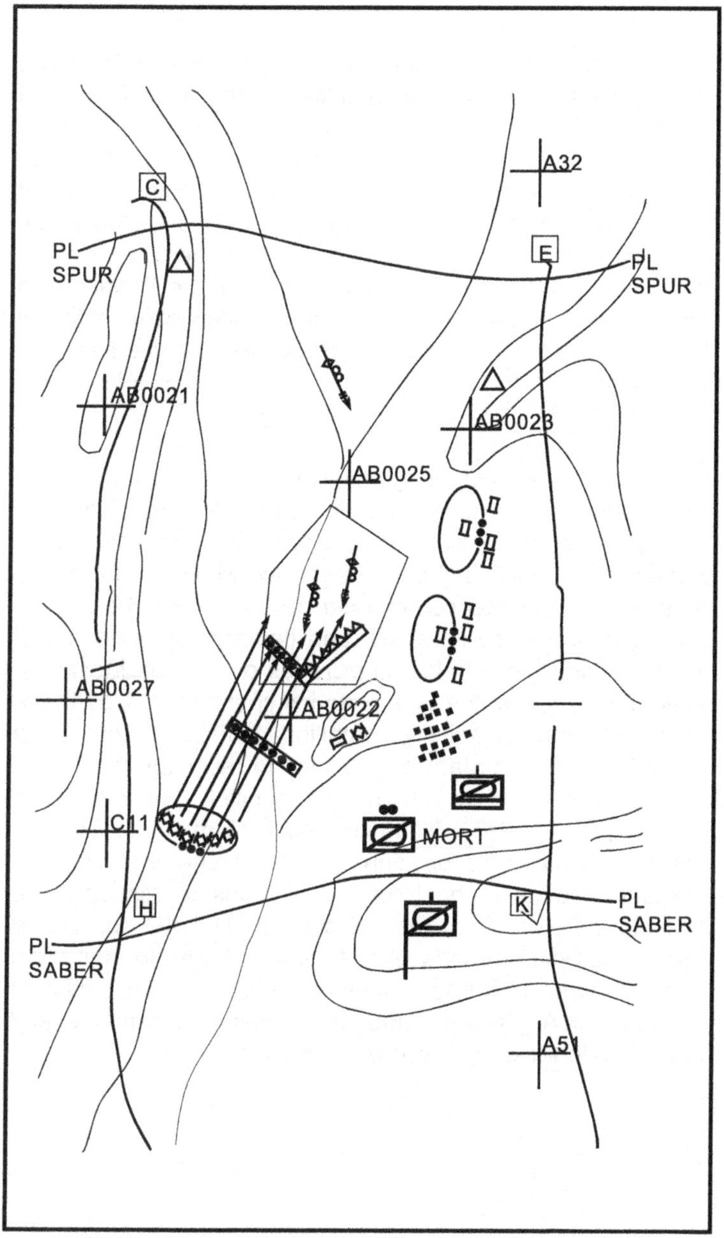

Figure 6-4. Defend in sector (part three).

The threat continues to advance through EA BEAR, piling into a minefield and tank ditch hidden on the reverse slope of a long, shallow draw. Several threat vehicles are caught in the obstacle. As the FSO observes this development, he concentrates all available fires on group targets that cover the obstacle. With the advance disrupted and stalled at the obstacle, the troop commander orders the 2d and 4th platoons to attack by fire into the flanks of the threat's lead echelon (see Figure 6-5). The concentrated cannon fire of both tank platoons quickly destroys most of the two lead companies.

The threat's third company comes into view in attack formation. The troop commander orders the 3d platoon to engage this force with long-range TOW missile fires. Survivors of this battalion begin to withdraw from the battlefield. The troop commander quickly orders the 2d and 4th platoons to counterattack and destroy the remnants of the battalion. The 3d platoon provides overwatch. Moving to alternate positions, the tank platoons counterattack by fire to finish off the remaining battalion vehicles. The troop commander then orders all platoons to reoccupy their initial hide positions and redistribute ammunition. He anticipates the second echelon will arrive within 30 minutes. The 1st platoon maintains its positions on the screen line. The first sergeant moves to each of the platoons to resupply. He collects the PWs from the 1st platoon. The XO collects all routine logistics reports and forwards them to squadron headquarters. The troop commander checks the status of leaders within the troop and designates replacements and cross-levels within the troop as necessary.

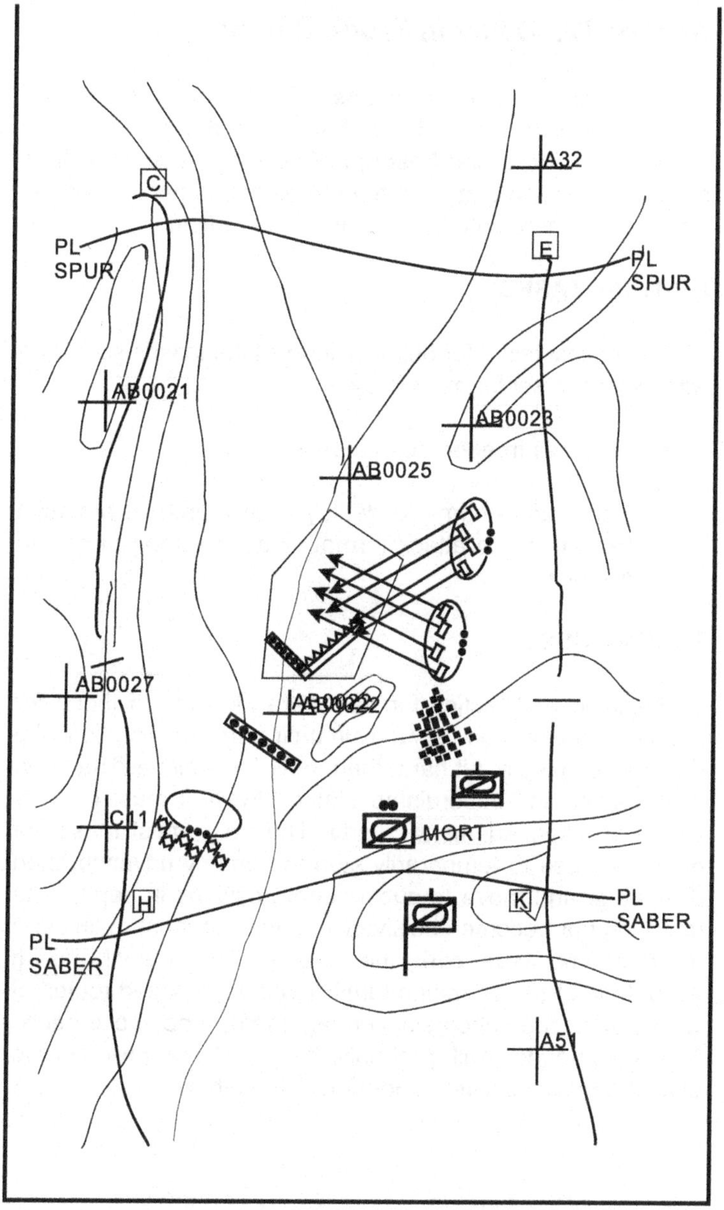

Figure 6-5. Defend in sector (part four).

Section IV. Delay in Troop Sector

Delay is a continuous series of defensive actions over successive positions in depth that trades the enemy space for time while retaining freedom of action. It is an economy-of-force operation that buys time to permit something else to happen at a more critical place on the battlefield.

CRITICAL TASKS

The critical tasks for delay include all the tasks associated with defend in sector as well as—

- Preserve freedom to maneuver.
- Cause the enemy to deploy from march or prebattle formation into attack formation as the troop moves to the rear.

TECHNIQUES

Planning and tactics for delay are identical to those for defend in sector, and vary only in their purpose. The flow of a delay resembles a "hit hard, then move" technique. The troop commander and subordinate platoon leaders must be very aware of disengagement criteria. The troop must mass the effects of fires to temporarily stop the enemy advance, then disengage and move to subsequent positions in depth. The troop cannot become decisively engaged. It must maintain a mobility advantage over the enemy. This means taking advantage of terrain, being familiar with high-speed routes of withdrawal, and rehearsing engagements and movements. The commander will probably have to use one or two platoons to assist disengagement of the others.

Chapter 7

Other Tactical Operations

Several combat operations are routinely associated with successfully accomplishing the missions described in Chapters 3 through 6. These operations require special planning and training considerations and techniques because of their complexity. At troop level, these operations are based on standing operating procedures (SOP) to ensure they can be conducted quickly and efficiently.

CONTENTS

		Page
Section I.	Tactical Road Marches	7-1
Section II.	Assembly Areas	7-9
Section III.	Relief in Place	7-15
Section IV.	Battle Handover/Passage of Lines	7-21
Section V.	Hasty Water Crossing	7-31
Section VI.	In-Stride Breach	7-35
Section VII.	NBC Defensive Operations	7-42

Section I. Tactical Road Marches

Troops not in contact with the enemy often travel long distances to position themselves for future operations. A successful tactical road march depends on unit discipline and the ability to execute the plan strictly in accordance with the SOP.

CRITICAL TASKS

Accomplish the following critical tasks when planning, executing, and supervising a tactical road march:

- Establish the readiness condition (REDCON).
- Issue a warning order.
- Reconnoiter the route to the start point.
- Conduct quartering party operations.
- Issue a movement order.
- Execute a tactical road march in accordance with movement order and SOP.
  - Cross and report the start point, checkpoints, and release point on time.
  - Conduct actions at halts.
  - Conduct actions on contact.
- Maintain security throughout movement and during halts.
- Operate a trail party.
- Clear the release point.

TECHNIQUES

The march discipline necessary to execute a road march with routine precision is attained only by strict adherence to SOP. Ensure the tactical road march portion of the troop SOP addresses, as a minimum, the following:

- Order of march.
- March speed.

- Distance between vehicles.
- Actions on contact (air, ground, and indirect fire).
- Actions at halts.
- Security.
- Contingency plans for vehicle breakdowns, breaks in column, and lost vehicles.
- Quartering party.
- Trail party.

The basic considerations in planning any road march are listed below.

- Time available.
- Distance of the move.
- Current situation.
- Availability and condition of routes.
- Size of the unit.
- Types, numbers, and characteristics of vehicles that must move.

The troop will most often move as part of a squadron operation, and the move will be based on the squadron order. The troop commander must, however, be familiar with planning considerations so he can plan and execute an independent troop move, if required.

March Planning Sequence. If time permits, follow this sequence when preparing for a road march:

- Prepare and issue a warning order, giving the troop's movement and any follow-on missions. Give them time to prepare for the operation.

- Analyze the situation to determine if any of the movement factors (order of march, rate of march, or interval) specified in the troop SOP must be altered to meet mission requirements. If the troop must reconnoiter two assembly areas, the two scout platoons might have to be first in the order of march.

- Conduct a map reconnaissance of the route (if assigned) or determine the best available route.

- Organize and dispatch reconnaissance and quartering parties (if occupying an assembly area).

- Prepare detailed movement plans based on mission requirements and reconnaissance information.

- Prepare and issue the march order to the orders group.

Route Reconnaissance. Reconnaissance is essential to movement planning. It reveals accurate, up-to-date information about the route. A route reconnaissance determines travel times, and identifies capacities of underpasses and bridges, locations of fords and bypasses, and critical points and obstacles. This lets the commander plan his movement and avoid areas or situations that could disrupt it.

Conduct the route reconnaissance with one of the scout platoons. Before the reconnaissance, let members know what information is needed and when and where they should submit it. Get them out as early as possible; troop plans are based on what they see and report.

If no other traffic control assets are available or if the troop is moving independent of the squadron, use the scouts from the platoon conducting the reconnaissance as traffic control teams at critical points along the route. Determine through map reconnaissance any built-up areas or key intersections that could disrupt the troop's movement, and position traffic control teams there. Based on their reconnaissance, the scouts may need to refine these positions.

If the movement is being conducted as part of the squadron, one troop will usually reconnoiter the route. The troop must, however, reconnoiter the route from its present location to the start point to determine if the route is suitable and how long it will take to reach the start point.

Quartering Party. A quartering party is used to reconnoiter and prepare a position before the troop main body arrives. The troop quartering party will often move as part of the squadron quartering party. Organize it with a scout platoon leader or the first sergeant as the party leader, with guides from each platoon, and with any additional personnel needed to clear the area.

Before the quartering party leaves, it must know the troop's route, order of march, estimated time of arrival, and any specifics on establishing the assembly area. Before the main body arrives, the quartering party reconnoiters the area, marks routes, and prepares to guide the main body as it enters the new area.

Trail Party. The troop trail party is made up of personnel and equipment (normally the troop trains) to handle emergency vehicle repair and recovery, medical aid and evacuation, and emergency refueling. The trail party moves just forward of the last maneuver platoon in the main body. The troop motor sergeant or first sergeant is in charge.

Main Body. The troop normally moves as a single march unit in column formation when conducting a tactical road march. The organization of the troop and any attached elements should be standardized during movement. Alter this organization to meet specific mission requirements as needed.

The troop commander has no prescribed place in the column. He positions himself where he can best control the movement of the troop. He will usually be well forward in the column, behind the lead platoon, to respond to contingencies while on the move. As a security measure, the troop command post should be positioned farther back in the column to disperse troop command and control.

The troop's column organization must provide adequate security against air and ground threats, while on the move and during halts. See Figure 7-1 for one way to organize a troop march column.

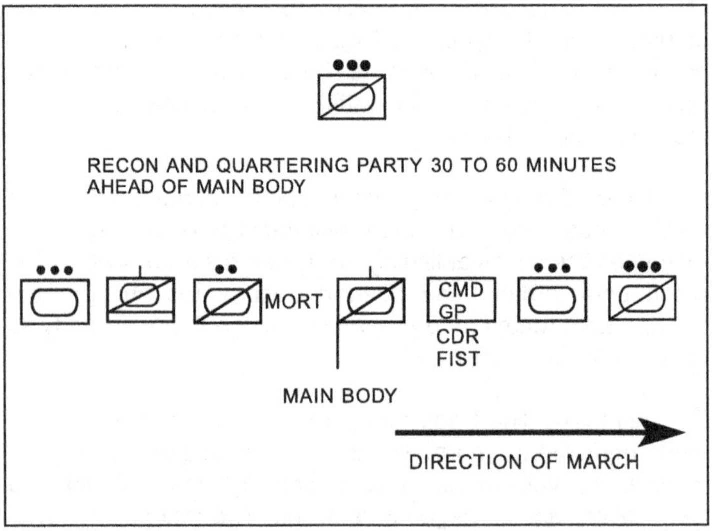

Figure 7-1. Troop column organization.

The 3d platoon, the first sergeant, and the quartering party move ahead of the troop. The troop order of march is 1st platoon, command group, 2d platoon, command post, mortars, trains, and 4th platoon. This order of march provides 360-degree security, disperses the command and control assets of the troop, and provides reconnaissance forward of the main body. Also, vehicle commanders assign sectors of observation to their crews, who search for air and ground threats (see Figure 7-2).

The troop performs the march in open or close column, depending on the situation. Use close column during limited visibility conditions. Vehicles are spaced 25 to 50 meters apart. This method takes advantage of the traffic capacity of the route, but provides little dispersion. Vehicle density is 15 to 30 vehicles per kilometer along the route of march. Use open column to provide greater dispersion and, thus, greater security. The distance between vehicles varies from 50 to 100 meters. Open column is normally used in daylight conditions. Base the troop's march speed on the slowest vehicle in the column. The troop trains will usually limit the rate of march.

Halts. Halts are used to rest personnel, provide personal comfort and relief, facilitate mess operations, refuel vehicles, maintain and inspect equipment, adjust the schedule, and allow other traffic to pass. The squadron march plan or troop SOP will specify the frequency and duration of halts, and will prioritize work to be done. For long movements, plan halts into the troop march table and ensure subordinate platoon leaders understand what actions must occur at the halt. When unscheduled halts occur, find the reason for the halt and let subordinates know how long it will last. Provide for security during halts, and establish OPs to provide early warning of enemy forces during long halts.

Vehicle commanders assign sectors of observation to their personnel so there is 360-degree observation. Each vehicle commander designates an air guard to provide air security, or specific vehicles may be designated air guard vehicles. In that case, the crews concentrate on only vehicle air observation rather than air and ground observation.

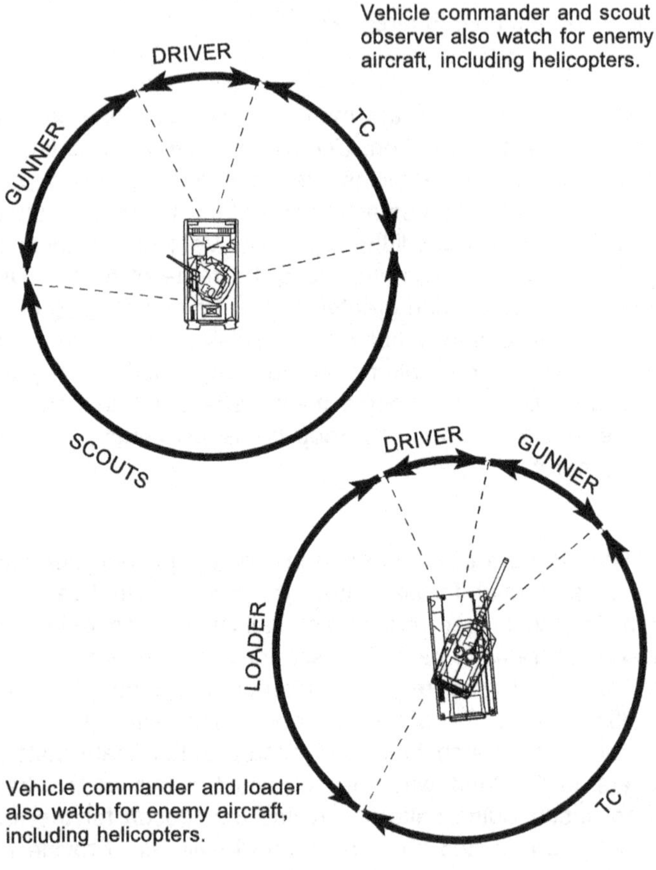

Figure 7-2. Maintain 360-degree security.

Vehicles that become disabled during movement must not obstruct traffic. The crew of the disabled vehicle must move the vehicle off the route, post guides to direct traffic, and find the problem. If the vehicle can be fixed, it rejoins the rear of the column. It does not return to its original position until the column has halted. If the vehicle cannot be readily repaired, the trail party recovers it.

Section II. Assembly Areas

An assembly area serves as a place where the troop gathers to prepare for future operations. In the assembly area, the troop prepares and issues orders, repairs and maintains vehicles, conducts resupply operations, and rests. As a minimum, assembly areas are positioned out of range of enemy light artillery.

TASKS

The following tasks are associated with assembly area operations:

- Plan the occupation of an assembly area.
- Conduct quartering party operations.
- Occupy an assembly area.
- Maintain security.
- Establish communications.
- Prepare for future combat operations.
- Depart an assembly area.

TECHNIQUES

The troop will normally be assigned a specific assembly area location. Within the area available to the troop, conduct a map reconnaissance, and if time is available, a ground reconnaissance. Look for an area that provides overhead concealment. This is extremely important if the troop will remain in the area for any length of time. Select an area that also provides—

- Cover from direct fire.
- Good drainage and a surface that will support troop vehicles.
- Good entrances and exits and an adequate internal road or trail network.
- Space for dispersion of vehicles, personnel, and equipment.

Quartering Party. The troop will often occupy an assembly area at the end of a road march. Before the quartering party leaves the troop's present location, tell them how to organize the assembly area. They must know any special requirements, such as a site for a logistics package (LOGPAC), so they can prepare the position. The quartering party must be organized to provide their own security during this operation. When the quartering party arrives at the forward assembly area, they must—

- Reconnoiter the area. If the area is not suitable, the leader of the quartering party must report immediately and provide a recommendation for another area.
- Organize the area. The leader of the quartering party selects locations for the platoons, mortars, command post, trains, and any attached elements, based on the commander's instructions. He may need to deviate from the commander's guidance to position an element in a suitable location.
- Improve and mark entrances, exits, and internal routes.

- Mark vehicle locations. Platoon representatives in the quartering party select general locations for vehicles in their platoon. Vehicle commanders and the chain of command refine these positions when they arrive.
- Perform guide duties. A platoon representative guides his platoon into position after the platoon clears the release point.

Occupation. Each platoon is guided from the release point into the assembly area by its quartering party member. Color-coded lights can be used to link up guides and lead vehicles (see FKSM 17-97-3). When the troop arrives at the assembly area, all elements move off the route without halting or slowing to keep the route of march clear. Keep this in mind when selecting routes, organizing the order of march for the road march, and allocating space in the assembly area. Once platoons have cleared the route and moved into their areas, they can adjust their positions without slowing the remainder of the troop.

Positioning troop elements in the assembly area is based on the size of the area, the terrain and avenues of approach, the length of time the troop will occupy the area, and any special requirements, such as resupply. Some considerations in allocating space for the troop are shown in Figure 7-3 and in the following list:

- If possible, space vehicles at least 100 meters apart to decrease exposure to enemy observation and fire.
- Keep vehicles in hide positions and establish OPs on terrain that provides good observation of approaches into the assembly areas for early warning of enemy movement. The vehicles can be moved up to a fighting position if necessary.
- Position platoons based on available terrain in the assembly area. Tank platoons may need more space to move vehicles in and out of position; scout platoons may occupy terrain that is more restrictive.

- Position the troop command post in the center of the assembly area. This makes establishing wire communications and issuing orders easier.
- Position troop trains where they have a good road network and space to conduct their maintenance and resupply operations. They may be used to secure the perimeter of the assembly area, but this will reduce their support ability.
- Emplace the mortars in a position from which they can support the troop with indirect fires. Overhead cover will reduce or eliminate their ability to fire.

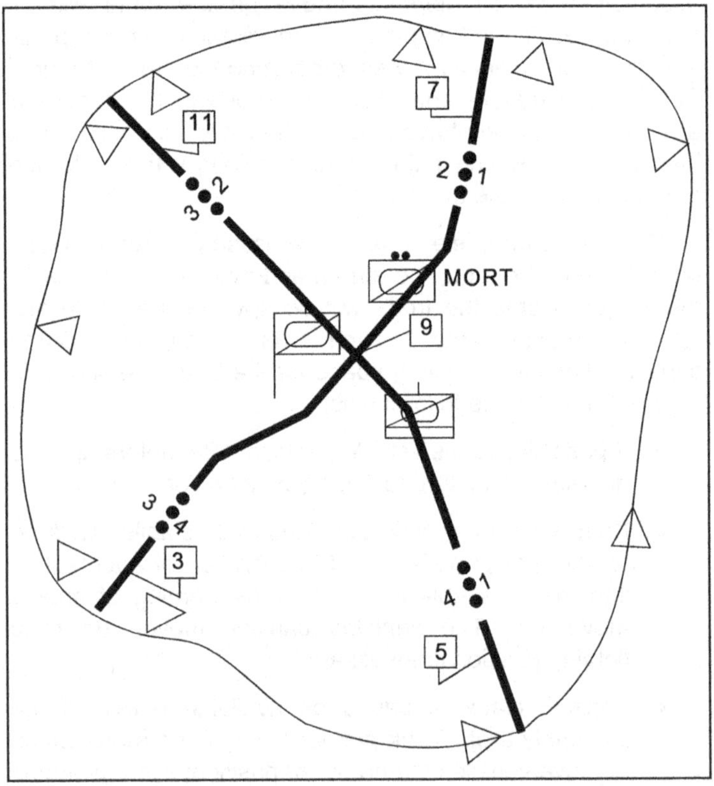

Figure 7-3. Troop assembly area.

Security. Although the assembly area is not a defensive position, the troop must be able to see and defeat enemy ground attacks. The best defense against air attacks is to remain hidden. Post guards at all entrances and exits to stop traffic that tries to enter the area. Establish OPs to observe key terrain features and likely avenues of approach for early warning of enemy approach. Each platoon must provide overlapping observation and fires within its platoon and with the platoons on its flanks. Establish a dismounted patrol plan so platoons make physical contact with their adjacent platoons. Ensure the platoons provide 360-degree coverage of the assembly area. Camouflage vehicles and equipment to prevent enemy detection from the ground and air. Place PEWS (platoon early warning system) in heavily vegetated areas or dead space to provide early warning of enemy movement in and around the position. Emplace NBC alarms upwind and about 150 meters from the troop's positions to provide early warning of an NBC attack. Assign the mortars an azimuth of fire on the most dangerous approach into the assembly area. Give the fire support officer guidance in preparing an indirect-fire plan.

Communications. Messenger and wire are the primary means of communications from the command post to each platoon and to each OP. If practicable, lay wire from the squadron to the troop. Plan on providing a messenger to the squadron command post. Use radio when no other means of communication can be used.

Preparation for Future Operations. Several tasks are routinely accomplished in an assembly area. These tasks are listed in the troop SOP (FKSM 17-97-3) under priority of tasks upon arrival in an assembly area, and include—

- Position vehicles.
- Establish local security.
- Establish OPs.
- Prepare fire plan.

- Establish wire communications.
- Maintain radio watch and man turret weapons.
- Camouflage positions.
- Prepare obstacles/mine plan.
- Select alternate and supplementary positions.
- Reconnoiter routes of withdrawal.
- Perform PMCS.
- Emplace NBC alarms and PEWS.
- Continue to improve positions.
- Conduct logistics resupply (Classes I, III, and V).
- Rest in accordance with REDCON status.

Modify this task listing to accomplish specific tasks (such as conduct rehearsals, test-fire weapons, and conduct inspections) in preparation for future operations. Ensure subordinates know how long the troop will remain in the assembly area and are told of any special requirements. Occupation of the assembly area will often be conducted and supervised by troop NCOs, while the commander and platoon leaders plan for upcoming operations.

Departure. Maintain the appropriate REDCON. Each REDCON level indicates critical tasks and time available to prepare for future operations.

- REDCON 1—be prepared to move immediately.
  - All personnel alert and ready for action.
  - Vehicles loaded and secured, and weapons manned.
  - Vehicle engines running and OPs not manned.

- REDCON 2—be prepared to move in 15 minutes.
  - All personnel alert.
  - OPs and wire pulled in.
- REDCON 3—be prepared to move in half an hour.
  - Fifty percent of each crew/squad stand down for rest, feeding, and maintenance.
  - Remaining 50 percent man vehicles, OPs, weapons, and monitor radios/phones.
- REDCON 4—be prepared to move in one hour.
  - Two men per platoon make dismounted checks of platoon area.
  - One man per vehicle monitors radios/phones and mans turret weapon.

All personnel remain at 100-percent alert until the prioritized work is complete after entering the assembly area. Initiate the appropriate REDCON when the work is finished. As the time for execution of a mission nears, increase the REDCON in accordance with guidance from squadron, assigning REDCON 1 just before the troop must move.

Section III. Relief in Place

A relief in place is an operation in which a unit in combat is replaced by another unit. Responsibilities for the combat mission and the assigned sector, battle position, or zone of the relieved unit are assumed by the relieving unit. A relief in place may be accomplished during offensive and defensive operations and may be conducted during any weather and light conditions.

The primary purpose of a relief in place is to maintain the combat effectiveness of committed units. A relief in place is conducted to replace a committed unit to give it the opportunity to reconstitute, rest, decontaminate, or perform a change in mission.

TASKS

The following tasks are associated with the relief in place:

- Plan and coordinate a relief in place.
- Establish communications.
- Establish liaison.
- Conduct reconnaissance.
- Initiate movement.
- Occupy positions.
- Maintain operations security.

TECHNIQUES

A troop will often perform a relief in place mission as part of a squadron operation to relieve or be relieved by a brigade. The troop will usually conduct a relief in place in conjunction with a battalion. The troop may relieve a battalion and operate in an economy-of-force role or be relieved by a battalion to conduct a change in mission.

To reduce confusion and maintain security, the following factors must be considered when planning a relief in place:

- The time that responsibility for the sector, battle position, or zone is to pass.
- Operations security.
- Deception plans.

- Time, method, and sequence of relief.
- Routes and critical control measures.
- Concept of subsequent missions.
- Plans for additional positions.
- Contingency plans.
- Location of obstacles and transfer of responsibility.
- Transfer of ammunition; wire lines; petroleum, oils, and lubricants (POL); and materiel to the relieving unit.

Communications. At a specified time or upon receipt of an order, the relieving unit minimizes radio traffic or begins to operate at radio-listening silence. The relieving unit command post monitors the command net of the unit to be relieved while continuing to operate on its internal command net.

Reconnaissance and Liaison. The orders group of the relieving unit moves to the command post of the unit being relieved for face-to-face coordination.

In preparation for the relief in place, the relieving unit moves to a forward assembly area under the command of the XO, if necessary.

If the situation permits, the relieving unit conducts a reconnaissance of the area when it completes coordination. The units establish a contact point behind the positions into which they will move; a guide from the relieved unit will meet them there to make any final coordination and guide them into position.

The commanders complete their plans and issue orders to their respective units after the reconnaissance and liaison are complete.

Methods. The relief in place can be conducted in several ways. Use the method appropriate to the troop's situation.

- One unit at a time. This is the slowest method, but the most secure.

- All units simultaneously. This is the quickest method, but the least secure.

- Center units first, followed by flank units.

- Flank units first, followed by relief of the center unit.

- Occupy adjacent or in-depth positions that cover the area of responsibility.

Passage of Command. The troop command post and battalion TOC collocate to facilitate command and control of the operation. The relieving unit departs its location in the prescribed sequence and lines up at the contact points with the guides from the relieved unit. The relieving unit moves into hide positions behind the relieved unit. The relieving unit moves into position after all pertinent information has been exchanged and the relieved unit moves out of position. The relieved unit moves to a rally point or assembly area to link up with the rest of the unit. When each element has completed the relief, the relieving unit assumes responsibility for the area and reports by messenger, wire, or radio to the command post. The relieved unit moves along its designated route to the new location. See Figures 7-4 and 7-5 for illustration of the relief in place.

The higher headquarters will often specify a time the relieving unit assumes responsibility for the sector, battle position, or zone. The commanders conducting the relief coordinate to ensure the relief is conducted on time. The relieving unit normally assumes responsibility when the relieved unit departs its positions. The commanders of the two units are usually collocated and responsibility passed face to face.

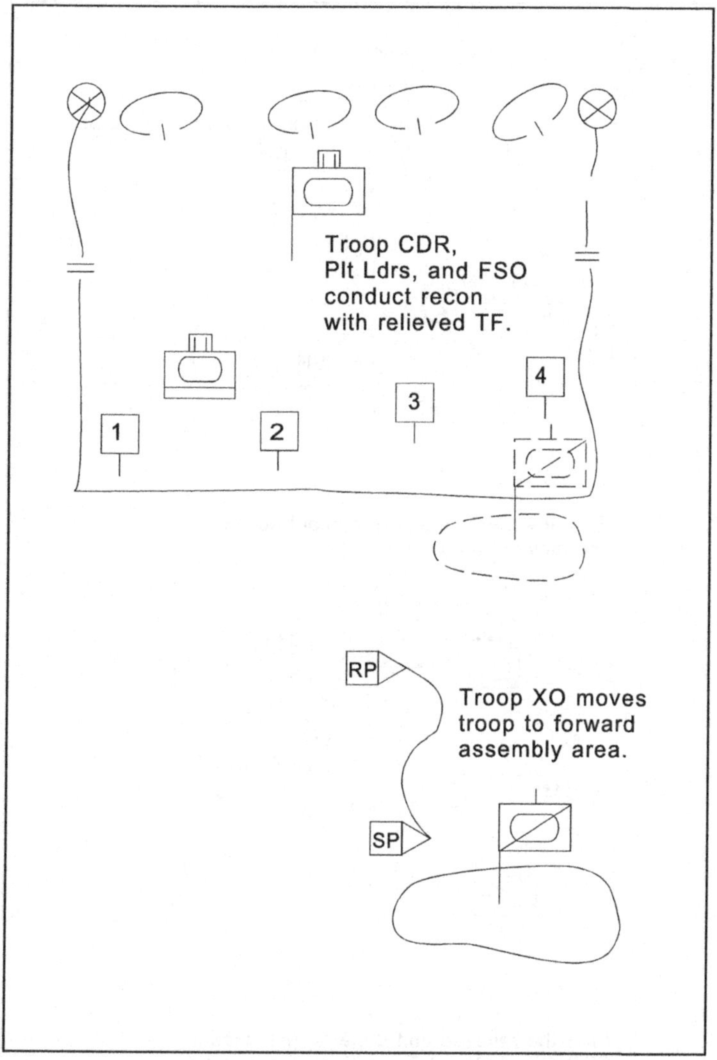

Figure 7-4. Troop prepares to conduct relief in place with task force.

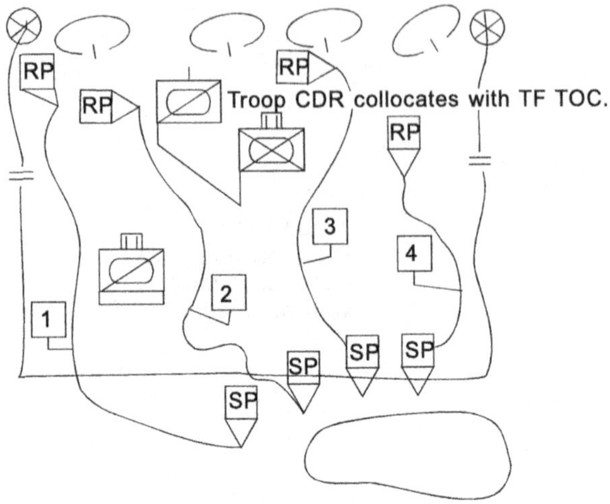

Platoons meet guides at contact points and move into position.

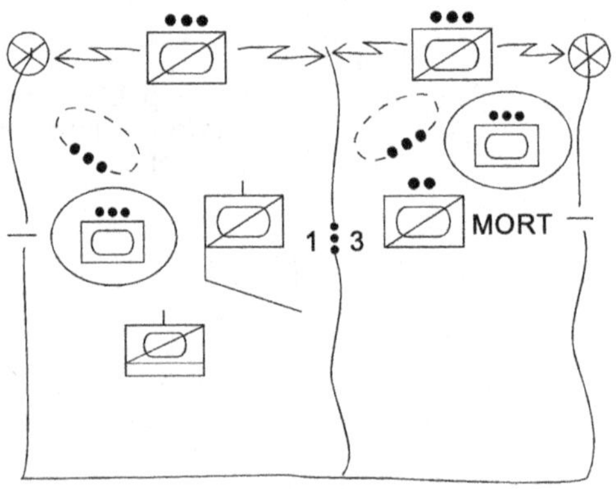

After the relieved unit departs, the troop CDR adjusts his PLT positions if necessary.

Figure 7-5. Troop completes relief in place.

Fire Support Assets. The troop fire support officer coordinates with the squadron fire support officer to ensure there is fire support throughout the operation. The fire support assets of the relieved unit remain in position throughout the relief of the maneuver forces, and are prepared to support both units. Fire support assets of the relieving unit position themselves as quickly as possible to provide additional support.

Enemy Contact. If either unit gains direct-fire contact with an enemy force, it immediately notifies the other unit and the higher headquarters directing the relief. If responsibility has not passed, the relieving unit becomes OPCON to the relieved unit. The relieving unit's mortars fire missions as directed by the relieved commander through his fire support officer. If responsibility has passed, the relieved unit becomes OPCON to the relieving unit. The collocation of the commanders and command posts facilitates rapid coordination and action in this situation.

Section IV. Battle Handover and Passage of Lines

Battle handover is an operation that transfers responsibility for fighting an enemy force from one unit to another in the close-in battle. Conducted by stationary and passing units, battle handover is designed to sustain continuity of the combined arms fight, and to prevent the enemy from getting a "free ride" anywhere on the battlefield as one force picks up the fight from another. It is also designed to preserve the fighting capabilities of both forces as they execute the operation. Battle handover is associated with almost all Army combat operations.

Passage of lines is a tactical event associated with battle handover. It is the controlled movement of one unit through the positions of another stationary unit that does not interfere with either unit's scheme of maneuver. A passage of lines is often used because the combat situation does not permit one unit to bypass another unit's positions. The passing unit must move through the positions of the stationary unit.

Passage of lines is often an integral part of the following operations:

- Deliberate attacks or counterattacks across the FLOT.
- Exploitation and pursuit.
- Route, zone, or area reconnaissance.
- Raids.
- Movement to contact.
- Defend, delay, or withdrawal.
- Defensive or offensive cover.
- Screen or guard.

Although a cavalry troop does not normally perform all of these operations, it may conduct them as part of a larger force.

CRITICAL TASKS

The three players in battle handover and passage of lines are: the *stationary unit*, the *passing unit*, and the *common commander* of both units. Each has critical tasks to perform to achieve smooth and efficient execution of the operation.

Critical Tasks of the Common Commander. The commander exercising command authority over both the stationary unit and the passing unit must accomplish three critical tasks.

- Establish where battle handover will occur by designating a phase line forward of the FEBA as the battle handover line (BHL). The line should be where combat maneuver forces of the stationary unit along the FEBA can effectively overwatch and protect the passing unit as it withdraws behind the FEBA or advances forward of the FEBA or FLOT. The distance forward of the FEBA or FLOT is limited to available fields of fire and the effective range of weapons of the stationary unit.

- Designate contact points just forward of the battle handover line at which stationary and passing units are required to conduct physical coordination (only in defensive operations).
- Ensure the passing unit is provided indirect-fire support while its artillery is displacing during battle handover and passage of lines.

The stationary unit normally recommends battle handover line and contact point locations to the common commander. It remains the responsibility of the common commander to establish the line and contact points. These control measures must be reflected graphically on an overlay and identified in the appropriate operation plan (OPLAN), OPORD, or FRAGO issued to subordinate units. The battle handover line, in effect, establishes a boundary between the ground owned and controlled by the stationary unit commander and the ground controlled by the passing unit commander. The stationary unit commander controls the ground forward of the FEBA up to the battle handover line. He can place security forces, obstacles, and direct and indirect fires into this area to support his scheme of maneuver.

While the battle handover line defines the point at which the battle handover should ideally occur, events normally preclude this from happening. The moving and stationary force commanders and their common commander should understand that the actual handover may occur in a "zone" centered on the battle handover line. Both the moving and stationary force must remain active and responsive enough to perform the battle handover anywhere in this zone, and should not dogmatically "stick to the plan."

Battle handover begins on the order of the common commander. To sustain unity of command in the passing operation, the passing unit is usually placed OPCON to the stationary unit to execute battle handover and passage of lines. OPCON by the stationary unit is limited to those actions necessary to get the passing unit through the stationary unit's area as quickly as possible. In defensive operations, battle handover from the passing unit to the stationary unit is not complete until the combat maneuver forces of the stationary unit along the FEBA have visual contact with the enemy, and the stationary unit commander indicates readiness to assume responsibility for the battle. In offensive operations, battle handover from the stationary unit to the passing unit is not complete until the combat maneuver elements of the passing unit have crossed the FEBA or FLOT, deployed for combat, and maneuvered across the battle handover line.

Passing Unit Critical Tasks. The passing unit must accomplish the following critical tasks to perform battle handover and passage of lines:

- Immediately establish communications with the stationary unit. Enter the command, operations and intelligence (OI), and fire support nets.
- Collocate a command post with the TAC CP or TOC of the stationary unit as soon as possible to enhance communications and unity of effort.
- Continuously report to the stationary unit the location, size, and composition of all enemy forces. Report the enemy's current activity. If the enemy is attacking, report his direction of movement, movement formation, and estimated rate of advance. If he is defending, report his locations, orientation, composition, fire sacks, reserves (if known), obstacle system, or flanks.
- Continuously report to the stationary unit the location, size, and activity of all subordinate elements to include combat support, combat service support, and command and control facilities.
- Given the current disposition of subordinate units, coordinate with the stationary unit to determine

contact points at which each subordinate company-size unit will physically coordinate handover and passage of lines with representatives of the stationary unit. Once contact points are determined, send a FRAGO to each subordinate unit specifying where they will physically coordinate passage with the stationary unit. Confirm recognition signals that must be displayed during passage (defensive operations).

- Once each subordinate unit acknowledges where it must physically coordinate passage, each unit will dispatch representatives to assigned contact points and coordinate passage for its unit. At the contact point, confirm recognition signals and exchange required information (defensive operations).
- Maintain visual contact with all enemy units and delay back to the battle handover line, avoiding decisive engagement (defensive operations).
- Display correct recognition signals and use correct challenge and password as specified in the CEOI during passage.
- Maintain proper weapons orientation.

Stationary Unit Critical Tasks. The stationary unit must accomplish the following critical tasks when ordered to conduct battle handover and passage of lines:

- Establish communications with the passing unit. Coordinate and direct the passing unit to contact points based on current dispositions of the subordinate units (defensive operations).
- Ensure contact points are manned and subordinate commanders have personal communications with their representatives (defensive operations).
- Ensure representatives at the contact points assign each passing unit a passage point into the area of operations and a route that extends from the passage points to the rear boundary or to an assembly area (defensive operations).

- Ensure representatives at the contact points exchange required information with the passing unit as outlined in FKSM 17-97-3.
- If security forces are employed, position them along the battle handover line to observe enemy avenues of approach. Adjust as needed for low visibility conditions (defensive operations).
- If obstacles are emplaced between the FEBA and the battle handover line, ensure routes through the obstacle system are clearly marked and physically controlled by guides, or provide an escort to the passing unit. Ensure that reserve targets on obligated routes are manned by soldiers in direct communication with their commanders.
- Ensure that all routes of withdrawal obligated to the passing unit are unobstructed and facilitate rapid movement to the release point (defensive operations).
- Ensure obligated routes of advance, attack positions, and routes to the battle handover line are unobstructed and permit rapid movement (offensive operations).

GRAPHIC CONTROL MEASURES

Figure 7-6 shows the graphic control measures that support battle handover and rearward passage of lines.

- Battle handover line. This line is established by the common commander in consultation with both commanders. The stationary commander has the major determination in locating the BHL, as his force must be able to overwatch the BHL with direct fires.
- Contact points. These are established on identifiable terrain and normally in the vicinity of the passage lanes. For rearward passage of lines, the contact points are established forward at the BHL. For forward passage of lines, the contact points are established in the stationary unit's rear area rearward of the passage lanes.

- Passage points. The passage point is that point on the passage lane where the moving unit moves through, and responsibility for the battle is passed to the stationary unit. It is usually placed where the passage lane begins.
- Passage lanes. The stationary unit establishes passage lanes to move the passing unit quickly through defending unit positions. This could include passing through gaps in friendly obstacles and moving near or through friendly engagement areas and battle positions. Lanes are restrictive; however, they should ideally be wide enough to allow the passing unit to move in a tactical formation. The passage lane begins at the passage point and ends at the rear of the stationary unit BPs. The passage is considered complete when the moving unit exits the lane.
- Routes. Routes are used to move the passing unit through the stationary unit rear area. The number of routes designated will vary based on METT-T, but as a general rule, multiple lanes/routes should be planned to facilitate rapid passage of moving units and to avoid unnecessary massing of units. The stationary unit may escort or guide the passing unit along the lane/route.
- Assembly area. An assembly area in the rear area of the stationary unit allows the passing unit to conduct hasty reorganization and emergency CSS actions. This assembly area is temporary in nature.
- Infiltration points. Units should plan infiltration points and lanes for personnel unable to complete the passage with the unit. Passing unit liaison officers may remain located with stationary unit CPs to serve as a point of contact for infiltrating personnel/equipment. Personnel who infiltrate must have some way of contacting the stationary unit before crossing into friendly territory.

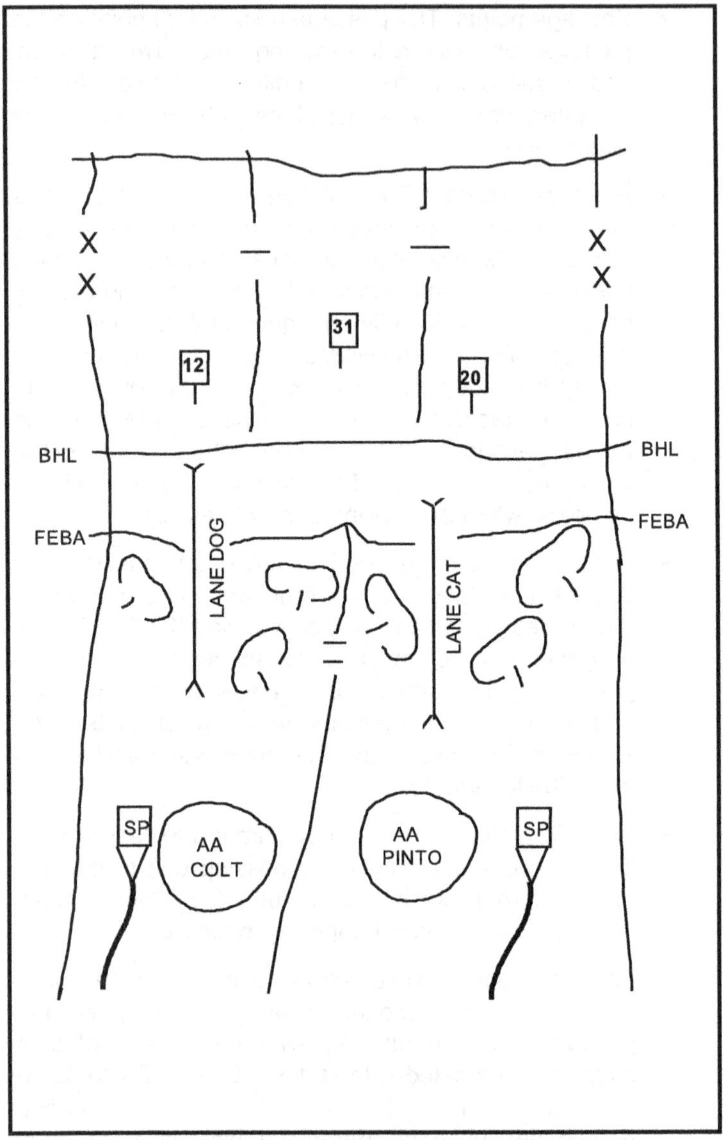

Figure 7-6. Rearward passage of lines.

TECHNIQUES

At troop level, the passage of lines will usually be performed as part of a squadron operation. The passage may be forward, such as to pass through a defending unit to conduct a counterattack, or rearward, such as when a covering force unit withdraws through units in the main battle area.

The troop is particularly vulnerable during a passage of lines. The unit may be concentrated and the fires of the stationary unit may be temporarily masked. Thorough reconnaissance and detailed coordination are critical to ensuring the operation is successful.

A troop will often perform a battle handover and passage of lines through a single battalion while the squadron conducts the operation through a brigade. Ideally, the boundaries will correspond to the battalion boundaries so that coordination for the battle handover is conducted through a single headquarters.

The troop commander has numerous considerations to ensure a successful passage of lines. During reconnaissance, he must confirm—

- The disposition of the stationary force through which the troop will pass.
- The location of contact points where both units are required to make physical contact at a predetermined time.
- The location of passage points on the battle handover line through which friendly forces will pass.
- The location of passage lanes that provide a clear route through a friendly position, and also facilitate a smooth and continuous passage. Areas selected for passage should be unoccupied or on the flanks of units in position. If possible, use multiple routes to reduce vulnerability during the operation.

- The location of an attack position (for forward passage) or assembly area. This position should provide cover and concealment and be located where the passing unit will not interfere with the stationary unit.
- The initial location for combat support and combat service support elements.

Based on the reconnaissance, the troop commander coordinates and plans for—

- Supporting fires. The stationary force supports the passing unit with direct and indirect fires up to the battle handover line. In forward passage, the stationary force supports the passing unit's move through the passage and until it crosses the battle handover line. In a rearward passage, the stationary unit supports the passing unit's move back across the battle handover line and through the passage of lines.

- Time of transfer of responsibility for control of the sector and handover of the enemy.

- Troop density. The passing troop commander should plan for multiple routes of passage to ensure rapid movement and to avoid congestion.

- Traffic control. Guides from the stationary unit pick up passing elements at each contact point and guide them through the position. The passing unit commander tells the stationary unit the type, number, and order of vehicles passing through each contact point.

- Communications. The leaders exchange CEOI information and mutually agreed upon recognition signals.

- CSS. The troop commander must coordinate the evacuation of casualties, PWs, vehicles, and resupply of fuel and ammunition. The stationary unit usually provides emergency service only. The passing unit supports itself.

- Liaison officers. The troop commander should designate a representative to perform the critical duties of a liaison officer. Commanders normally coordinate a forward passage of lines and the XO coordinates a rearward passage. Liaisons are normally located at critical points during the passage. If the commander or XO is not available, a scout platoon leader should perform liaison duties. Ensure he is thoroughly briefed on the situation and follows the checklist in the troop SOP.

Section V. Hasty Water Crossing

A hasty water crossing is the movement across an inland waterway using a crossing means at hand or readily available without significant delay once the waterway is reached. It is preplanned and conducted as a continuation of the operation underway. Although the crossing is termed hasty, detailed planning assures that fire support and crossing means are available on arrival at the water obstacle.

CRITICAL TASKS

The following critical tasks are associated with the successful conduct of a hasty water crossing:

- Plan the water crossing.
- Conduct reconnaissance.
- Establish security.
- Cross the obstacle.
- Provide continuous fire support.
- Continue the mission.

TECHNIQUES

Plan. A troop may perform a hasty water crossing independently or as part of a squadron. A hasty water crossing is performed as an extension of the ongoing operation. It gives the troop the ability to sustain the momentum of an operation, and is often associated with a movement to contact and a zone reconnaissance. The hasty water crossing is characterized by speed, surprise, minimum loss of momentum, and minimum concentration of forces.

The crossing should be preplanned to ensure fire support and crossing means are at the crossing site. The need for a water crossing should be determined in the mission analysis and added to the commander's concept of the operation. This ensures the troop's critical assets are positioned to support the water crossing. Seizing bridges intact before the enemy can destroy them is the quickest and most economical means of crossing, and is used when possible.

Reconnaissance. Scout platoons will be the first to encounter a water obstacle. When this occurs, they report it and reconnoiter it to obtain the following information, which is forwarded to the troop command post:

- Width and depth of the waterway.
- Water velocity.
- Possible entry and exit points and their conditions.
- Enemy situation on the far bank.

Scouts reconnoiter multiple crossing sites to prevent the enemy from discovering the true site and to reduce the concentration of vehicles at the site. If follow-on forces will cross at the same sites, scouts mark the entrances and exits for their use.

The troop commander, with the fire support officer, positions himself forward during the reconnaissance of the water obstacle so he can personally supervise the operation. The fire support officer will support the scouts with indirect and direct fires, if needed.

Security. A crossing can be performed with or without opposition. To avoid enemy fire while crossing, prevent him from knowing where the crossing will be.

Before the main body crosses, the scouts must secure the far bank by fording or swimming across the water obstacle and establishing positions. They should cross on both sides of the site and secure positions on the dominant terrain of the far side to provide early warning of threat activity. After securing the far side of the water obstacle, the scouts establish positions to the flanks of the crossing site to provide local security (see Figure 7-7).

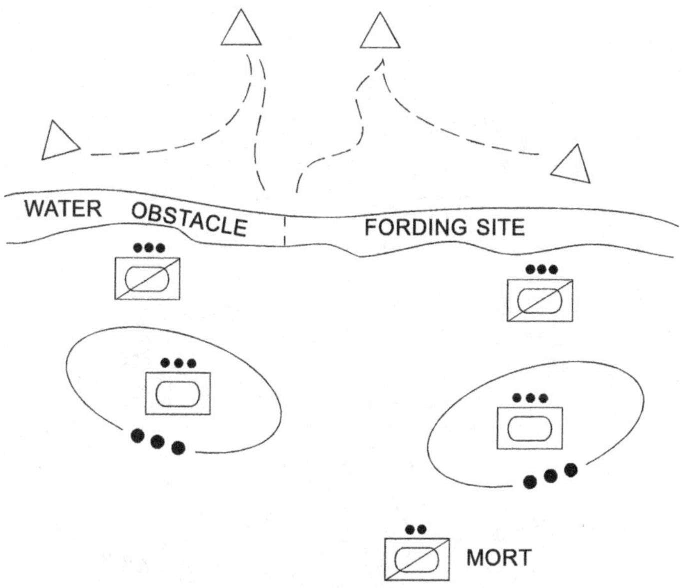

The scouts encounter a water obstacle, find fording sites, and secure the far side. Position the tank platoons and mortars to support with direct and indirect fires.

Figure 7-7. Secure the far bank.

The rest of the troop should not bunch up at the crossing site. Position them a terrain feature short of the water obstacle, so they have quick access to the crossing sites but are not exposed to enemy observation and fires.

Cross the Obstacle. Once the scouts have established positions on the far bank, the troop main body is ready to cross the obstacle. Time the movement to the crossing site so that no vehicles have to pause on the near side, but all can move directly through the crossing site and to their positions on the far side to continue the operation. Once the troop starts the crossing, complete it as quickly as possible (see Figure 7-8).

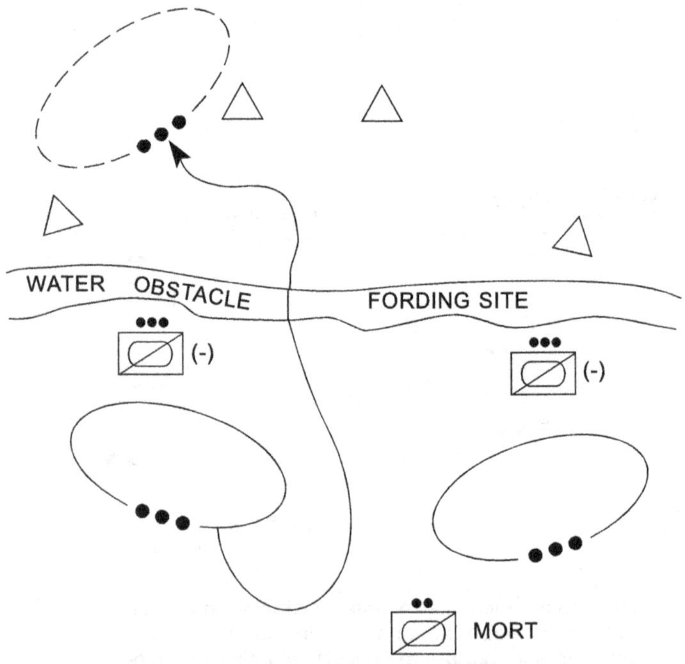

After the far side is secure, rapidly move platoons across the obstacle and continue the mission.

Figure 7-8. Cross the obstacle.

Fire Support. Fire support is used during the hasty water crossing to suppress known and suspected enemy positions at the crossing site. Smoke is employed to the front or flanks to screen the reconnaissance and the crossing. The fire support officer must time the movement of the mortars so they are in position to support the water crossing operation with continuous fire support.

Section VI. In-Stride Breach

When obstacles are encountered, an in-stride breach will maintain the tempo of offensive operations. Treat obstacles with caution and expect them to be covered by enemy observation and fires. The troop's ability to breach obstacles using available assets is important to ensure mission accomplishment. The size of the enemy force covering an obstacle with direct fire will, however, affect the troop's ability to conduct a hasty breach.

TASKS

The following combat tasks are associated with an in-stride breach:

- Detect the obstacle, reconnoiter it, and search for a bypass.
- Suppress all enemy positions with direct and indirect fires.
- Obscure enemy observation with smoke.
- Secure the near side of the obstacle.
- Breach or neutralize the obstacle.
- Move forces across the obstacle.
- Continue the mission.

TECHNIQUES

The enemy will reinforce natural obstacles with man-made ones to slow, disorganize, and canalize the troop. These obstacles may consist of—

- Minefields.
- Log obstacles, such as abatis, log cribs, stumps, and posts.
- Antitank ditches.
- Wire entanglements.

When scouts encounter an obstacle, they must follow the procedures for actions on enemy contact described in Chapter 3.

- Deploy and report.
- Develop the situation.
- Choose a course of action.
- Recommend a course of action.

As they develop the situation, the scouts should rapidly reconnoiter the obstacle to determine if bypass routes exist and whether or not they are covered by enemy fires, if breaches through the obstacle already exist, if the obstacle is part of an occupied defensive position, what the enemy strength is, and where he is located.

To conserve time and manpower, the troop should attempt to bypass all obstacles. Bypassing will not always be possible because available routes around the obstacle may lead into a fire sack. Forcing through an obstacle is the least desirable method of breaching because of the resulting loss of personnel and equipment. A hasty breach may be the only viable course of action.

Before deciding on a course of action, the troop commander should position himself to observe the situation and control the operation, if possible. Once the commander decides that a hasty breach is the only course of action, he must organize the troop into a breaching element and an overwatch element. When executing the hasty breach, the commander should avoid concentrating the troop in any one area, making it vulnerable to enemy fires.

During zone reconnaissance or movement to contact, the troop operates over a 6- to 10-kilometer-wide zone. During the hasty breach, the commander may not be able to use all the platoons in the troop. In this situation, until the breach operation is complete, give the scout platoon not involved in the breach operation a limit of advance so it does not move too far ahead of the rest of the troop. It can provide the troop with an early warning of enemy activity.

The Breaching Element. The breaching element is composed of scouts and engineers attached or in support of the troop. Their mission is to create lanes through the enemy obstacle system to allow the troop to pass.

The Overwatch Element. The overwatch element is the tank/AT platoons, scouts not in the breaching force, and any additional engineers. Its mission is to provide close, continuous overwatching fires in support of the breaching element. After the breaching element has cleared lanes through the obstacle, part of the overwatch element will assault through the lane to defeat any enemy in the area of the breach. The fire support officer coordinates the fires of the mortars and any supporting artillery to suppress enemy fires with HE and to obscure his observation with smoke.

Sequence of Events in a Hasty Breach. After encountering an obstacle and the commander chooses a hasty breach as the course of action, the sequence of events is as follows:

- Scouts.
  - Reconnoiter the near side of the obstacle to determine its front edge and lateral limits, enemy dispositions, if the obstacle is part of a defensive position, and if a partial or complete breach exists.
  - Secure flanks in support of the breaching operation.

- Overwatch element.
  - Occupy overwatch positions and provide direct and indirect suppressive fires on the enemy for elements moving to and through the obstacle.
  - Use smoke to obscure enemy observation of the obstacle (see Figure 7-9).

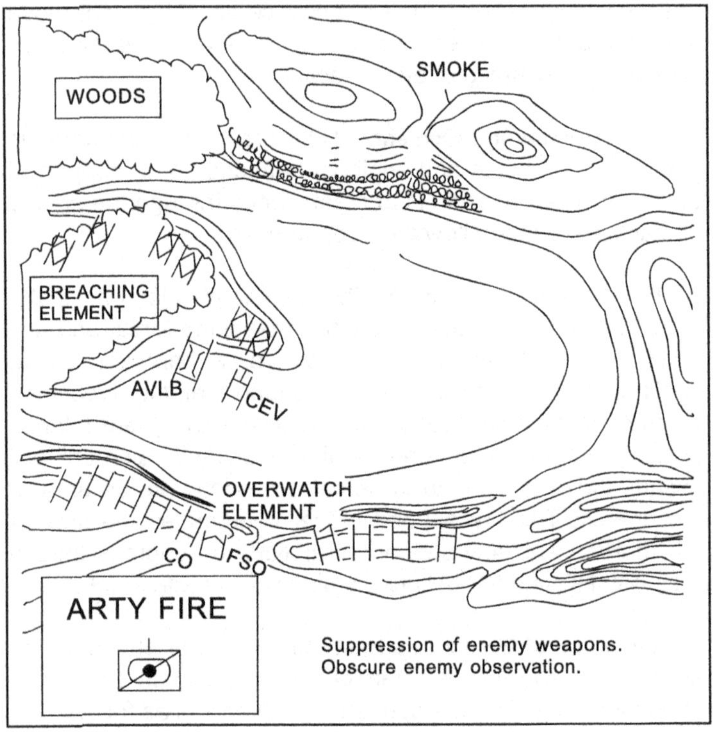

Figure 7-9. Obscure enemy observation.

- Breaching element.
  - Occupy covered and concealed positions to coordinate activities and prepare equipment, demolitions, and routes to the obstacles.
  - After enemy fire has been suppressed, breach the obstacle.
  - Secure fighting positions on the near and far sides of the obstacle (see Figure 7-10).

- Overwatch element.
  - As the breaching element breaches the obstacle, a tank platoon from the overwatch element prepares to attack through the obstacle.
  - Once the breaching element secures positions on the far side of the obstacle, the tank/AT or scout platoon attacks through the obstacle and destroys enemy elements that can fire directly on it. This assault element either continues to advance as the troop lead element or establishes hasty defensive positions while the remainder of the troop passes through the obstacle (see Figure 7-11).
- Troop. Continue the mission.

Neutralizing and Breaching Minefields. Minefields will differ in layout and composition, depending on the availability of mines and the nature of the avenue of approach. When possible, mines should be detonated in place. The objective in breaching the obstacle is to make a safe route to the far side. Multiple routes across the obstacle reduce vulnerability to enemy fires, but may be very time consuming and beyond the capabilities of the troop.

Use the following methods to neutralize and breach minefields:

- **Foot Lanes.** Establish foot lanes through the obstacle when a mounted assault breach is not at first practicable. This lane is normally two meters wide, and can later be widened into vehicle lanes. Foot lanes allow the troop to move forces through the obstacle to secure the far side.

- **The Bangalore Torpedo.** This device will clear foot lanes through mines and wire obstacles. It will clear a path three to four meters wide through wire entanglements and a narrow footpath through a minefield.

- **Vehicle Lanes.** After the first breach is made, foot lanes may be widened to one-way vehicle lanes at least eight meters wide. Vehicle lanes may also be breached separately from foot lanes. Use existing roads when possible. Vehicle-pushed rollers, if available, should be used to proof the lane cleared by line charges.

- **The M173 Rocket-Projected Line Charge.** This is an AT minefield clearing device. When the line charge explodes, it clears a vehicle lane about six meters wide.

Section VII. NBC Defensive Operations

OPERATIONS

NBC defense operations reduce casualties and damage to equipment and materiel, and minimize confusion and interruption of the troop's mission in the event of enemy NBC attacks. These operations are performed concurrently with all combat operations to preserve the fighting strength of the troop.

TASKS

The following combat tasks are associated with the conduct of troop NBC defense operations:

- Prepare for a nuclear attack.
- Respond to the initial effects of a nuclear attack.
- Respond to the residual effects of a nuclear attack.
- Cross a radiologically contaminated area.
- Conduct radiological reconnaissance.
- Perform radiological decontamination.
- Prepare for a chemical agent attack.
- Respond to a chemical agent attack.
- Cross a chemically contaminated area.
- Conduct a chemical reconnaissance.
- Perform hasty decontamination.
- Coordinate for deliberate decontamination of equipment.
- Exchange protective clothing.

ORGANIZATION

The troop's NBC defense system is composed of a trained NBC officer (usually the XO), an NBC NCO, and an enlisted alternate from troop headquarters. Troop soldiers are also designated and trained to operate all assigned NBC equipment and to assist decontamination teams.

The NBC officer supervises troop NBC defense activities and assists the commander in training NBC equipment operators and decontamination teams. The NBC NCO and his alternate directly supervise radiological monitoring, chemical detection, and decontamination operations. During combat operations, the NBC NCO is located in the command post where he—

- Receives, prepares, evaluates, and disseminates information and reports enemy and friendly NBC attacks.
- Supervises employment of detection, monitoring and survey, and decontamination teams.
- Maintains unit radiation exposure status records.
- Assists the troop commander in analyzing guidance from squadron, mission, threat, and weather as they affect NBC operations. Recommends appropriate MOPP level based on this information.

To facilitate operations in an NBC environment, designate the NBC teams from all platoons and sections by vehicle crews so vehicle commanders have responsibility for their crews and the execution of the NBC defense team mission. Each section in the troop has a chemical agent detection team and each platoon has a radiological monitoring and survey team.

Chemical Agent Detection Team. Each troop has twelve chemical agent detection teams. There are three in each scout platoon and one in each tank/AT platoon, one in the mortar section, one in the command post, and one in each combat trains and field trains. Each team has an M256 detector kit, a chemical agent alarm, detector paper, 400 meters of WD-1/TT for remoting the alarm, and enough batteries and supplies for seven days of continuous operations.

Radiological Monitoring and Survey Team. Each troop has seven teams, each equipped with an IM 174/PD dose rate instrument and an IM 93/147 dosimeter. One detection team from each platoon serves as a radiological monitoring and survey team. Also, the mortar section, maintenance section, and troop headquarters each have a radiological monitoring and survey team.

Deliberate Decontamination Team. Deliberate decontamination teams are organized by the vehicle crew. One crew from each scout and tank/AT platoon is designated as a decontamination team. This team is used to support the deliberate decontamination of the troop. One crew from each platoon should receive special training from the squadron chemical section to execute this task.

TECHNIQUES

Proficiency in NBC defense operations is attained only by strict adherence to standardized procedures as outlined in the troop SOP. The troop must be proficient in the following three fundamentals of NBC defense to survive and remain an effective fighting unit:

- Contamination avoidance (before, during, and after an attack).
- Protection (before, during, and after an attack).
- Decontamination (after an attack).

Actions Before the Attack. The best defense against a chemical or nuclear attack is to avoid being detected and targeted. To avoid becoming a lucrative target, implement and strictly enforce the following passive avoidance measures during all operations:

- **Dispersion.** Maintain dispersion between vehicles whether stationary or moving; vehicles bunched together are much easier to find. This is especially important when occupying assembly areas, conducting resupply operations, or when crossing obstacles and waterways.
- **Concealment.** Concealed positions prevent the enemy from observing vehicles and personnel from the ground or air.
- **Camouflage.** When concealment is not available or adequate, use camouflage to hide vehicles and positions.
- **Communications security (COMSEC)/signals security (SIGSEC).** The enemy can easily locate positions with direction-finding equipment if the troop does not adhere to proper communications procedures. Ensure all troop elements properly encode messages sent over unsecured nets.
- **OPSEC.** Take all possible actions to deceive the enemy about troop movement, positioning, intentions, and size. Mask movement by using folds in the terrain. Use hide positions when not moving or engaging the enemy. Enforce noise and light discipline. To deceive the enemy about the number of troop vehicles, start all vehicles simultaneously, and cover all track marks that lead into positions.

Nuclear/Chemical Vulnerability Analysis. Based on the NBC threat and the troop mission, determine the troop's vulnerability and incorporate those factors into the commander's planning.

Take the following actions to reduce the effects of a chemical or nuclear attack:

- **Improve positions.** Use available natural and man-made cover, such as folds in the earth, buildings, and dug-in positions, to protect personnel, equipment, materials, and supplies.
- **Alert personnel.** Inform soldiers to assume the appropriate MOPP level (see Table 7-1), based on the possibility of enemy attack and the MOPP analysis. Inspect unit and individual NBC equipment and prepare for operations.

Table 7-1. Mission-oriented protection posture levels.

MOPP LEVEL	OVERGARMENT	BOOTIES	MASK/HOOD**	GLOVES
0	Stored Nearby*	Stored Nearby*	Carried	Stored Nearby*
MASK ONLY	Carried	Carried	Worn	Carried
1	Worn, open or closed based on temperature	Carried	Carried	Carried
2	Same as MOPP 1	Worn	Carried	Carried
3	Same as MOPP 1	Worn	Worn, hood open or closed based on temperature	Carried
4	Worn, closed	Worn	Worn	Worn

*On combat vehicle or in fighting position.
**Includes M258A1 kit and detector paper.

- **Protect equipment and supplies.** Protect against chemical contamination inside a vehicle since it is very hard to decontaminate. Keep all hatches closed, if possible, and turn on vehicle overpressure system. Cover equipment and supplies stored outdoors or in tank bustles. Nonporous plastic sheets make good covers. Use tarpaulins if plastic is not available.

- **Position alarms.** Activate and emplace automatic alarms to provide the earliest possible warning of an attack. Position the alarms upwind and at least 150 meters away from stationary positions. Keep adjacent alarms about 300 meters apart.

- **Alert NBC personnel and teams.** Ensure NBC personnel and teams are available and prepared to perform their assigned missions.

Actions in Response to an Attack. Every soldier must be proficient at individual NBC basic skills. Masking is conducted immediately in response to these conditions:

- Chemical alarm sounds.

- Positive reading on detector paper.

- Individuals exhibiting symptoms of chemical or biological agent poisoning.

All soldiers mask automatically when an attack is possible, after the initiation of chemical/biological warfare, in response to artillery shells exploding in the area, and when aircraft drop bomblets or spray a mist or fog. In response to these conditions, all soldiers assume MOPP 4 and remain at maximum protection until the absence of contamination is confirmed or the commander has completed the analysis and determined a lower level of MOPP is appropriate.

Actions in response to a chemical or nuclear attack are standardized and included in the troop SOP.

- Initiate immediate masking and sound alarm (visual, audible, or voice).

- Continue mission.

- Assume MOPP 4 as rapidly as possible.
- Treat casualties.
- Send NBC-1 report and update it as more information becomes available.
- Identify the agent.
- Initiate decontamination operations.

These procedures must be rehearsed, and all individuals must adhere to them.

Actions After an Attack. Warn all subordinate elements immediately so they can take appropriate actions. In addition, alert squadron and all adjacent units, giving priority to the unit downwind from the attack. The troop and squadron are extremely vulnerable at the beginning of a chemical attack. Continue the troop mission while preparing for the enemy; he will take advantage of the chemical attack by conducting tactical operations.

In accordance with the troop SOP, prepare and send an NBC-1 report immediately. The personnel observing the attack send the report to the troop XO and NBC NCO in the command post. The XO forwards the report to the squadron TOC. Send the report immediately, and update it as more information is available. Treat any casualties of the attack using self-aid and buddy-aid, and report their status through the troop command post to the squadron S1. Operating in MOPP 4 will, over time, reduce the troop's effectiveness in combat operations. Soldiers become fatigued and lose mental and physical dexterity. Maintaining command and control becomes very difficult during extended MOPP 4 operations.

The troop must determine the type of agent used in the attack and the extent of contamination. Initiate tests with the M256 detector kit to identify the type of agent. The M256 kit detects and identifies field concentrations of nerve, blister, or blood agents. This test will help determine when unmasking may be safe after a chemical attack. Update the NBC-1 report with any information provided by the test. Decontaminate if required, and unmask as soon as possible.

When the troop is contaminated with persistent or semipersistent chemical agents, decontamination is necessary before reducing the MOPP level. Implement the following decontamination techniques:

- Start the skin decontamination within one minute of contamination.
- Apply decontaminating agents to equipment and vehicles to limit the spread of contamination and shorten the duration of the hazard. Use the M258A1 kit for personnel wipe down, and the M11 or M13 decontamination apparatus for vehicle/equipment spray down.

Decontaminate as far forward as possible. After leaving the contaminated area, perform hasty decontamination as soon as the situation permits. Hasty decontamination consists of exchanging MOPP gear and washing vehicles with hot, soapy water or high-pressure water to remove gross contamination. This operation is conducted at platoon or troop level, assisted by the decontamination team and other assets provided by the squadron. Select a site that is uncontaminated, near a water source, and concealed. Hasty decontamination may allow temporary relief from MOPP 4 for eating, drinking, and personal hygiene. Deliberate decontamination is required later when time and additional decontamination assets are available.

Initiate unmasking procedures as soon as possible after a chemical attack, to reduce the performance degradation that occurs while soldiers are in MOPP 4. Unmasking procedures differ based on the availability of detection equipment. When nonpersistent agents are used in the attack and an M256 detector kit is available, use the kit to determine when the agent is no longer present. Instruct two or three individuals to unmask for 5 minutes, remask, and sit in a shady area for 10 minutes. Check the individuals for chemical agent symptoms. If no symptoms appear, it is safe to unmask.

When no detector kit is available, direct two or three soldiers to perform the following procedures. Have them keep their eyes open, take a deep breath, hold it, and break the seal on their mask for 15 seconds. Reseal and clear the mask, and wait 10 minutes in a shady area. Check them for chemical agent symptoms. Have them break the seal and take three or four breaths and reseal the mask. Wait 10 minutes and check them for symptoms. Have them unmask for 5 minutes and remask for 10. Check for symptoms. If no symptoms appear, assume it is safe for all to unmask.

Following a persistent or semipersistent attack, begin unmasking procedures after hasty decontamination is complete and the M256 kit indicates no agent is present.

OPERATIONS IN A NUCLEAR ENVIRONMENT

Radiological Monitoring. This is the detection of radiation and the measurement of dose rate with radiac instruments. Monitoring finds a hazard that would otherwise go unmeasured or undetected. It is conducted by the troop radiological monitoring and survey team, which monitors—

- Total dose using the IM 93/147.
- Dose rate using the IM 174/PD.
- Food, water, and personnel contamination using the AN/PDR-27.

Conduct periodic monitoring when directed, or after nuclear war breaks out. Monitor a specific point in the area at least once an hour. Start continuous monitoring after receiving a fallout warning; when conducting a movement; when a nuclear burst is reported, seen, or heard; after detecting radiation above 1 cGy/hr during periodic monitoring; or when the squadron commander orders. Stop continuous monitoring when the squadron commander orders or, except while on the move, when the dose rate falls below 1 cGy/hr.

Based on NBC-4 monitoring reports, the troop may conduct a radiological survey to determine the degree and extent of contamination. This information is forwarded to squadron in an NBC-4 report and on DA Form 1971-R.

Exposure Control. While operating in a nuclear environment, the troop must control exposure to nuclear radiation. An operational exposure guide is a method of determining the maximum radiation dosage to which platoons can be exposed and still accomplish a mission. Determine radiation exposure by the cumulative dose or the platoon's radiation history.

Platoons record and report their total dose radiation daily, per SOP, to the troop command post where the NBC NCO compiles the data. The dose is determined by averaging the readings from both platoon dosimeters and rounding to the nearest 10. The NBC NCO totals each platoon's average dosage, and coverts it to a radiation status (see Table 7-2). He maintains a record of the radiation exposure status of each platoon and submits a daily radiation exposure status report to the squadron (see Table 7-3).

Table 7-2. Radiation dose categories.

RADIATION DOSE CATEGORIES

THE TABLE INDICATES THAT THE NUMERICAL VALUE OF cGys RECEIVED BY EACH PLATOON IS CONVERTED TO A RADIATION EXPOSURE STATUS (RES) FOR EACH PLATOON. THE CONVERSION FACTORS ARE SHOWN BELOW.

RADIATION STATUS CATEGORY	TOTAL CUMULATIVE DOSE (cGy)
RES-0	0 (NO EXPOSURE)
RES-1	GREATER THAN 0 BUT LESS THAN OR EQUAL TO 70
RES-2	GREATER THAN 70 BUT LESS THAN OR EQUAL TO 150
RES-3	GREATER THAN 150

The regimental and squadron commanders may move units on the battlefield, or relieve a particular unit that has been exposed to a high level of radiation. The squadron commander specifies the degree of exposure that units will not exceed for each operation. There are three degrees of radiation exposure risk—negligible, moderate, and emergency. Each level of risk is correlated to unit radiation status (see Table 7-4).

Table 7-4. Unit radiation status.

RADIATION STATUS (RS) CATEGORY	TOTAL PAST CUMULATIVE	EXPOSURE CRITERIA FOR A SINGLE OPERATION (cGy)
RS-0 UNITS	NO EXPOSURE	NEG RISK: 50 MOD RISK: 70 EMERG RISK: 150
RS-1 UNITS	GREATER THAN 0 BUT NOT MORE THAN 70	NEG RISK: 50-PAST DOSE MOD RISK: 70-PAST DOSE EMERG RISK: 150-PAST DOSE
RS-2 UNITS	GREATER THAN 70, NOT MORE THAN 150	ANY FURTHER EXPOSURE IS CONSIDERED EMERG RISK
RS-3 UNITS	GREATER THAN 150	ALL FURTHER EXPOSURE WILL EXCEED EMERG RISK

1. RADIATION EXPOSURE STATUS CATEGORIES ARE BASED ON PREVIOUS EXPOSURE TO RADIATION.

2. RECLASSIFICATION OF UNITS FROM ONE RADIATION EXPOSURE STATUS CATEGORY TO A LESS SERIOUS ONE IS MADE BY THE COMMANDER UPON ADVICE OF THE MEDICAL PLATOON LEADER AFTER AMPLE OBSERVATION OF ACTUAL STATE OF HEALTH OF THE EXPOSED PERSONNEL.

3. ALL EXPOSURES TO RADIATION ARE CONSIDERED TOTAL BODY AND SIMPLY ADDITIVE. NO ALLOWANCE IS MADE FOR BODY RECOVERY FROM RADIATION INJURY.

4. NEGLIGIBLE RISK WILL CAUSE NO CASUALTIES AND A LESS THAN 2.5 PERCENT INCIDENCE OF VOMITING. MODERATE RISK WILL RESULT IN A LESS THAN 5 PERCENT INCIDENCE OF VOMITING AND HEADACHES, BUT WILL CAUSE NO CASUALTIES. EMERGENCY RISK WILL RESULT IN LESS THAN 5 PERCENT CASUALTIES AND A HIGHER PERCENTAGE OF NAUSEA, VOMITING, AND HEADACHES.

Minimize the Effects of Friendly Nuclear and Chemical Attacks. A NUCWARN/CHEMWARN message is normally not sent below squadron level. Squadron alerts the troops using a code word prescribed in the OPORD. Squadron will also issue a FRAGO containing relocation instructions, the level of protection required, and the time of the friendly strike. To assure maximum reaction time, each soldier in the troop must be informed. Each crew immediately prepares its vehicles, equipment, and position for the attack as is appropriate to its location within the attack area. Crews prepare their vehicles by taking the following actions:

- Move if necessary.
- Position vehicle behind best available cover with front of vehicle toward the blast.
- Point gun away from blast.
- Lock brakes.
- Secure loose equipment in vehicle.
- Secure inside the vehicle all exterior components that could be damaged by the blast.
- Close and lock all hatches, to include ballistic shields.
- Wear helmets and protect eyes.
- Turn off and disconnect all radios but one, which is needed to retain communications with squadron.

Complete these tasks within five minutes of receiving notification. After the friendly strike, continue radiological monitoring and report if necessary. Quickly prepare to continue operations.

If the troop is contaminated with fallout, the level of contamination is monitored by the monitor and survey teams. Decontamination is accomplished by brushing or wiping the dust off the equipment. Brush off or shake out MOPP gear and wash down the equipment and vehicles to reduce radiological contamination. Use the AN/PDR-27 to determine decontamination requirements and to determine if supplies of food and water must be destroyed.

Chapter 8

Combat Support

The effective integration of combat support (CS) will spell the difference between success and failure on the battlefield. The commander must know the capabilities and limitations of CS assets, and how to employ them properly.

CONTENTS

		Page
Section I.	The Fire Support Team	8-1
Section II.	Mortar Support	8-9
Section III.	Artillery Support	8-12
Section IV.	Engineer Support	8-14
Section V.	Ground Surveillance Radar Support	8-20
Section VI.	Air Defense	8-23
Section VII.	Army Aviation Support	8-26
Section VIII.	Close Air Support	8-28

Section I. The Fire Support Team

CAPABILITIES AND LIMITATIONS

The FIST is organized, equipped, and trained to provide—

- A fire support advisor/coordinator—the FSO.
- An observation/laser designation capability.
- A communications link to all available fire support.

The FIST has secure FM communications on four radios, and digital communications on two nets through the use of digital message devices (AN/PSG-5 and AN/PSG-2). These capabilities enable the FIST to maintain communications on the following nets (see Figure 8-1):

- Troop command net.
- Troop fire support net.
- Supporting artillery fire net (digital).
- Squadron fire support net.

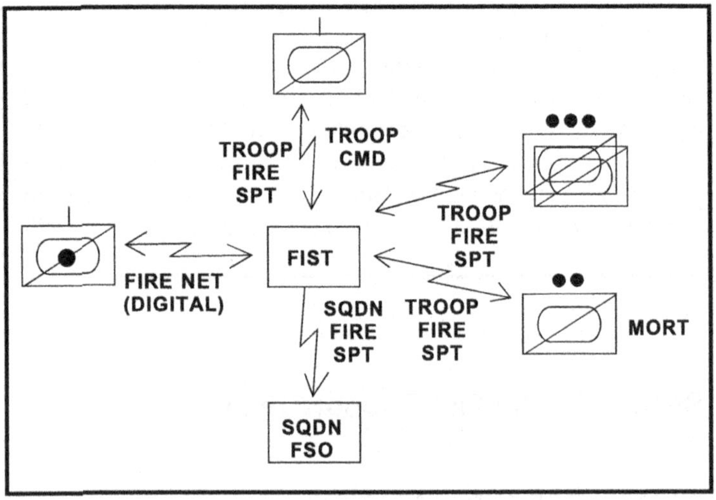

Figure 8-1. Fire support team communication nets.

The FIST is also equipped with a target laser locator designator set to designate targets for terminally guided munitions and to determine target locations accurately at ranges up to 10 kilometers (limited by line of sight). The target designator set is equipped with a thermal sight to use under limited visibility conditions.

The digital message devices allow the FIST to transmit and receive high-speed digital messages from the howitzer battery computer system, the direct support artillery battalion TACFIRE, OH-58D aircraft, and the squadron FSO's variable format message entry devices, among other digital message devices. Digital message devices can quickly transmit standard fire requests, fire adjustments, and plain text messages. The FIST can store nine fire requests on the digital message devices for rapid transmission in support of troop operations.

The FIST's abilities may be constrained by its communications with the supporting field artillery (FA). To coordinate any fire support for the troop, the FIST must be positioned where it can maintain at least a voice communications link to the supporting artillery. The FIST may not be able to see the battlefield and use the laser designator capability in all situations. When the troop is operating in a large sector, the FIST will be able to see only a small portion of the area. The scouts, therefore, must be proficient in requesting indirect-fire support. The scouts' calls for fire will be relayed through the FIST to the appropriate fire support unit.

EMPLOYMENT TECHNIQUES

Planning and Coordination. Fire support planning and coordination begin on receipt of a mission, and continue throughout its planning and execution. The troop FSO should always accompany the commander to mission briefings at squadron, so that he is fully aware of the squadron scheme of maneuver and the availability of fire support assets. He and the squadron FSO can also begin coordinating any specific requirements for fire support, such as preparations, groups, and series.

The troop commander's guidance should include the following to allow the troop FSO to plan the operation:

- What is the purpose of FA and mortar fires?
- How will FA fires support the troop's maneuver?
- What are the attack criteria? Will the commander use FA for immediate suppression, screening, or destruction?
- Are the attack and engagement criteria for mortars different from FA?
- On what and where does the commander want to mass FA and mortar fires?

Plan indirect fires on known or suspected enemy positions; at choke points; on the troop's objective or flanks; or on any other areas that would support the scheme of maneuver. These targets can be fired at a specific time, on call, or when a particular event takes place.

The FSO incorporates his approved fire plan into a target list and forwards it to the squadron FSO. The squadron FSO compiles the lists from all the troop FSOs, eliminates any duplicate targets, forwards the list to the supporting artillery, and informs the troop FSOs of any changes to their target lists.

Once he has completed his plan, the FSO will brief the commander on planned targets, scheduled fires, and fire support coordination measures that will be in effect. The commander will make the final decision on the fire support plan. Ensure it complements the scheme of maneuver and maximizes the available firepower for the troop. Once approved, the fire plan is continuously modified and updated to meet changing situations.

Silence is Consent. Once the FSO has briefed the commander and the commander approves the troop fire support plan, calls for fire are executed during the conduct of the operation by the silence is consent method. This method allows subordinate leaders within the troop to request indirect fires in support of their maneuver within the framework of the commander's guidance and the troop fire support plan. All initial requests for fires will go over the troop command net and the commander will simply acknowledge the request with "ROGER—OUT." The troop FSO will process the call for fire through the appropriate indirect-fire system based on the commander's guidance. All adjustments and corrections for the fires will be communicated on the troop fire direction net directly between the observer and the troop FSO. If the request for fire that is submitted by the subordinate leader is not within the commander's fire support guidance, the troop commander will deny fires or modify his guidance based on the situation.

Execution of the Fire Support Plan. During the operation, the FSO is responsible for—

- Keeping the commander advised of all fire support matters.
- Resolving any fire support conflicts that may arise during the planning and execution of the operation.
- Attacking targets with the most suitable fire support means available to support the scheme of maneuver.

Also, the commander may make the FSO responsible for positioning the troop mortars to best support the operation.

The FSO operates on the four radio nets mentioned previously. The scouts in the troop normally serve as the observers calling for indirect-fire support. The FIST monitors all spot reports on the troop command net and, as necessary, prepares calls for fire. The FSO will allocate fires from mortars and artillery, based on the commander's guidance. The commander may give the FSO additional guidance concerning a fire mission as it is requested (see *Silence is Consent* paragraph). Further information about the fire mission and all other calls for fire are passed on the troop fire support net. Keep the troop command net free for other traffic.

Both the FIST and the mortars monitor the troop fire support net. When scouts forward adjustments or corrections for indirect-fires, they use the troop fire support net. If the FSO forwards the mission to supporting artillery, he sends the message "LIGHTNING UP" on the troop fire support net. This tells the scouts calling for fire and the mortars that the mission is being fired by supporting artillery and that all adjustments and the "end of mission" report must be relayed through the FSO. If the mortars will fire the mission, the FSO sends the message "THUNDER DOWN" on the troop fire support net. This tells the mortar squad to fire the mission, and tells the scouts to send all adjustments and the "end of mission" report to the mortar squad (see Figures 8-2 and 8-3).

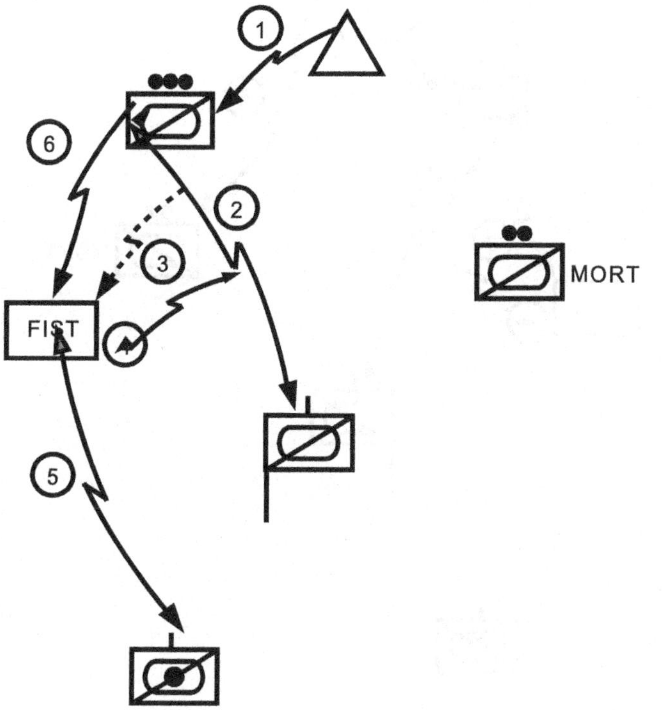

(1) SCOUTS SEND SPOT REPORT TO PLATOON LEADER (PLATOON NET).

(2) PLATOON LEADER FORWARDS THE SPOT REPORT TO THE CP (TROOP COMMAND NET) WITH REQUEST FOR FIRE.

(3) FIST MONITORS SPOT REPORT, RECEIVES PERMISSION TO FIRE THE MISSION, AND DETERMINES WHAT ARTILLERY TO USE.

(4) FIST SENDS MESSAGE "LIGHTNING UP" ON TROOP COMMAND NET. SCOUTS AND MORTARS MONITOR.

(5) FIST SENDS DIGITAL CALL FOR FIRE TO DS ARTILLERY. FIST SENDS SHOT ON TROOP COMMAND NET.

(6) SCOUTS RELAY ALL ADJUSTMENTS TO FIST ON TROOP FSN.

Figure 8-2. Fire request channels (artillery).

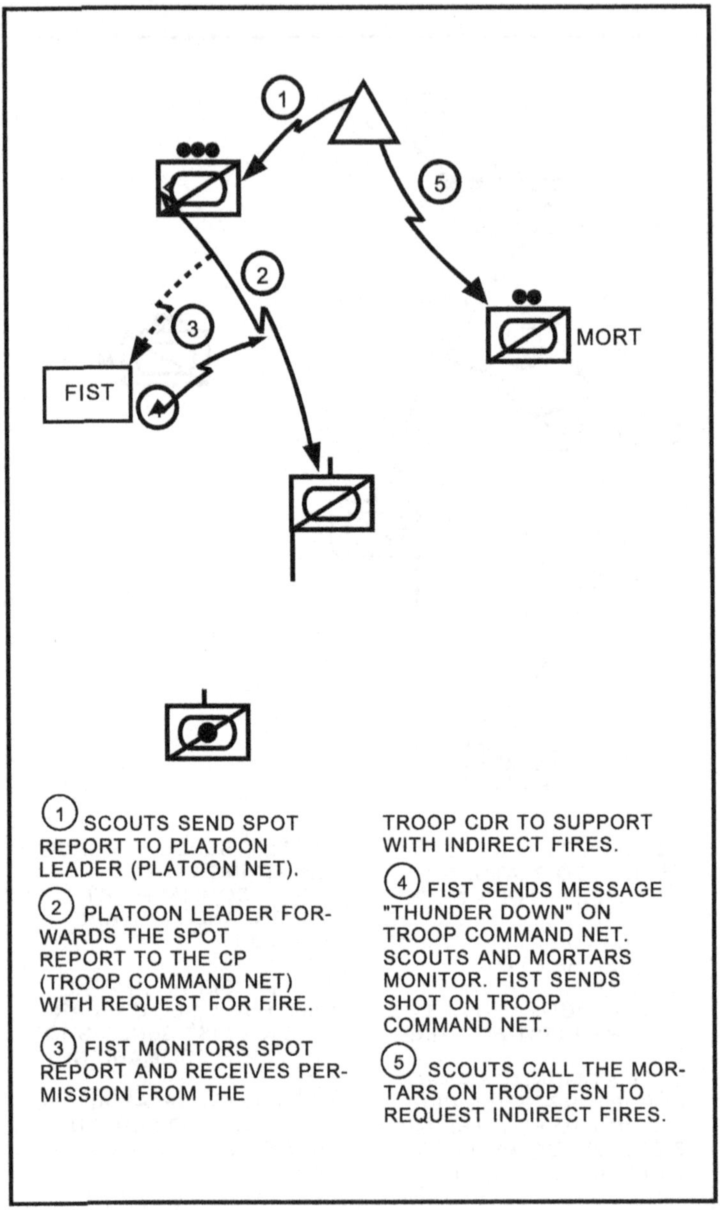

Figure 8-3. Fire request channels (mortar).

Section II. Mortar Support

CAPABILITIES AND LIMITATIONS

The mortar section provides a heavy volume of accurate, sustained indirect fires in response to the needs of the troop. They are ideal weapons for attacking targets on reverse slopes, in narrow ravines, built-up areas and in other areas that are difficult to strike with low-angle fires. Mortars are most effective for—

- **Suppression.** Suppression forces the enemy to button up or move to less advantageous positions. High explosive (HE) ammunition with a variable-time fuze or quick fuze does this best.

- **Screening.** Obscuration smoke is placed directly on or just in front of enemy positions to obscure vision. Screening smoke is placed between troop positions and the enemy to conceal the troop's movements. Smoke will degrade the effectiveness of laser range finders, so the tank platoons must adjust their gunnery techniques when engaging the enemy.

- **Illumination.** The use of illumination permits daytime engagements during some limited visibility conditions by improving the effectiveness of image intensification devices. Illumination rounds are used to mark targets for close air support and attack helicopters, and to allow observers to adjust artillery onto a target. Illumination rounds are usually adjusted so they illuminate above or behind the enemy location to silhouette him.

The height of the illumination should be adjusted to burn out about 200 meters above the ground. This gives maximum illumination time, about 90 seconds for one round, and will not degrade the effectiveness of image intensification devices. Because illumination rounds light the battlefield for both sides, it should be used when night-vision devices are not available in sufficient quantities or when ambient light levels are very low.

EMPLOYMENT TECHNIQUES

The troop mortar section is employed as a separate element during operations. It moves independently of the platoons, and provides its own security. The troop commander normally designates positions for the section sergeant and provides guidance for support forward of the scouts. The commander may find that having the FIST position the mortars is easiest and most effective. This technique allows the mortar section sergeant the ability to concentrate on providing fires and computing fire data and the commander to focus on maneuvering the troop, not one section.

The mortars can operate on two nets. If the FIST is responsible for moving the section, both nets may be on the troop fire support net. The FSO will keep them updated on the situation. However, since many spot reports over the troop net become fire missions, the mortars can eavesdrop and provide more responsive fires if they stay abreast of the situation themselves. If the commander gives the mortar section sergeant authority to select positions and to move the section, he must operate on the troop command net to stay abreast of the situation and on the troop fire support net.

Offense. During offensive operations the movement of the mortars will be based on the progress of the scout platoons. While the section is on the move, it must be prepared to provide immediate fires using direct lay, direct alignment, or hip shoot. Plan the moves of the section so it is in position to support the troop at critical times, such as during river crossings and counterattacks, and when clearing defiles.

Defense. Considerations for using mortars in the defense are similar to those for using them in the offense. Plan on prestocking ammunition at subsequent positions to reduce resupply problems during a defensive operation.

Reconnaissance and Security. During reconnaissance and security missions, the troop will often be operating over a large frontage that cannot be completely covered by the section. The commander must decide whether to position the mortars to cover the most dangerous area, or to move them to a position where they can cover a part of different areas and adjust as necessary. Knowing what other artillery assets are available will help in making that decision.

Commander's Guidance. A list of considerations the commander can use to help plan mortar section employment follows:

- Ammunition constraints by type and quantity.
- Priority of fire to a designated platoon.
- Anticipated changes in mortar employment.
- Communications constraints.
- General designation of positions.
- Movement guidance.
- Coordination requirements.
- Resupply.

Section III. Artillery Support

ORGANIZATION AND SUPPORT RELATIONSHIP

The howitzer battery organic to the regimental cavalry squadron consists of eight 155-mm self-propelled howitzers. The squadron will often be supported by an artillery battalion. The artillery battalion usually has three batteries of 155-mm self-propelled howitzers, and provides direct support to the squadron.

CAPABILITIES AND LIMITATIONS

Field artillery provides the troop its main fire support. It gives accurate fires with a wide variety of munitions. Field artillery adds a powerful dimension to troop direct-fire and maneuver capabilities.

- Artillery can—
  - Provide fire support under all weather conditions and terrain types.
  - Shift and mass fires rapidly without the requirement to displace.
  - Support the battle in depth with long-range fires.
  - Provide a variety of conventional shell/fuze combinations.
  - Provide continuous fire support by careful displacement.
  - Be as mobile as the supported unit.

- Artillery has—
  - Limited capability against moving targets.
  - Limited self-defense capability against air and ground attack.
  - Limited capability to destroy point targets without considerable ammunition expenditure. (Copperhead is the exception. It can be used to engage point targets.)
  - Vulnerability to detection by enemy target acquisition systems because of firing signatures.

Field artillery has a wide variety of munitions tailored for the engagement of different types of targets. The ammunition types include—

AMMUNITION	**TARGET**
High explosive	Personnel, field fortifications, vehicles
HC smoke	Obscuration and screening
White phosphorus	Obscuration
Illumination	
Cannon-launched guided projectiles (Copperhead)	Armored vehicles or high-payoff targets requiring precision target engagement

In addition, the troop must become acquainted with the characteristics of improved conventional munitions (ICM) and scatterable mines. ICMs include antipersonnel (APERS) and dual-purpose ICM. The commander must consider the danger to friendly troops in areas in which APERS is fired. Also, the high ICM dud rate makes maneuver hazardous in the area of an ICM field. Scatterable mines are area denial munitions for use against personnel and remote antiarmor mines for use against armored vehicles.

EMPLOYMENT TECHNIQUES

Artillery units normally provide direct support to the squadron, and are positioned by the artillery commander in close coordination with the squadron S3 and FSO. The troop FSO is the troop's link to the supporting artillery (see Section I, The Fire Support Team, and Chapters 1 and 2).

Section IV. Engineer Support

ORGANIZATION AND SUPPORT RELATIONSHIP

A combat engineer company is organic to the heavy and light armored cavalry regiment. Additional engineer assets are often provided from divisional and corps engineer units to support the regiment. Up to a platoon of engineers is frequently attached to or placed in direct support of a cavalry troop for combat operations. A combat engineer platoon from the heavy regimental engineer company is organized for combat as shown in Figure 8-4.

HEAVY ENGINEER CAPABILITIES AND LIMITATIONS

A combat engineer platoon is uniquely equipped and trained to conduct mobility, countermobility, and survivability operations in support of troop operations. The platoon headquarters has two M9 armored combat earthmovers, which are highly mobile, armored, and amphibious (see Figure 8-5).

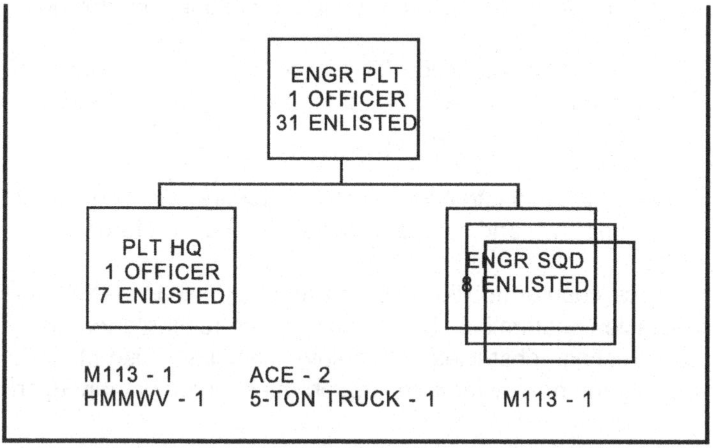

Figure 8-4. Regimental engineer company (heavy) organization.

Figure 8-5. Armored combat earthmover, M9.

In mobility operations, the engineer platoon can provide—

- Obstacle reduction. The engineers can reduce or negate the effects of obstacles to improve the troop's ability to maneuver.

- Route construction. The engineers can construct, improve, and maintain roads, bridges, and fords.

In a countermobility role, the engineers can assist with obstacle construction. The engineers can reinforce terrain by constructing obstacles to delay, canalize, disrupt, and destroy the enemy force to support the scheme of maneuver.

In survivability operations, the engineers can improve positions by constructing berms, dug-in positions, and overhead protection to reduce the effectiveness of enemy weapons.

A cavalry troop is frequently supported by additional engineer assets, such as an armored vehicle launched bridge (AVLB) and combat engineer vehicles (CEV) from the engineer company. The AVLB section of headquarters and headquarters troop of the armored cavalry squadron consists of three AVLBs that could be available during combat operations.

An AVLB is a tank chassis modified to transport, launch, and retrieve a 60-foot class-60 bridge. The bridge is capable of carrying military load class-60 track vehicles across a 17-meter gap and military load class-70 track loads across a 15-meter gap (see Figure 8-6).

Figure 8-6. Armored vehicle launched bridge.

The CEV is an M60A1 tank mounted with a hydraulically operated dozer blade, a 165-mm turret-mounted demolition gun, a retractable boom, and a winch (see Figure 8-7).

Figure 8-7. Combat engineer vehicle.

The small emplacement excavator (SEE) is a small highly mobile truck with a front-end loader and a backhoe (see Figure 8-8).

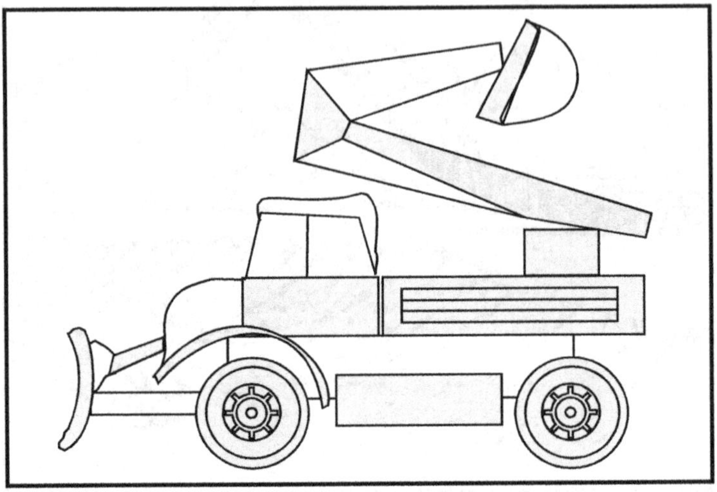

Figure 8-8. Small emplacement excavator.

LIGHT ENGINEER CAPABILITIES AND LIMITATIONS

The engineer platoon from the light regimental engineer company is equipped with five HMMWVs. The platoon can accomplish most of the same tasks that a heavy platoon can, but the time to complete most jobs is greater. The platoon can provide the troop with the following:

- In mobility operations.
  - Obstacle reduction. Engineers can reduce or negate the effects of obstacles to improve the troop's maneuver ability. To improve the platoon's ability, they may be reinforced with an M9 ACE, SEE, MICLIC, and Bangalor torpedo. The engineer company from the light regiment does not provide the regiment (troop) with any bridging capability.
  - Route reconnaissance. Engineers can conduct route reconnaissance.

- In a countermobility role, the engineers are usually left under squadron control to complete priorities of work in accordance with the squadron obstacle plan. The engineers can reinforce terrain by constructing obstacles to delay, canalize, disrupt, and destroy the enemy force to support the scheme of maneuver.
- In survivability operations, the engineers can improve positions by constructing berms, dug-in positions, and overhead protection to reduce the effectiveness of enemy weapons (see Figure 8-9).

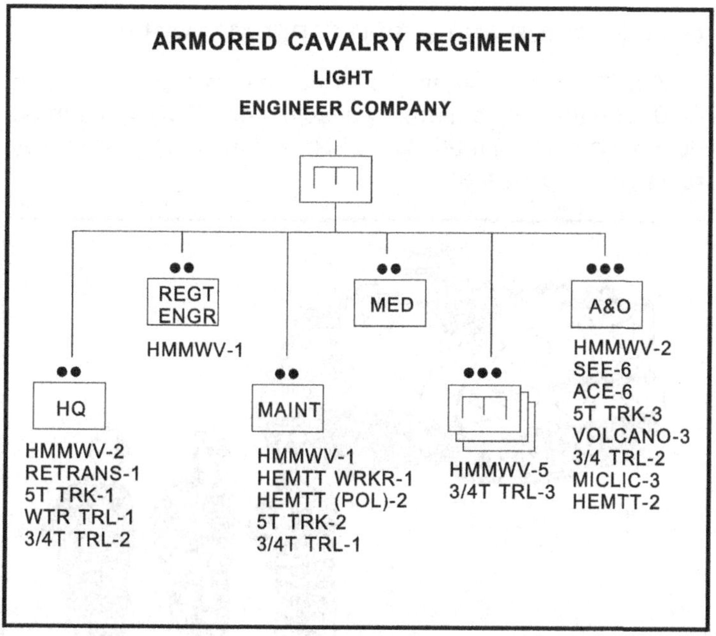

Figure 8-9. Regimental engineer company (light) organization.

EMPLOYMENT TECHNIQUES

The troop has several options in the task organization and employment of engineer assets to support its operation. Normally, the troop will keep the entire platoon under troop control and assign missions just as with a maneuver platoon. The commander can also task organize the engineer platoon, and assign squads to scout platoons to support the operation. Whatever method is chosen, the engineer platoon leader needs to be involved in the troop planning process.

Section V. Ground Surveillance Radar Support

ORGANIZATION AND SUPPORT RELATIONSHIP

A GSR team is organized for combat as shown in Figure 8-10. The team(s) is normally attached or OPCON to a troop during combat operations. GSR is found only in heavy regiments and divisions.

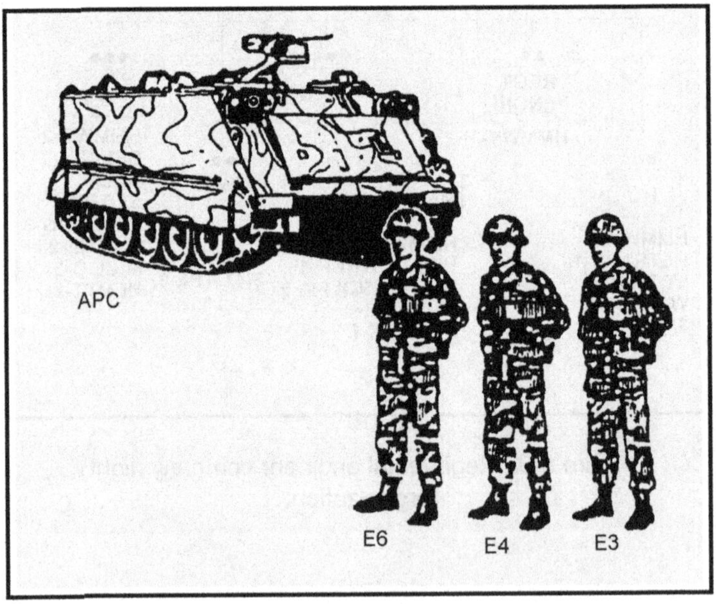

Figure 8-10. Ground surveillance radar team.

CAPABILITIES AND LIMITATIONS

GSRs provide mobile, all-weather battlefield surveillance. GSRs employed in pairs can provide observation from a given vantage point 24 hours a day, detecting targets and providing accurate range and azimuth readings.

The AN/PPS-5 has a line-of-sight range of 10,000 meters against vehicles and 6,000 meters against personnel. A GSR can detect targets through light camouflage, smoke, haze, light snow and rain, darkness, and light foliage. Heavy rain and snow seriously restrict its radar detection capability.

The GSR is generally ineffective against air targets unless the aircraft is flying close to the ground. The GSR is designed to detect targets moving in the presence of a background. It is vulnerable to enemy direction finding and jamming.

EMPLOYMENT TECHNIQUES

The troop will normally employ supporting GSR teams to augment the surveillance capabilities of its scout platoons. The commander should assign each GSR team a specific sector of surveillance and a frequency of coverage, or give the scout platoon leader guidance for employing the GSR team. The enemy can detect radar signals, so the GSR cannot conduct continuous surveillance. The tasks assigned GSR surveillance teams include—

- Searching avenues of approach or possible enemy positions on a time-schedule basis or randomly to determine location, size, composition, and nature of enemy activity.

- Monitoring point targets such as bridges, defiles, and road junctions and reporting quantity, type, and direction of enemy movement through the target.

- Extending the observation capabilities of scouts by enabling them to survey distant points and areas of special interest.
- Vectoring patrols to keep them oriented during times of limited visibility.

Position the GSRs in an area that is free of ground clutter and in one that affords them long-range observation and a wide field of view. Normally, they should be assigned a general area by the troop commander or scout platoon leader and the GSR team leader will select the specific location. The team should be prepared for rapid displacement, and have several alternate positions selected and reconnoitered to avoid enemy suppressive fires.

Offensive Operations. Employ GSRs in the offense to augment reconnaissance and security efforts. Fast-moving operations may preclude the continuous, effective use of GSRs.

In a movement to contact or a zone reconnaissance, employ GSRs on high-speed avenues of approach to the front or flanks of the troop to provide information on enemy movement. A GSR team can provide long-range observation while the scouts conduct a detailed reconnaissance of the local area. Employ GSR teams on the flanks of the troop's movement in pairs so they can bound forward. In this manner, the GSR can provide continuous early warning and keep pace with the operation.

Defensive Operations. In defensive operations, employ GSR teams to augment scouts. Position them to provide long-range observation on expected avenues of approach or to maintain surveillance of flanks.

Position the GSR teams behind the screen line if the terrain affords them long-range observation. Use GSR to assist scouts in maintaining contact with the enemy as the scouts deploy to subsequent screen lines. This also reduces their vulnerability to enemy fires.

Section VI. Air Defense

SUPPORT RELATIONSHIP

An air defense artillery (ADA) battery is organic to the regiment. For combat operations, these assets are normally attached or direct support to the squadron. The squadron commander may provide a slice of ADA to the troop.

Air defense assets are scarce, so the troop cannot plan on dedicated air defense protection. It is the troop's responsibility to protect itself from enemy air attack. The commander must take measures to avoid enemy air attack, to limit the damage if attacked and, if necessary, to fight back.

AIR DEFENSE EMPLOYMENT

Include air defense planning in all tactical operations. Air defense includes passive and active air defense and the employment of ADA assets.

Passive Air Defense

Passive air defense measures are the troop's first line of defense against enemy air attack. They include all measures, other than active defense, taken to minimize the effects of hostile air action. There are two types of passive air defense—attack avoidance and damage limiting measures.

Attack Avoidance. If the enemy pilot cannot find the troop, he cannot attack it. Use concealment, camouflage, obscuration, and deception to hide from the enemy.

Select positions that provide good concealment. When adequate concealment is not available, camouflage the vehicles to blend with their surroundings. Cover all track marks leading into the troop's position. Cover all shiny objects that could reflect light and attract attention. Take all actions necessary to reduce the enemy's ability to find the troop.

Damage Limiting Measures. Damage limiting measures are taken to reduce the effects of enemy air attack. Disperse vehicles while occupying static positions such as assembly areas or when preparing to cross a water obstacle or a breached obstacle. Dispersion reduces the effects of munitions. While on the move, air guards must be alert for enemy air attacks. When an enemy air attack is identified, quickly disperse, go to a concealed position, if possible, and then stop moving. A stationary vehicle is more difficult to see than a moving vehicle, and dispersion will reduce the effect of the air attack.

Use natural or man-made cover to reduce the effects of enemy air attack. Use folds in the earth, depressions, buildings, and sandbagged positions as damage limiting cover. These methods to reduce the effects of air attack must be used scrupulously by troop trains and command posts to reduce their vulnerability on the battlefield.

Active Air Defense

Although passive measures are the first line of defense against air attack, the troop must be prepared to fight back if necessary. The troop gains two things by fighting back. It may kill or drive off the attacker, and the morale and spirit of the troops improve by fighting back.

The decision to fight an air threat is based on the situation and the capabilities of the troop's weapon systems. Generally the troop may defend itself against direct attack, but unless attacked, it does not engage aircraft unless directed to do so by the squadron commander.

Stinger/Avenger Employment

Stingers/Avengers are used most effectively in point defense. However, mobility during tactical operations requires that Stinger crews move with the troop during maneuvering to provide the best possible air defense.

The troop commander must determine the air defense priorities for each phase of the operation and brief the Stinger/Avenger section chief accordingly. In order for the section chief to design an effective troop air defense plan for the commander, the commander must furnish the section chief the mission, objective, routes of march, intended scheme of maneuver, battle formations, and intended response to hostile air activity.

Unit air defense falls under two major categories: air defense of a fixed asset (assembly area or defensive position), and air defense of a mobile asset (convoy or maneuvering troop).

Fixed Asset. The air defense section leader places his crews from 2 to 4 kilometers apart near the troop's position, which provides overlapping and mutually supporting fires for troop defense. He provides command and control to his section, and air defense early warning to the troop unit and his deployed crews. The commander must provide logistical support for the section, indirect protective fires for their defense, and security within the troop defensive perimeter during hours of darkness.

Mobile Asset. While the unit moves, either in a tactical road march or during movement to contact, the air defense crews must move with each element to provide uninterrupted air defense protection.

Convoy operations require the air defense section leader to weight his crews near the front and rear of the troop's column(s), and to spread the remainder of his crews evenly throughout the column(s). While traveling, he provides command and control to his crews and air defense early warning to the troop and his crews.

Section VII. Army Aviation Support

SUPPORT RELATIONSHIP

Army aviation units are seldom OPCON to the troop; however, the troop frequently conducts joint operations with air cavalry or attack helicopter units within the troop area of operations. Air and ground units work together to make an effective team in conducting cavalry operations.

AVIATION TROOP COMMANDER

Air operations are controlled and coordinated by one man, the aviation troop commander. He coordinates directly with the ground troop commander, on the command net, when air operations extend into the troop's area of operations. When time is available, coordinate face-to-face to prepare for an operation. Let him know—

- Situation.
  - Enemy. Ground and ADA units by type and location.
  - Friendly. Locations and FLOT to include supporting artillery.
- Mission.
- Execution.
  - Scheme of maneuver.
  - Fire support.

The aviation troop commander tells the ground troop commander—

- His capabilities.
  - Number and type of aircraft.
  - Armament.
  - Time available.

- Execution. Scheme of maneuver to include approach direction into area.

- Command and signal.

  - Aviation troop commander, frequency, and call sign.

  - Succession of command and designated platoon leader in the troop commander's absence.

The aviation troop commander reports all spot reports and fire requests through the troop command post or FIST while operating in the ground troop's area of operations.

Reconnaissance. The capabilities of air and ground scouts complement each other. Air scouts can quickly reconnoiter a large area while ground scouts are conducting a detailed terrain reconnaissance. Employ the air scouts forward to provide information about areas where the ground scouts need to concentrate their efforts, such as at possible fording sites or route restrictions. The ground scouts can focus their efforts where the air scouts have directed them, instead of conducting a complete and time-consuming reconnaissance. The air scouts can also provide early warning of enemy movement to the front or flanks of the troop.

Security. During screen missions, employ the air assets to the front or flanks of the troop to observe areas between OPs and those areas that are difficult to observe from the ground. The air scouts can place effective indirect fires on the enemy and in maintaining contact with the enemy when the ground scouts move to subsequent screen lines. In offensive and defensive operations, Army aviation will normally work for the squadron to complement the squadron scheme of maneuver.

Section VIII. Close Air Support

CAS sorties are usually allocated to the regiment by the corps commander to meet preplanned requests submitted by the regimental fire support element. CAS for the troop is obtained by submitting an immediate CAS request.

CAS is requested by submitting an immediate request to the squadron air liaison officer. When CAS is on station, the tactical air control party may talk directly to the troop commander to determine enemy strengths, descriptions, and locations, as well as friendly unit locations. A forward air controller (FAC) normally directs the air strike. In the absence of a FAC, the troop FSO is trained to direct air strikes.

The ground troop may be required to mark the target for CAS aircraft. Use the mortar section for this. Fire white phosphorus or illumination rounds set for a ground burst. The FIST can designate targets for aircraft equipped with laser-spot trackers, such as Pave Penny or laser-guided bombs.

CAS capabilities are listed below.

- High speed and long range.
- Versatile weapon and ammunition mix.
- Accurate delivery.
- Excellent air-ground communications in A-10 and A-16 aircraft.
- Can locate and strike moving targets.

CAS limitations include—

- Limited resources.
- Delivery restrictions caused by limited visibility and weather or the proximity of friendly forces.
- Flight restrictions imposed by enemy air defense.
- Delayed response and short time on station.

Chapter 9

Combat Service Support

CONTENTS

Section I. Organization
Section II. Logistics
Section III. Personnel Service Support
Section IV. Prisoners of War

Mission accomplishment depends on the troop's ability to obtain, transport, and distribute fighting resources such as fuel, ammunition, replacement personnel, and rations. Dependable ways to treat and evacuate wounded soldiers are crucial to good morale. Evacuation and repair of combat equipment sustain troop combat power and readiness to fight.

Section I. Organization

Troop trains are organized as combat trains and field trains. The combat trains are made up of immediate battlefield service support. The rest of the troop CSS elements are collocated with squadron CSS assets in the squadron field trains and at the unit maintenance collection point (UMCP). The troop CSS team consists of the troop XO, first sergeant, supply sergeant, communications sergeant, maintenance sergeant, and senior troop aidman.

There is no dedicated service support radio net at troop level. All logistical reports and initial requests for logistical support are conducted on the troop command net. Additional coordination is conducted on the platoon radio nets. Routine reports are sent before and after combat operations, and are delivered by messenger when possible. All service support coordination with squadron is conducted on the squadron administration and logistics (A/L) net. The first sergeant operates on this net. The command post does not routinely operate on the A/L net, but uses it when necessary to forward reports and to conduct logistical coordination.

COMBAT TRAINS

The troop combat trains provide CSS for the troop during combat operations. They are organized for combat as shown in Figures 9-1 and 9-2.

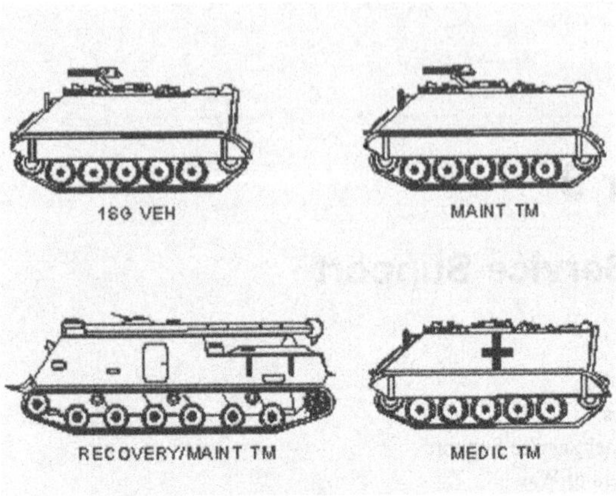

Figure 9-1. Heavy troop combat trains.

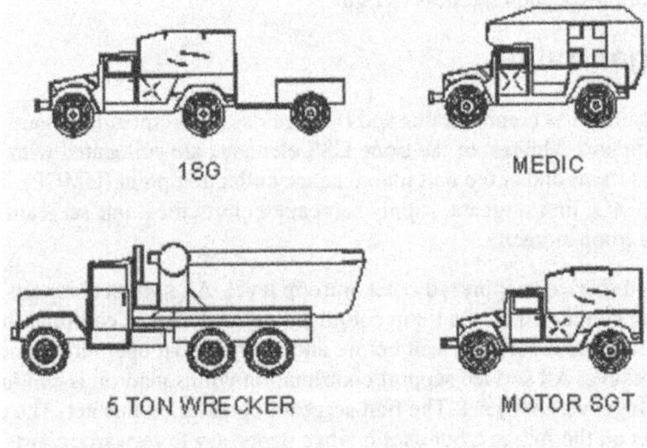

Figure 9-2. Light troop combat trains.

The first sergeant, assisted by the maintenance sergeant, directs movement and employment of the combat trains. He receives logistical reports directly from the platoon sergeants or through the troop command post, coordinates logistical matters with the XO, and assists the XO in ensuring the troop is logistically ready for battle. During operations, the first sergeant directs aidman teams and recovery teams forward as needed, leads the troop LOGPAC forward from the logistics release point (LRP) to resupply the troop and then leads it back to the LRP, and organizes the LOGPAC to conduct troop resupply operations. He also directs the medical evacuation of soldiers who are wounded in action. Additionally, he directs the evacuation of soldiers killed in action, enemy prisoners of war, and disabled vehicles.

The troop combat trains can be consolidated and moved under the control of the first sergeant, or they can be split and moved. If the combat trains are split, the motor sergeant and first sergeant will divide the medics, mechanics, special tools, and high-use PLL items into equal portions and move parallel and behind troop combat elements to provide responsive medical evacuation and battle damage assessment across the troop zone. Recovery assets will continue to move center of zone to facilitate recovery. This technique may be used if the troop zone is wide and split operations are necessary to provide proper support.

Position the combat trains where they can support the troop, but will not be exposed to enemy direct fires. This location is usually from 2 to 5 kilometers to the rear.

Offensive and Reconnaissance Operations.During such operations as a movement to contact or a zone reconnaissance, position the trains in the center of the troop zone and about 2 kilometers, or one terrain feature, behind the tank platoons. Use the established road network, if possible, to move the trains. Bound the trains forward to successive concealed positions, based on the movement of the troop, to keep them in position to support the troop.

Defensive and Security Operations.During defensive and security operations, position the combat trains about 3 kilometers behind the tank platoon positions in the center of the troop sector. If possible, hide the trains in a small built-up area, which provides cover and concealment and a road network to facilitate the movement of the trains. The trains must be ready to move to support the troop or displace to a new location. The XO and first sergeant plan subsequent locations, and as the situation requires, the first sergeant or maintenance sergeant moves the trains to the next location.

Security.Trains must provide their own local security. They do not have the manpower or equipment to clear large areas, so they should move into areas that have already been cleared by scouts. The trains' best defense is to avoid detection. The trains should disperse their vehicles and camouflage the positions by using natural concealment like a wooded area or a small built-up area. A built-up area is best, because there are a limited number of approaches and the thermal signature of the trains is concealed. Enforce strict noise and light discipline. Hide vehicles in barns or garages, if available. The buildings in a built-up area and the trees in a wooded area provide some protection against the effects of indirect fires and air-delivered munitions. Dispersion also limits the damaging effects of these weapons.

Establish OPs around the trains to provide early warning of enemy movement toward the position. Put them where they can cover major avenues of approach. An OP has at least two soldiers, a crew-served weapon (if available), a map, binoculars, a night observation device, and a radio or field phone. The OP must immediately report, by radio or phone, enemy ground and air attacks so the trains can take appropriate actions.

Passive defense is the trains' best defense against air attack. The steps taken to locate them properly, disperse vehicles, camouflage positions, and use the cover and concealment of built-up or wooded areas help prevent the enemy from detecting and attacking, and limit the effects of enemy munitions if attacked.

Air attacks cannot always be avoided. If attacked, take active air defense measures. The first sergeant or senior soldier present must coordinate the small arms fires of the trains for them to be effective against an air threat. Refer to Chapter 8, Section VI, Air Defense, for a complete discussion of passive and active air defense.

FIELD TRAINS

Troop CSS assets not in the combat trains are collocated with the squadron field trains or the UMCP, and sustain the troop's fighting capability by moving rations, ammunition, fuel, repair parts, and replacements from the squadron rear area to the troop on the battlefield. Troop field trains are controlled primarily by the supply sergeant, and consist of the personnel and equipment shown in Figures 9-3 and 9-4. Troop maintenance assets in the field trains may be consolidated under the control of the squadron maintenance officer.

The troop supply sergeant is the troop representative in the squadron field trains. The supply sergeant is responsible for the following tasks:

- Maintaining records of troop clothing and equipment.
- Processing or forwarding requests for all Class I, II, III, IV, V, VI, and VII. He also assists the maintenance section in ordering Class IX repair parts and the medics in resupplying Class VIII (medical).
- Assembling and leading all vehicles in the troop LOGPAC from the field trains to the LRP. If the squadron is conducting resupply operations, the supply sergeant may move under the control and direction of one of the following squadron personnel: squadron support platoon leader or platoon sergeant, S4, HHT commander, or XO.
- Assisting the first sergeant in organizing the LOGPAC site for troop resupply and in evacuating soldiers who are killed in action, enemy prisoners of war, and disabled vehicles; and in medically evacuating wounded and injured soldiers.
- Delivering mail, when available, via the troop LOGPAC.

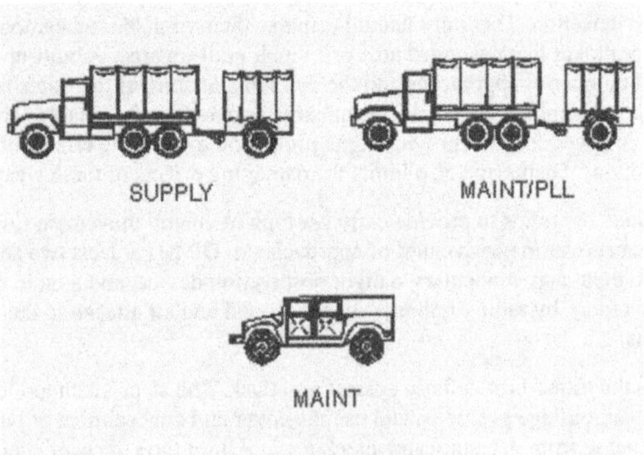

Figure 9-3. Heavy troop field trains.

Figure 9-4. Light troop field trains.

At the squadron field trains, the supply sergeant coordinates with the S4 and support platoon leader to ensure the LOGPAC meets the troop's CSS requirements.

The supply sergeant will coordinate with the troop PLL clerk for parts and requisitions and with the squadron personnel and administration center (PAC) section for mail, promotions, awards, Standardization Installation/Division Personnel System (SIDPERS), and other personnel actions. He will coordinate with squadron maintenance and communication for repaired vehicles and equipment and with the squadron food service sergeant for rations. The supply sergeant must frequently check with the HHT command post in the field trains for any additional troop requests reported on the A/L net.

The squadron field trains are responsible for their own local security. The supply sergeant will be tasked to assist the HHT commander in maintaining security of the field trains. He must execute and supervise security operations in accordance with the HHT commander's plan. During the movement through nonsecure areas, the supply sergeant should ensure that all crew-served and individual weapons are manned and ready. He should also ensure that all members of the troop field trains are briefed on immediate action drills for enemy contact or vehicle breakdowns.

If available, the troop should provide the supply sergeant with a radio to enhance communications between the supply sergeant and the first sergeant.

SQUADRON COMBAT SERVICE SUPPORT

In addition to organic troop CSS, one or two medical/evacuation teams are attached to the troop for combat operations from the squadron medical platoon. During resupply operations, the troop normally receives two fuel and two ammunition carriers from the transportation section of the support platoon. These carriers are not usually attached, but are provided to the troop as needed or as part of a standardized LOGPAC. The LOGPAC is discussed in Section II, Logistics.

To ensure responsive CSS for the squadron, the S4 and squadron maintenance officer organize the squadron support into echeloned trains and establish a main supply route (MSR), a UMCP, and an LRP. These trains are made up of combat trains (which provide immediate recovery, maintenance, medical, and emergency resupply support), and field trains (which provide the remainder of the squadron service support and limited direct-support maintenance). The UMCP is established to provide maintenance support for combat operations. It is positioned near or collocated with the squadron combat trains, and is the closest point to which damaged or failed equipment and systems are recovered. The UMCP is the focal point of the squadron maintenance effort. The MSR links troop combat trains to the squadron combat trains or UMCP and the field trains. LRPs are established along the MSR to facilitate the handover of LOGPACs and equipment between troops and squadron. An example of a squadron support plan is shown in Figure 9-5.

Section II. Logistics

SUPPLY OPERATIONS

The supply sergeant is responsible for getting supplies and delivering them to the troop. He delivers small items, but is largely dependent on support platoon assets to deliver bulky or high expenditure items. The commander establishes priorities for delivery, but the demands of combat normally dictate Classes I, III, V, VIII, and IX as most critical.

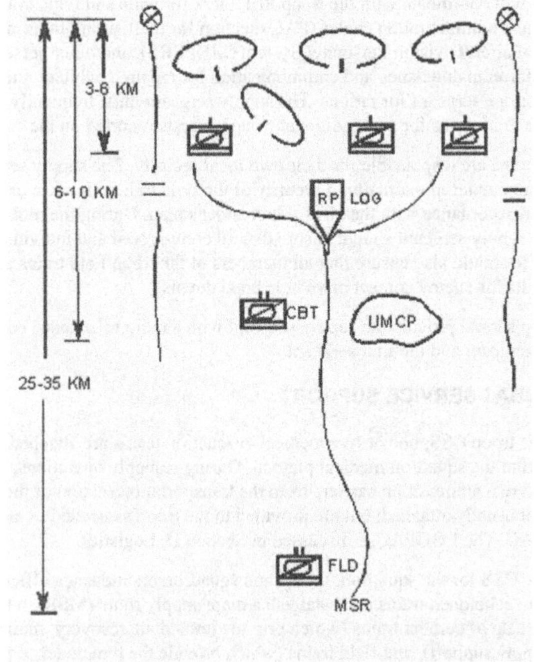

Figure 9-5. Squadron combat service support plan.

Class I (Rations).Meals ready to eat (MRE), C rations, are stocked on board each troop vehicle in a basic load prescribed by SOP (three-day supply). Class I is delivered daily by the supply sergeant as part of the

LOGPAC from the squadron field trains. Hot meals (B or tray-pack rations) are served when possible, but the normal ration cycle is C-C-C during combat operations. This requires resupply of at least 34 cases of MRE per day per troop. Water is a critical item, and must be resupplied daily. Soldiers may require up to 5 gallons of water per soldier per day, or more than 600 gallons for the entire troop. The troop can transport only 400 gallons of water on the water trailer. Soldiers should top off their water cans when possible. The supply sergeant should carry 5-gallon cans of water on his truck to supplement the needs of the troop. When he resupplies the troop, he exchanges the full cans for empty ones.

Class II (Supplies and Equipment).Class II items are requisitioned through the S4 and delivered by the supply sergeant as needed. The supply sergeant also maintains the following supplies, as a minimum, in his troop supply:

- One set of NBC overgarments and filter per soldier.
- One case of BA-30s.
- One mile of WD-1 wire.
- Additional batteries as necessary for night-vision devices and other equipment.
- Five sets of TA-50, to include personal clothing of various sizes, stored in duffel bags.
- Fifteen body bags.
- Map cases, grease pencils, permanent markers, etc.

Class III (POL).Class III is delivered by support platoon assets both as part of the troop LOGPAC and on an as-needed basis. Platoon sergeants report their Class III status to the command post daily, when status is requested, or when any Class III item is below a certain percentage, usually 59 percent.

Each troop vehicle carries a small basic load of packaged products as prescribed by SOP, and the fuel carriers normally have only small amounts of packaged products on board. Requests for large quantities of packaged products must be requested through the S4.

Class IV (Construction Materials). Class IV supplies are requested through the CTCP (combat trains command post) and delivered by the supply sergeant with the LOGPAC. Class IV supplies needed at troop level may include concertina wire, sandbags, and lumber.

Class V (Ammunition).The troop deploys with a basic load of Class V as prescribed by the unit for each type of vehicle. Ammunition status is reported by platoon sergeants to the command post daily, upon completion of enemy contact, or as needed. Normal resupply of Class V is delivered by the support platoon vehicles with the LOGPAC. Emergency resupply is requested through the first sergeant to the S4, and is delivered from the combat trains if it is available. Emergency resupply of Class V will normally consist of ammunition for major weapon systems, such as tank main guns, TOW missiles, and 25-mm guns in the heavy troop; and TOW missiles and caliber .50, 40-mm grenades, 7.62-mm, and individual weapons ammunition in the light troop.

Class VI (Personal Demand Items).Class VI items (such as soap, toothpaste, and cigarettes) are requested through the S4 by the first sergeant. They are usually picked up by the supply sergeant in the field trains and delivered as part of the LOGPAC.

Class VII (Major End Items).Class VII items, such as vehicles and night-vision devices, are automatically requested based on equipment shortages and battle losses. Items are delivered to the S4, who notifies the troop of the availability of the equipment. The equipment is delivered with a LOGPAC or brought forward immediately to an LRP where the first sergeant meets it and delivers it to the troop.

Class VIII (Medical Supplies).Class VIII is provided by the squadron medical platoon. The troop senior medic requests supplies through the squadron aid station. During intense combat, supplies may be pushed forward. They are then delivered to the combat trains where the troop medic picks them up from the squadron aid station, or are brought forward on ambulances.

Class IX (Repair Parts).The troop PLL clerk, who is in the field trains or the UMCP, works under the control of the squadron maintenance officer and the squadron maintenance technician. The PLL clerk requisitions, receives, and maintains records for all Class IX items. All troop PLLs are consolidated under the squadron maintenance officer during combat operations, and one or more PLL trucks remain in the UMCP at all times to provide a supply of Class IX for forward area support. When the troop maintenance sergeant needs a repair part, the request is sent through the squadron maintenance officer. The squadron maintenance officer determines if the part is available through the PLL, and fills the request if possible. If the part is not in the PLL, he then requests it through the squadron maintenance team in the field trains. They have the troop PLL clerk requisition the part for immediate use or replenish the part used from the PLL. The troop PLL clerk submits the requisition through squadron maintenance to the supporting maintenance unit. The part is delivered through squadron maintenance to the troop when the requisition is filled. The PLL clerk updates the records to reflect the requisition.

Maps.Maps are requested through the troop command post to the S4. As maps are available, the supply sergeant picks them up in the field trains and delivers them to the troop as part of a LOGPAC.

RESUPPLY OPERATIONS

Resupply of combat resources is accomplished using standardized procedures to rearm, refuel, and refit the troop as fast as possible to sustain its combat potential. There are two types of resupply operations—routine and supplementary. Methods of resupply are tailgate issue and service station.

Routine Resupply

Routine resupply operations include daily resupply of Classes I, III, V, and IX, and mail and other items needed by the troop. Routine resupply takes place when the troop is not in heavy contact, and may be conducted in an assembly area or behind troop positions when the troop is deployed in sector or zone. Class III is required more than once a day when the vehicles are in continuous operation.

Routine resupply is conducted using the LOGPAC from the field trains and the troop combat trains. The LOGPAC is organized in the field trains. Its composition is based on the troop's needs as reported to the S4, on requisitions, and on the availability of supplies. The troop supply sergeant reports to the S1 and S4 sections, squadron mess section, and squadron maintenance section to ensure all available supplies are picked up and loaded on trucks for troop resupply. The S4, support platoon leader, and troop supply sergeants assemble the LOGPACs for each troop by adding the troop supply trucks to the troop slice of Class III and V trucks from the support platoon. The support platoon leader leads the LOGPACs to the LRP, where the first sergeants meet them. Each first sergeant leads his LOGPAC to the troop resupply site. When it arrives, members of the troop combat trains guide the LOGPAC vehicles into position. Once the LOGPAC is established, the first sergeant reports to the command post that he is set. The troop commander or the XO coordinates resupply operations, and ensures all platoons and sections are resupplied. Resupply is conducted by one of two methods—tailgate issue or service station.

Tailgate Issue Method.The tailgate method is used in static positions such as assembly areas. Class III and V supply vehicles and other bulk-issue vehicles move from vehicle to vehicle to conduct resupply. The rest of the service support vehicles are centrally located in the troop area. Little or no movement is required by the combat vehicles. Personnel move to a centralized location to receive supplies, Class I, and mail. This method provides 360-degree security throughout the resupply operation; however, it is very time-consuming and requires an adequate road network for the wheeled supply vehicles to reach each vehicle. An example of this method is shown in Figure 9-6.

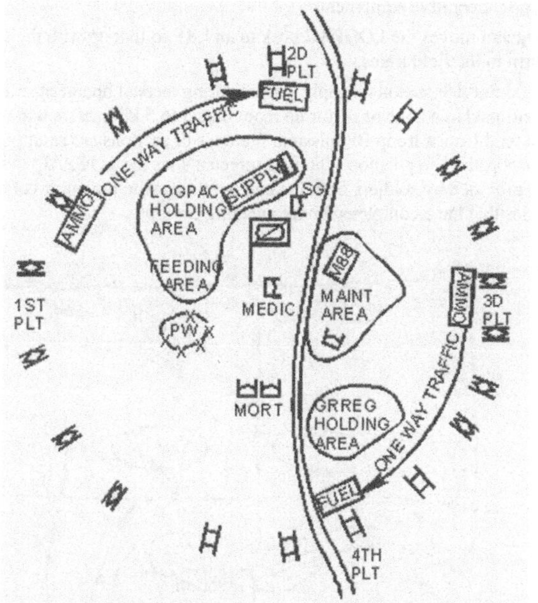

Figure 9-6. Tailgate issue method in a troop assembly area.

The following takes place during tailgate resupply:

- Combat vehicles remain in place. POL and ammunition trucks travel in a clockwise direction around the assembly area to each vehicle position, in turn, to conduct resupply.
- Crewmen rotate through the feeding area and pick up supplies, water, and mail.
- The first sergeant and platoon sergeants arrange for pickup of those killed in action (KIA) and their personal effects. The KIA are brought to a holding area near the medical aid station, but kept out of view.
- Ambulances pick up, treat, and evacuate seriously wounded soldiers. Other wounded soldiers are

carried to the ambulance or walk to it for emergency medical treatment.

- Prisoners are kept together and guarded. As soon as possible, they are moved to the squadron trains on a returning supply vehicle.
- The troop armorer, the radio repairman, and the organizational mechanics repair known problems and spot-check other vehicles.
- Vehicles needing maintenance are brought to the maintenance area.
- The first sergeant and platoon sergeants closely monitor the resupply operation.
- Empty LOGPAC vehicles are moved to a holding area, where they are loaded with KIA, prisoners of war (PW) and inoperative equipment.
- The supply sergeant moves the LOGPAC back to an LRP to link up with the support platoon leader and return to the field trains.

Service Station Method.Service station resupply is used during tactical operations. It is most effective when the troop is positioned in a zone or sector no more than 3 to 5 kilometers wide, such as is found in a defend in sector or defend from a troop BP mission. Platoons or sections are resupplied at the LOGPAC while the rest of the troop stays in position. The first sergeant sites the LOGPAC as shown in Figure 9-7. LOGPAC security is provided by soldiers from the combat trains who are not involved in the resupply, and by platoon vehicles that have completed or are awaiting resupply.

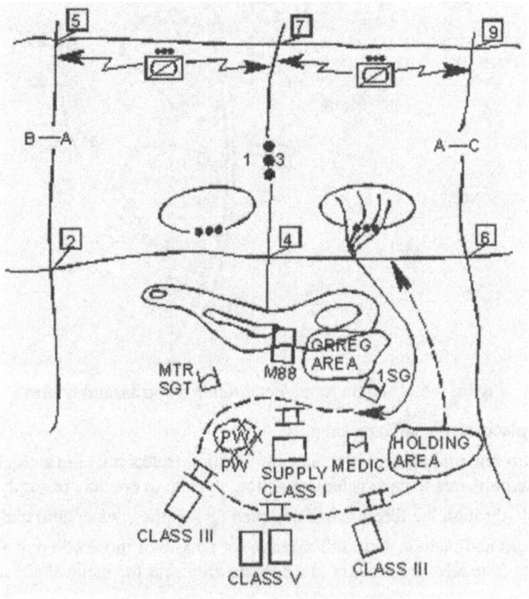

Figure 9-7. Service station method.

The following takes place during service station resupply:

- Vehicles of the first platoon or section enter the LOGPAC at a designated location and follow one-way traffic flow.
- The vehicle carrying the KIA in body bags moves to the holding area, out of view of the troop, and leaves the remains and personal effects.
- Soldiers requiring medical attention are delivered to medics and treated or prepared for evacuation.
- The platoon sergeant supervises the operation and coordinates face-to-face with the first sergeant for any special requirements.
- The platoon leader dismounts his vehicle and uses the first sergeant's vehicle and radio to contact the troop commander for orders and situation reports.
- Crews requiring unit-level maintenance remain in the maintenance holding area.
- Vehicles rotate through stations.
- Unit-level mechanics, the armorer, and the radio repairman repair known problems and spot-check other vehicles.
- Crews rotate to the supply truck to pick up mail, supplies, and Class I.
- The platoon leader and platoon sergeant conduct precombat inspection.
- When the platoon or section has completed resupply, it moves to its designated position.
- The rest of the platoons rotate individually through the LOGPAC for resupply.

At the end of the resupply operation, the troop LOGPAC returns to the LRP where it links up with the support platoon leader and returns to the field trains. In the field trains, the supply sergeant returns the fuel and ammunition carriers to the support platoon, returns ration-serving equipment to the mess team, delivers the KIA to the graves registration (GRREG) collection point, requests any additional supplies from the squadron supply sergeant, and returns to his position in the field trains.

Supplementary Resupply

In the heavy troop, daily routine resupply will not sustain the combat power of the M1/M3 fleet when operating without a break for 8 to 10 hours. The tanks will consume so much fuel that they will have to supplement routine resupply of Class III one or two times daily. The troop may also need to resupply Class V. Supplementary resupply can be conducted in a couple of ways depending on the mission and if the troop is in contact with the enemy. Regardless of the technique used, the commander must push the supplies forward to his platoons; the tempo of combat operations cannot be disrupted in order to resupply. Conduct emergency resupply when in heavy contact with the enemy, and prestock resupply when not. Once the supplies are brought forward, either the tailgate issue or service station methods are used.

Supplementary resupply must be planned when continuous operations are expected. Resupply must be requested through the S4 several hours in advance to ensure the service support assets are prepared to support the operation. The troop must quickly resupply once fuel trucks are available, and then return them to the field trains. There are few fuel carriers and drivers available in the squadron and these valuable assets cannot be wasted.

Platoon Prestock.When operating over wide frontages, such as during screen, zone reconnaissance, or movement to contact operations, break the supplies down into platoon packages and push the supplies forward to each platoon position. Position the vehicles behind the platoon and resupply individual vehicles or sections at a time using the service station method. If the situation allows, the resupply vehicles can move through the platoon position and resupply the platoon in place using the tailgate issue method.

Controlling platoon prestock is difficult because resupply vehicles are moving to several locations rather than remaining under the centralized control of the first sergeant. The platoon leader must also provide security for their own prestock (see Figure 9-8).

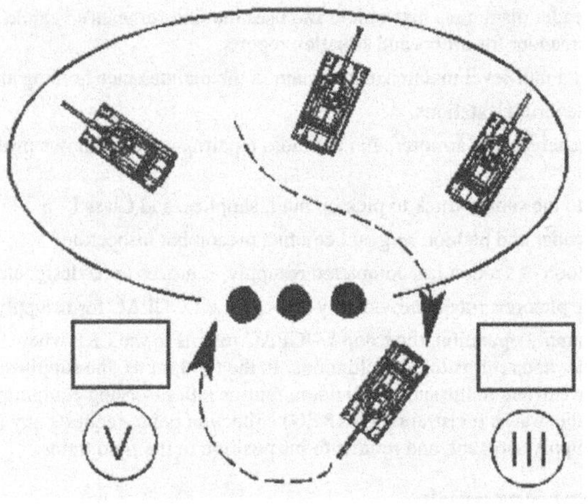

Figure 9-8. Platoon prestock.

Troop Prestock.Troop prestock is used to supplement routine resupply of Classes III and V. Troop prestock is used when the troop is operating in a narrow zone or sector. The resupplies are pushed forward to the troop position using either the service station or tailgate method. Each platoon rotates through the resupply site, using the service station method, while the rest of the troop remains in position; or each vehicle is resupplied in position using the tailgate method (see Figure 9-9).

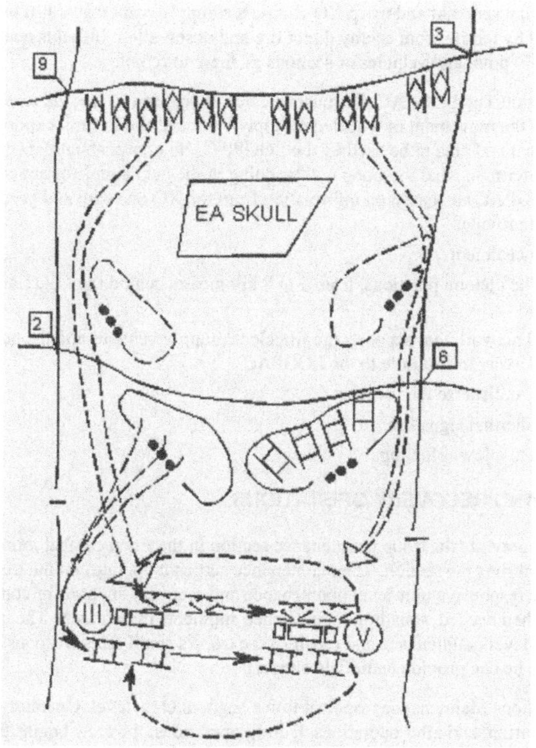

Figure 9-9. Troop prestock.

Variations of these methods can be used to meet the troop's situation. The troop must, however, rehearse these techniques to ensure it can quickly and efficiently conduct resupply operations and continue the mission.

Emergency Resupply

Emergency resupply normally involves only fuel and ammunition and is conducted while in contact with the enemy. The resupply begins at section and platoon level by redistributing ammunition between vehicles to cross-level loads. The platoon sergeant reports his need for emergency resupply to the first sergeant, who relays the request to the S4. The squadron combat trains maintain a small load of Class III and V for these situations. The S4 or support platoon leader coordinates a linkup between the squadron combat trains and the troop first sergeant. The first sergeant meets the resupply trucks and moves back to the troop area. The first sergeant and troop XO choose a resupply point that is just behind the troop position and masked by terrain from enemy direct fire and observation. If fuel is needed, the fuel truck is moved to the resupply point and vehicles or sections go there to refuel.

Resupply Site Selection.The LOGPAC site must be carefully chosen to provide responsive support for the troop, to support the movement of wheeled resupply vehicles, and to limit exposure to enemy fires. The XO selects the general area to be used by the LOGPAC. He knows about the current tactical situation, and can determine what the troop will be doing in the next couple of hours. The first sergeant selects the exact LOGPAC site based on information from the XO and map and ground reconnaissance. He selects a site that provides—

- Cover and concealment.
- Proximity to the platoon positions, from 3 to 8 kilometers behind the FLOT and center of the sector.
- A road or trail network that supports the wheeled resupply vehicles and the heaviest troop vehicles and allows one-way traffic flow to the LOGPAC.
- Enough room to disperse the vehicles.
- Reduction of thermal signature.
- Level enough to allow refueling.

MAINTENANCE AND RECOVERY OPERATIONS

The first sergeant supervises the troop maintenance section in the troop combat trains and the maintenance sergeant runs the section. The maintenance sergeant operates on the troop command net and stands by to provide responsive unit-level maintenance and recovery support for combat vehicles and radio equipment. When needed, squadron maintenance augments troop assets. The troop armorer is responsible for unit-level maintenance and repair of the troop's small arms weapons. Position him in the combat trains where he can provide battlefield support.

Maintenance Operations.Maintenance responsibilities begin at crew level. Operator-level PMCS must be performed before, during, and after operations. During operations, the crew begins the maintenance process when it identifies the fault. Once a problem develops, the vehicle commander determines support requirements, to include self-recovery, assistance from another vehicle, or assistance from troop or squadron maintenance, and takes the following actions:

- Reports the situation to the platoon sergeant, providing condition(s), location, and circumstances.
- Attempts to self-recover, if mired.
- Uses another vehicle to move to a secure location if exposed to enemy observation and fire. Uses smoke to screen the area between the mired vehicle and the enemy.

When repairs are beyond the capability of the crew, the platoon sergeant uses the troop command net to notify the first sergeant of the situation and to request assistance. The crew must maintain radio contact (if the radio is operational) on the platoon net and maintain local security. The maintenance sergeant monitors the radio traffic. He switches his radio to the platoon frequency to coordinate maintenance support and dispatches the appropriate assets to the "down" vehicle. The maintenance team checks whether the problem can be corrected in place within the time criteria (30 minutes in defensive situations and 2 hours in offensive situations). They move the vehicle to a more secure location, if necessary, and fix the vehicle. If repairs will take too long, the maintenance sergeant uses the squadron A/L net to notify the squadron maintenance officer who will coordinate for a handover of the equipment from troop to squadron at the LRP or along the MSR. The squadron maintenance officer needs the following information:

- Identification of the troop.
- Equipment involved.
- Location of equipment.
- Nature of damage.
- Pertinent information about the tactical situation, and any other necessary information.

Based on the information above, the squadron maintenance officer will designate a time and place for equipment handover.

The troop does not have the personnel, equipment, or time to create a holding area in the troop combat trains. If repairs will take more than the allowable time, evacuate the equipment to squadron.

Recovery Operations.When a vehicle cannot be repaired within the allowable time or is damaged beyond repair, the maintenance section must recover it and pull it to the MSR where it is turned over to squadron. Squadron maintenance will pull it to the UMCP or the field trains.

If a vehicle is catastrophic, or not able to be recovered based on the enemy situation, the commander may authorize the destruction of the vehicle. Before using thermite grenades to destroy a vehicle, remove as many valuable items as time permits, such as—

- Classified equipment and documents.
- Communications equipment.
- Weapons.
- Ammunition.

FIELD SERVICES

GRREG is performed by the field services platoon from corps, located in the regimental support area. The initial collection, identification, safeguarding of personal effects, and evacuation of the dead is the troop's responsibility.

When remains are discovered, be careful to preserve all items that may be used for identification. If metal identification tags (dog tags) are on the remains, do not remove them. Secure all personal effects in a bag or poncho and tie it to the remains. Place each casualty in a body bag, poncho, or shelter half and evacuate with the first available means of transportation, such as LOGPAC vehicles or disabled vehicles, to the squadron field trains. The supply sergeant receives all remains, and is responsible for turning them over to the GRREG collection point.

If the tactical and logistical situation makes evacuation impossible, emergency on-site burial is performed. On-site burial requires the permission of the squadron commander. If an on-site burial is performed, do the following:

- Complete two copies of DD Forms 551 and 1077 for each body. These forms are available through the squadron PAC. The first sergeant and supply sergeant should carry them.

- If there are several remains, engineer support may be needed to dig trenches.
- Dig a trench 6-1/2 feet wide and 3-1/2 feet deep. The length is determined by the number of remains.
- Remove one identification tag from each body, and string the tags on a wire in the order in which the remains are buried.
- Place the remains in the grave shoulder to shoulder.
- Bury all personal effects with the remains.
- Mark the ends of each row with a stake. Tag each stake to identify the marker as a grave. Indicate the length of the grave.
- Prepare an overlay that shows the location of the grave site.
- Cover the grave with earth removed from the trench.
- Forward the completed forms, overlay, and identification tags to squadron.

If the remains are contaminated, the grave site must be clearly marked and separated from noncontaminated grave sites. This must also be indicated on the grave-site overlay.

BATH AND LAUNDRY SERVICES

Bath and laundry services are provided by supply and service units from the corps support command. When available, these services are coordinated through the S4.

Section III. Personnel Service Support

PERSONNEL AND ADMINISTRATIVE

All troop administrative and personnel actions are handled by the squadron PAC in the field trains. The troop chain of command is responsible for ensuring that soldiers receive passes, leaves, promotions, awards, mail, legal assistance, financial services, and other personnel and welfare services on a fair and prompt basis. The first sergeant interfaces with PAC daily through the supply sergeant, at the administrative/logistics operations center with the S1/S4, or at the LRP.

Casualties and missing personnel are reported on DA Forms 1155 and 1156. These forms are initiated by the individual who witnessed the incident and are forwarded to the first sergeant. This action is the basis for notification of the next of kin and awarding benefits such as Serviceman's Group Life Insurance (SGLI), so accuracy is a must.

The commander should continuously manage the troop personnel situation to ensure trained personnel are manning key positions, crew shortages are filled with available personnel, and replacement personnel are trained and incorporated into crews. During combat, platoon leaders and platoon sergeants must balance their crews after casualties are evacuated. They must fill key positions on vehicles with the most qualified soldier. If necessary the commander may direct cross-leveling of personnel within the troop.

The troop submits daily personnel reports to the squadron as prescribed by SOP. The personnel reports are important because they allow the squadron commander to make tactical decisions for the employment of the troop. The reports also allow the S1, with the squadron commander's guidance, to properly distribute personnel replacements. Replacement personnel are received by the supply sergeant in the field trains. The supply sergeant in-processes the soldier into the troop, ensures he has the proper equipment, then delivers him to the first sergeant during the troop LOGPAC.

combat health support

First Aid.First aid is the responsibility of all soldiers; they use first aid, self-aid, buddy-aid, and combat lifesavers. All must be trained to take action after a soldier is wounded to keep him breathing, stop the bleeding, prevent shock, and dress the wound until medical personnel are available to treat the soldier. Training must include treatment of NBC casualties and crew evacuation drills to get soldiers out of a vehicle without creating further injury. Selected soldiers from each platoon will be trained in advanced medical skills in the combat lifesaver program.

Treatment.The first sergeant should position troop medical aid and evacuation teams on the battlefield where they can be most responsive. They will usually operate under the control and direction of the first sergeant in the troop combat trains. If the troop zone is pushing 10 kilometers wide, the troop may operate split combat trains with the first sergeant controlling medical evacuation on one side of the zone and the motor sergeant controlling the other half. The medics must know the locations of and routes to each platoon, the troop combat trains, the squadron combat trains, and each collection point.

Combat Stress Control.The psychological effects of combat on soldiers influence their ability to execute their missions. Individual self-control and self-discipline in the face of danger are maintained through unit discipline and firm leadership. The commander and his subordinate leaders are key to the mental toughness of soldiers. Commanders should be visible to soldiers and share their hardships with them. They should talk to their soldiers individually and as a troop to keep them informed. The commander's tone of voice on the radio must indicate that he has the situation under control. He should transmit the successful accomplishments of the troop and tell the soldiers they are doing a good job. Keep details about any casualties in the troop off the radio and ensure that soldiers who are killed in action are covered up and kept out of view of the troop. The chaplain can be a great asset in maintaining troop morale. Do not neglect the mental fitness of soldiers, because uncontrolled fear is contagious and can quickly lead to chaos.

Evacuation.To receive medical assistance, a platoon leader or platoon sergeant calls the first sergeant on the troop command net. The medic team monitors the call and goes to the platoon frequency to make any other necessary coordination. The medic team moves to the reported location to treat the soldier(s). The medics transport as many wounded soldiers as possible in their armored ambulance. Do not transport the dead with the wounded; evacuate them separately. If the medics determine the soldiers are seriously wounded, they either transport them directly to the squadron combat trains, or coordinate to hand them over to squadron at a collection point. Soldiers with slight wounds should not be evacuated unless it is necessary. They should be treated and returned to duty. Secure the seriously wounded soldiers on their litters to prevent any further injury during the evacuation.

As a general rule, do not use combat vehicles to evacuate the wounded; this reduces combat strength. Use combat vehicles only if absolutely necessary, and no other transportation is available.

Weapons and military equipment (except NBC protective equipment) of personnel being evacuated are secured with the platoon sergeant, first sergeant, or supply sergeant. Ensure the soldier keeps his protective mask and overgarments. When the situation permits, all equipment that was secured by the first sergeant or platoon sergeant is turned over to the supply sergeant.

disease and nonbattle injuries

More soldiers are lost in combat to illness and disease than to combat-related wounds. Maintaining the health and fighting fitness of the troop is a leadership responsibility. Personal hygiene, field sanitation, and rest must be incorporated into all troop operations. If these are ignored, the combat effectiveness of the troop will decrease rapidly.

UNIT MINISTRY TEAM

The unit ministry team (UMT), consisting of an appropriate number of chaplains and chaplain assistants, is assigned to provide unit, area, and denominational coverage to all troops in the cavalry squadron. UMT provides the following eight subfunctions of worship opportunities:

- Administration of sacraments, rites, and ordinances.
- Pastoral care and counseling.
- Development and management of the UMT.
- Management of material resources.
- Advising the commander on matters of religion, morale, and morals as affected by religion.
- Ministry in support of soldiers suffering from battle fatigue.
- Development of programs that enhance the total well-being of the soldier.
- Development of activities to enhance unit cohesion.

PERSONAL HYGIENE

The personal hygiene of the troop's soldiers is a leader's responsibility. Rules of hygiene must be observed to ward off disease and improve troop morale. Ensure soldiers wash and change their socks and underwear daily, if possible. All soldiers should shower as time and resources are available. The senior aidman plays a big part in personal hygiene by educating soldiers and by checking for signs of trench foot or immersion foot and frostbite.

Disease can quickly spread through the troop and incapacitate its soldiers. Watch for symptoms and ensure the soldiers' immunizations are current.

REST

Rest is extremely important for both commanders and their soldiers. When the troop is at REDCON 3 or 4, troop leadership should ensure a sleep plan is planned and executed.

FIELD SANITATION

Field sanitation is vital to the prevention of the spread of disease. The medics must assist in this effort by checking troop water supplies to ensure they are potable, and by ensuring mess utensils and equipment are properly cleaned. In static situations, such as when in assembly areas, soldiers will use slit trenches or latrines; at other times soldiers will use cat holes. Slit trenches and cat holes must be covered up after use to prevent the spread of disease.

Section IV. Prisoners of War

CAPTURE AND HANDLING

Maps, military documents, letters, and diaries obtained on the battlefield, as well as PWs, are valuable sources of combat intelligence. Scout platoons usually obtain information while searching the battlefield in the conduct of reconnaissance operations or after contact with the enemy. They should rapidly report this information to the troop commander for evacuation instructions. The scouts may be ordered to take the documents to the troop command post or to rendezvous with the first sergeant at a specific location to turn over the documents. Proper handling and evacuation of captured documents and equipment are important. The materials must be immediately turned over to trained intelligence personnel. In most cases, captured documents lose their value over time.

EVACUATION TECHNIQUES

In most situations, returning supply vehicles, aircraft, or troop headquarters vehicles can be used to evacuate PWs from the troop combat trains to the squadron collection point. The troop is responsible for guarding PWs until they are turned over to the S1. Wounded PWs are treated through normal medical channels, but are kept separate from US soldiers.

When support vehicles are not readily available and the troop must continue its operation, secure the PWs in a holding area such as a basement or compound. Notify the first sergeant and the S1 of the PW location and continue the mission. The first sergeant will pick up the PWs and complete their evacuation to the collection point.

Appendix A

Fratricide Prevention

Fratricide is as old as warfare itself, a complex problem that defies simple solutions. Fratricide can be broadly defined as the employment of friendly weapons and munitions, with the intent to kill the enemy or destroy his equipment or facilities, that results in unforeseen and unintentional death or injury to friendly personnel. This appendix focuses on actions leaders can take with current resources to reduce the risk of fratricide.

CONTENTS

Section I. Magnitude of the Problem
Section II. Risk Identification and Preventive Measures
Section III. Risk Assessment
Section IV. Fratricide Reduction Measures
Section V. Fratricide Risk Considerations (OPORD Format)

Section I. Magnitude of the Problem

The modern battlefield is more lethal than any in history. The tempo of operations is rapid, and the nonlinear nature of the battlefield creates command and control challenges for all unit leaders.

The accuracy and lethality of modern weapons make it possible to engage and destroy targets at these extended acquisition ranges. At the same time, however, the ability of US forces to acquire targets using thermal imagery and other sophisticated sighting systems exceeds our ability to accurately identify these targets as friend or foe. As a result, friendly elements can be engaged unintentionally and destroyed in a matter of seconds.

Added to this is the problem of battlefield obscuration, which becomes a critical consideration whenever thermal sights are the primary source of target identification. Rain, dust, fog, smoke, and snow degrade identification capability by reducing the intensity and clarity of thermal images.

On the battlefield, positive visual identification cannot be the sole engagement criteria at ranges beyond 1,000 meters. Situational awareness is the key; it must be maintained throughout an operation.

Section II. Risk Identification and Preventive Measures

Reduction of fratricide risk begins during the planning phase of an operation and continues throughout preparation and execution. Risk identification must be conducted at all levels during each phase, and the results clearly communicated up and down the chain of command so risk assessment can begin. This section covers considerations that influence risk identification; it also focuses on measures the platoon leader can implement both to make the identification process more effective and to help prevent friendly fire incidents from occurring. Section III of this appendix covers the risk assessment process; Section IV lists additional fratricide reduction measures and guidelines.

Planning Phase

A plan that is thoroughly developed and understood helps to minimize fratricide risk. The following considerations help indicate the potential for fratricide in a given operation:

- The clarity of the enemy situation.
- The clarity of the friendly situation.
- The clarity of the commander's intent.
- The complexity of the operation.
- The planning time available at each level.

Graphics are a basic tool that commanders at all levels use to clarify their intent, add precision to their concept, and communicate their plan to subordinates. As such, graphics can be a very useful tool in reducing the risk of fratricide. Commanders at all levels must understand the definitions and purpose of operational graphics and the techniques of their employment.

Note. See FM 101-5-1 for the definitions of each type of graphic control measure.

Preparation Phase

The following factors may cause fratricide risks to become evident during rehearsals:

- Number and type of rehearsals.
- Training and proficiency levels of units and individuals.
- The habitual relationships between units conducting the operation.
- The physical readiness (endurance) of the troops conducting the operation.

Backbriefs and rehearsals are primary tools in identifying and reducing fratricide risk during the preparation phase. The following are some considerations for their use:

- Backbriefs ensure subordinates understand the commander's intent. They often highlight areas of confusion, complexity, or planning errors.
- The type of rehearsal conducted determines what types of risks are identified.
- Rehearsals should extend to all levels of command and involve all key players.

Execution Phase

During execution, in-stride risk assessment and reaction are necessary to overcome unforeseen fratricide risk situations. The following are factors to consider when assessing fratricide risks:

- Intervisibility between adjacent units.
- Amount of battlefield obscuration.
- Ability or inability to positively identify targets.
- Similarities and differences in equipment, vehicles, and uniforms among friendly and enemy forces.
- Vehicle density on the battlefield.
- The tempo of the battle.

Maintaining situational awareness at all levels and at all times is another key to fratricide reduction as an operation progresses. Units must develop and employ effective techniques and SOPs to aid leaders and crewmen in this process. These techniques include—

- Monitoring on the next higher net.
- Radio cross-talk between units.
- Accurate position reporting and navigation.
- Training and use/exchange of liaison officers (LO).

Section III. Risk Assessment

Risk assessment must be conducted whenever fratricide risk factors are identified. It must take place at all levels during each phase of operations. As with risk identification, the results of the assessment must be passed on to all levels of the chain of command so that fratricide reduction measures can be developed and implemented. Refer to Section IV for specific reduction measures.

Figure A-1 is a worksheet for evaluating fratricide risk in the context of mission requirements. The worksheet lists six mission-accomplishment factors that affect the risk of fratricide, along with related considerations for each factor. Assess the potential risk in each area as low, medium, or high, and assign a point value to each (one point for low risk, two for medium risk, three for high risk). Add the point values for the overall fratricide assessment score. Use the resulting score only as a guide, however. The final assessment must be based both on observable risk factors like those on the worksheet and on your "feel" for the intangible factors affecting the operation. Note that descriptive terms are listed only in the low- and high-risk columns of the worksheet. The assessment of each factor will determine whether the risk matches one of these extremes or lies somewhere between them as a medium risk.

FACTORS	LOW (1)	MEDIUM (2)	HIGH (3)
1. UNDERSTAND PLAN			
• Commander's Intent	Clear		Foggy
• Complexity	Simple		Complex
• Enemy Situation	Known		Unknown
• Friendly Situation	Clear		Unclear
• ROE	Clear		Unclear

2. ENVIRONMENT ● Intervisibility ● Obscuration ● Battle tempo ● Positive target ID	Favorable Clear Slow 100%		Unfavorable Obscured Fast 0%
3. CONTROL MEASURES ● Command relationships ● Audio ● Visual ● Graphic ● SOPs ● LOs ● Location/Navigation	Organic Loud/Clear Well Seen Standard Standard Proficient Sure		Joint/Combined Jammed Obscured Not understood Not used Untrained Unsure
4. EQUIPMENT (Compared to US) ● Friendly ● Enemy	Similar Different		Different Similar
5. TRAINING ● Individual proficiency ● Unit proficiency ● Rehearsal ● Habitual relationship ● Endurance	MOS Qual Trained Multiple Yes Alert		Untrained Untrained None No Fatigued
6. PLANNING TIME (1/3 - 2/3 Rule) ● Higher HQ ● Own HQ ● Lower HQ	Adequate Adequate Adequate		Inadequate Inadequate Inadequate
OVERALL FRATRICIDE ASSESSMENT	LOW 26-46%*	MEDIUM 42-62%*	HIGH 58-78%*
● Commander may use numbers as the situation dictates. ● Numbers alone may not give accurate fratricide risk.			

Figure A-1. Fratricide risk assessment worksheet.

Section IV. Fratricide Reduction Measures

The following measures are provided as a guide to actions that can reduce fratricide risk. They are not directive in nature, nor are they intended to restrict initiative. Apply them as appropriate based on the specific situation and METT-T factors.

- Identify and assess potential fratricide risks in the estimate of the situation. Express these risks in the OPORD or FRAGO.
- Maintain situational awareness, focusing on such areas as current intelligence; unit locations and dispositions; denial areas (minefields/FASCAM); contaminated areas, such as ICM and NBC; SITREPs; and METT-T factors.
- Ensure positive target identification. Review vehicle/weapon ID cards. Know at what ranges and under what conditions positive identification of friendly vehicles/weapons is possible.
- Establish a command climate that stresses fratricide prevention. Enforce fratricide prevention measures, emphasize the use of doctrinally sound tactics, techniques, and procedures. Ensure constant supervision in the execution of orders and the performance of all tasks and missions to standard.
- Recognize the signs of battlefield stress. Maintain unit cohesion by taking quick, effective action to alleviate it.
- Conduct individual, leader, and collective (unit) training covering fratricide awareness, target identification and recognition, and fire discipline.
- Develop a simple, decisive plan.
- Give complete and concise mission orders.
- Use SOPs that are consistent with doctrine to simplify mission orders. Periodically review and change SOPs as needed.
- Strive for maximum planning time for you and your subordinates.
- Use common language/vocabulary and doctrinally correct standard terminology and control measures, such as fire support coordination line, zone of engagement, and restrictive fire lines.
- Ensure thorough coordination is conducted.
- Plan for and establish effective communications.
- Plan for collocation of command posts whenever it is appropriate to the mission, such as during a passage of lines.
- Designate and employ LOs as appropriate.
- Ensure rules of engagement are clear.
- Include fratricide risk as a key factor in terrain analysis (OCOKA).
- Conduct rehearsals whenever the situation allows time to do so.
- Be in the right place at the right time. Use position location/navigation devices (GPS and POSNAV); know your location and the locations of adjacent units (left, right, leading, and follow-on); and synchronize tactical movement.
- Include discussion of fratricide incidents in after-action reports.

Section V. Fratricide Risk Considerations (OPORD Format)

This section, which parallels the five-paragraph OPORD, contains key factors and considerations in fratricide reduction. This is not a change to the OPORD format; rather, it should be used during OPORD development to ensure fratricide reduction measures are included in the order. It is not a strict guide. The factors and considerations are listed where they would likely appear in the OPORD, but they may warrant evaluation during preparation of other paragraphs.

1. Situation.

a. Enemy forces.

(1) Are there similarities among enemy and friendly equipment and uniforms that could lead to fratricide?
(2) What languages do enemy forces speak? Could these contribute to fratricide risk?

(3) What are the enemy's deception capabilities and his past record of deception activities?

(4) Do you know the locations of enemy forces?

b. Friendly forces.

(1) Among the allied forces, are there differences (or similarities with enemy forces) in language, uniform, and equipment that could increase fratricide risk during combined operations?
(2) Could differences in equipment and uniforms among US armed forces increase fratricide risk during joint operations?

(3) What differences in equipment and uniforms can be stressed to help prevent fratricide?

(4) What is the friendly deception plan?

(5) What are the locations of your unit and adjacent units (left, right, leading, follow-on)?
(6) What are the locations of neutrals and noncombatants?

c. Own forces.

(1) What is the status of training activities? What are the levels of individual, crew, and unit proficiency?

(2) Will fatigue be a factor for friendly forces during the operation? Has an effective sleep plan been developed?

(3) Are friendly forces acclimatized to the area of operations?

(4) What is the age (new, old, or mix) and condition of equipment in friendly units? What is the status of new equipment training?

(5) What are the expected MOPP requirements for the operation?

d. Attachments and detachments.

(1) Do attached elements understand pertinent information regarding enemy and friendly forces?
(2) Are detached elements supplied this pertinent information by their gaining units?

e. Weather.

(1) What are the expected visibility conditions (light data and precipitation) for the operation?
(2) What effect will heat and cold have on soldiers, weapons, and equipment?

f. Terrain.

(1) Do you know the topography and vegetation (such as urban, mountainous, hilly, rolling, flat, desert, swamp/marsh, prairie/steppe, jungle, dense forest, open woods) of the expected area of operations?
(2) Have you evaluated the terrain using the factors of OCOKA?

2. Mission. Is the mission, as well as all associated tasks and purposes, clearly understood?

3. Execution.

a. Task organization.

(1) Has the unit worked under this task organization before?
(2) Are SOPs compatible with the task organization (especially with attached units)?

(3) Are special markings or signals (for example, cats' eyes, chemlites, or panels) needed for positive identification of uniforms and equipment?

(4) What special weapons and/or equipment will be used? Do they look or sound like enemy weapons and/or equipment?

b. Concept of the operation.

(1) Maneuver. Are main and supporting efforts identified to ensure awareness of fratricide risks and prevention measures?

(2) Fires (direct and indirect).

(a) Are priorities of fires identified?

(b) Have target lists been developed?
(c) Has the fire execution matrix/overlay been developed?

(d) Have locations of denial areas (minefields, FASCAM) and contaminated areas (ICM, NBC) been identified?

(e) Are the locations of all supporting fire targets identified in the OPORD/OPLAN overlays?

(f) Are aviation and CAS targets clearly identified? Have signals been established to positively identify these targets for the aircraft? Have airspace coordination areas been developed? Have enemy air defense systems been suppressed?

(g) Has the direct-fire plan been developed and synchronized with the fire support plan?

(h) Have final protective fires been designated?

(i) Have you identified and verified sector limits?

(j) Have executors for each target been assigned and do they understand when and where to shoot? Do the shooters have "eyes on" the target?

(k) Are the observers surveyed in or are they using a map spot? Target location errors can cause big problems.

(l) Do all leaders and executors understand where the fire support coordination measures are and when they go into effect? Rehearsal is the key.

(m) Can the fire support officer hear what targets are being called on the maneuver nets?

(n) Have all targets been rehearsed with the executors and the field artillery battalion?
(o) Does the reinforcing or general support reinforcing field artillery have all the proper graphics and understand where they fit in? Did they attend the rehearsal?

(p) Have restrictions on specific munitions been established and does everyone know where they are planned and emplaced?

(3) Engineer tasks.

(a) Are friendly minefields, including FASCAM and ICM dud-contaminated areas, known?

(b) Are obstacles identified, along with the approximate time needed for reduction/breaching of each?

(4) Tasks to each subordinate unit. Are friendly forces identified, as appropriate, for each subordinate maneuver element?

(5) Tasks to CS/CSS units. Have locations of friendly forces been reported to CS/CSS units?

(6) Coordinating instructions.

(a) Will rehearsals be conducted? Are they necessary? Are direct and indirect fires included?

(b) Is a backbrief necessary?

(c) Are appropriate control measures clearly explained and illustrated in the OPORD and overlays? Have they been disseminated to everyone who has a need to know? What is the plan for using these control measures to synchronize the battle and prevent fratricide?

(d) Have target/vehicle identification drills been practiced?

(e) Do subordinate units know the immediate action, drill, or signal for "CEASE FIRE" or "I AM FRIENDLY" if they come under unknown or friendly fire? Is there a backup action?

(f) Is guidance in handling dud munitions, such as ICM and CBUs, included?

4. Service Support.

a. Are train locations and identification markings known by everyone?

b. Do medical and maintenance personnel know the routes between train units?

5. Command and Signal.

a. Command.

(1) What are the locations of the commander and key staff members?
(2) What is the chain of command and the succession of command?

b. Signal.

(1) Do instructions include backup code words and visual signals for all special and emergency events?

(2) Do instructions cover how to identify friendly forces to aircraft?

(3) Are SOI distributed to all units with a need to know, such as higher, lower, adjacent, leading, and follow-on elements?

Appendix B

Standing Operating Procedures

The intent of this appendix is to provide an outline as an aid for developing a squadron tactical SOP.

I. GENERAL

A. Purpose: This tactical SOP prescribes standard procedures for use during all combat situations and provides a comprehensive reference for conducting operations in a field environment.

B. Conformity: All assigned, attached and OPCON personnel will read and comply with the provisions of this tactical SOP.

II. BATTLE COMMAND

A. Command.

1. Organization.

a. Succession of command.

b. Cues for assuming command.

c. Operation of the command post.

(1) Shifts.

(2) Displacement/set-up/tear-down.

2. Troop-leading procedures.

a. Estimate input (checklist).

b. Precombat inspection (checklist).

c. Backbriefs.

d. Rehearsals.

e. Combat orders.

(1) Formats.

(2) Preparation.

(3) Reproduction.

(4) Dissemination.

3. Coordination with adjacent units (checklist).

4. Liaison with main body elements (checklist).

B. Control.

1. Combat graphics and symbols.

2. Control measures designation (numbering system).

3. Terrain index reference system (TIRS).

4. Operational terms.

5. Vehicle identification marking system.

6. Unit recognition signals.

7. Communication.

a. Net diagrams.

b. Fixed call signs.

c. Brevity codes/cue words.

d. Antijamming actions.

e. Alternate means.

8. Reports.

a. Battle.

b. Intelligence.

c. Logistical.

III. MANEUVER

A. Readiness condition (REDCON).

B. Quartering party.

C. Road marches.

D. Assembly areas.

E. Squadron organization for combat.

F. Formations.

G. Battle plays.

1. Actions on contact.

2. Fix and bypass.

3. Close assault.

4. In-stride breach.

5. Bridges/defile.

6. Formation changes.

7. Passage of lines.

8. Hasty defense.

9. Consolidate on the objective.

10. Ambush.

11. Other plays.

H. Reconnaissance operations.

1. Zone reconnaissance.

a. Graphics.

b. Critical tasks.

2. Area reconnaissance.

a. Graphics.

b. Critical tasks.

3. Route reconnaissance.

a. Graphics.

b. Critical tasks.

I. Security operations.

1. Screen.

a. Graphics.

b. Critical tasks.

2. Area security.

a. Route.

b. Convoy.

J. Offensive operations.

1. Movement to contact.

2. Hasty attack.

3. Offensive graphics.

K. Defensive operations (defensive graphics).

L. Limited visibility operations (checklist).

M. Break in action (checklist).

1. Redistribution of ammunition, personnel, and equipment.

2. Evacuation of casualties and PWs.

3. Redistribution of ammunition under fire.

N. Relief in place.

1. Relief in place graphics.

2. Critical tasks.

O. Air-ground team.

1. Command relationship.

2. Formations/techniques of movement.

IV. MOBILITY AND SURVIVABILITY

A. Mobility.

1. Standard tasks.

2. Standard priorities.

B. Countermobility.

1. Standard tasks.

2. Standard priorities.

3. Engineer target turnover (checklist).

C. Survivability.

1. Standard tasks.

2. Standard priorities.

3. Fighting position construction.

D. NBC defense.

1. NBC team organization and equipment.

2. Unit NBC equipment.

3. Defense against nuclear attack.

4. Defense against chemical/biological attack.

5. Decontamination.

6. NBC reconnaissance.

V. FIRE SUPPORT

A. Fire support request sequence.

B. Fire support planning and execution matrix format.

C. Graphics.

VI. AIR DEFENSE

A. Air defense warning and cue words.

B. Weapons control status and cue words.

VII. INTELLIGENCE/ELECTRONIC WARFARE

A. Intelligence.

1. Standard tasks.

2. Standard priorities.

B. Electronic warfare.

1. Standard tasks.

2. Standard priorities.

3. Countermeasures.

VIII. LOGISTICS

A. Resupply procedures.

1. LOGPAC procedures.

2. Battle loss actions.

B. Combat Health Support.

1. Emergency medical treatment.

2. Medical evacuation.

3. Field hygiene and sanitation.

4. Combat stress control.

5. Routine sick call.

C. Maintenance support.

1. Battle damage assessment and repair.

2. Exchange criteria.

3. Cannibalization criteria.

4. Destruction criteria.

5. Maintenance repair time guidelines.

D. Personnel.

1. Replacements.

2. Accountability.

3. Personnel actions.

E. Combat trains.

1. Layout.

2. Operations.

IX. SAFETY

Appendix C

Operations Order

Operations order format

A sample OPORD is outlined below.

1. Situation.

a. Enemy.

(1) Weather (note effects - include light data).

(a) Light data.
(b) Weather forecast for operation.
(c) Effects of weather and light data on operations.

- Trafficability.
- Visibility.
- Effect on lasers/thermals.
- Effect on air operations.

(2) Terrain.

(a) Obstacles, hills, valleys, road types and conditions, streams, rivers, bridges, towns.
(b) Avenues of approach.

- Size unit supported.
- Start and end point.
- Objective.

(c) Key terrain.

(d) Observation.
(e) Cover and concealment.
(f) Engagement areas.
(g) Effect of terrain on the operation.

(3) Enemy forces.

(a) Identification of enemy forces.

(b) Activity of enemy forces.
(c) Location of enemy units.
(d) Disposition of enemy forces.
(e) Strength of enemy forces.
(f) Composition of enemy forces, to include type of equipment.
(g) Other enemy information critical to the upcoming operation, to include the following:

- Chemical and nuclear capabilities.
- Air defense artillery (ADA).
- Aviation, including helicopters.
- Electronic warfare.

(h) Enemy courses of action.
(i) Probable enemy course of action.

b. Friendly.

(1) Mission of higher headquarters (squadron) and commander's intent.
(2) Mission of adjacent units (left, right, front, rear).
(3) Mission of other organic units in higher headquarters.
(4) Mission of reserves in higher headquarters.
(5) Mission of supporting units who are in direct support/reinforcing (DS/R) to higher headquarters (field artillery, engineer, ADA).
(6) Which element from higher headquarters has priority of fires.
(7) Reinforcing units with a reinforcing/general support (R/GS) role to supporting units.
(8) Close air support (CAS) and number of sorties allocated to higher headquarters.

c. Attachments and detachments to the troop.

2. Mission. The *who, what, when, where,* and *why* for troop/company. State the essential task or tasks to be accomplished by the entire unit, to include on-order missions. Clearly define the troop's objective.

3. Execution.

a. Concept of the Operation. This paragraph explains the commander's intent by stating the *purpose*, *method*, and *endstate* of the operation. The *purpose* tells the "why" of the operation. The *method* tells how the commander visualizes achieving success with respect to the troop as a whole and the utilization of any combat multipliers, in general terms. The *endstate* tells what the final disposition of forces will be and how the endstate will facilitate future operations. For example--

We will conduct the reconnaissance to determine the location, disposition, and composition of enemy main defensive belt in order to facilitate movement through zone, and penetration of the main defensive belt by 52d ID (Mech), allowing 10th Corps to continue offensive operations to the north. We will conduct a forward passage of lines, and perform zone reconnaissance in a troop vee with the engineer platoon conducting a detailed route reconnaissance of Route Cherry behind the scout platoons. Initially movement will be rapid and aggressive until we make contact with the enemy main defensive belt, then transition to a more deliberate, detailed reconnaissance. Indirect fires will be used to provide immediate suppression and assist in the development of the situation by scout platoons. Success is determining the disposition, composition, and location of the enemy main defensive belt then determining the axis and point for the 52d ID (Mech) penetration of the enemy's defense. Our endstate is the troop screening along the enemy main defensive belt ready to pass elements of the 52d ID (Mech) forward for their attack.

(1) Scheme of Maneuver. How major units (maneuver and combat support) will be employed; including movement techniques, bypass criteria (if any), and engagement/disengagement criteria. See the following example.

C Troop moves at 130430 from TAA Viper along Route Dog in troop column. Order of march on Route Dog is Red, Blue, Mortars, FIST, Engineers, Black 6, White, Green, TOC, and combat trains.

C Troop crosses LD at 0500 with Red on left (south), Blue on right (north), Mortars, Engineers, White, Green, TOC, and combat trains moving center of zone (troop vee).

Scout platoons reconnoiter in zone abreast, establishing contact at designated troop internal and flank contact points. Upon enemy contact, develop the situation at platoon level, initially with indirect fire, and report. Destroy squad- and section-sized light-armored and soft-skinned vehicles, within capability. Anything beyond capability, be prepared to assist hasty attack by White and/or Green. Bypass only enemy dismount elements squad-sized or smaller. Bypass all built-up areas larger than 1 kilometer square not on Route Cherry. Upon contact with enemy main defensive belt, conduct dismounted patrols to ascertain enemy disposition, composition, orientation, and location of obstacles.

Mortars cross LD and establish MFP 10 following Blue. Maneuver from MFP 11 to MFP 12 through MFP 25 in sequence. Maintain ability to range at least one-third maximum range forward of lead scouts. Key movement on scouts reporting crossing of phase lines throughout zone.

Engineers cross LD following the FIST and conduct a detailed route reconnaissance of Route Cherry from SP to RP. Maneuver on Route Cherry approximately 500 meters rearward of scouts. Upon identification of enemy's main defensive belt, be prepared to assist scouts with obstacle reconnaissance. White and Green cross LD in platoon columns, White leading. Deploy into platoon wedges forward of LD, terrain permitting. Maneuver in center of troop zone, Green following White by 500 meters. White and Green maneuver approximately one phase line behind Red and Blue through hide positions T-44 through T-52 in sequence. Key movement on scouts reporting crossing of phase lines throughout zone. Scouts will assist in positioning tanks for conduct of hasty attacks by fire or by fire and maneuver upon approval of the troop commander.
TOC and combat trains cross LD in column, following Green. TOC maneuvers center troop zone, establishing positions H-44 through H-48 in sequence. Remains approximately two phase lines rearward of lead scouts. Trains move center troop zone in column, establishing positions T-27 through T-32 in sequence. Remain approximately two phase lines rearward of lead scouts. Key movement on scouts reporting crossing of phase lines throughout zone.

(2) Fires.

(a) Purpose for field artillery and mortar fires. (How will indirect fires support the scheme of maneuver?)

(b) Preparation starting time, duration and description of any fires landing in the area of operations.
(c) Allocation of final protective fires (FPF).
(d) Which element will have priority of fires.
(e) Number of priority targets allocated and who will control them.
(f) Special fires, restrictions, allocation/use of smoke, illumination, or CAS.

(g) Description of scheduled fires (offensive).
(h) Reference to fire support or target annexes.
(3) Obstacles, mines, fortifications.

(a) Priority of engineer effort.

(b) Priority of engineer mission.
(c) Obstacle overlay.
(d) Obstacle list.
(e) Logistical constraints.
(f) On-order missions.
b. Specific tasks to subordinate units. List specific missions in "battle sequence" for each subordinate unit, including attached units. Include movement techniques, flank coordination requirements, other details, and any "be-prepared" missions.

c. Coordinating instructions.
(1) Movement instructions.

(2) Time schedule of events.

- Rehearsal.
- Backbrief times.
- Precombat inspection time.
- First movement time.

(3) Passage of lines.

- Contact points.
- Passage points.
- Lanes and identification.

(4) Mission-oriented protection posture (MOPP) level.

(5) Operational exposure guide (OEG).
(6) Actions on contact.
(7) Actions at danger areas.
(8) Priority intelligence requirements (PIR) and intelligence requirements (IR).
(9) Effective time of attachment or detachment.
(10) Rally points.
(11) Rules of engagement (ROE).
(12) Priority targets for direct-fire weapons.
(13) Uniform and equipment, to include weapon and ammunition.
(14) Any changes regarding battlecarry and battlesight ranges.
(15) Air defense posture and weapons control status.
(16) Any instructions not provided in concept of the operation or specific instructions.

4. Service Support.
a. General.

(1) Organization of trains (squadron and troop).

- Combat.
- Field.

(2) Location of trains (initial and subsequent).

(3) Movement of trains (movement instructions).

b. Materiel and services.
(1) Supply.

- Class I: Time and type.
- Class III: Time, location, and method.
- Class V: Time, amount, and type.
- LOGPAC instructions (include squadron logistics release point [LRP]).

(2) Transportation. Location of main supply route.

(3) Services.

- Mortuary affairs.
- Clothing exchange and bath.
- Maintenance.

c. Medical evacuation and treatment.
(1) Location of squadron aid station.

(2) Displacement of squadron aid station.
(3) Location of regimental clearing station.
(4) Aero-medical evacuation information.
(5) Location of ambulance exchange points.
(6) Handling of contaminated wounded personnel.

d. Personnel.
(1) Prisoner of war (PW) handling and disposition instructions.

(2) PW guard instructions.
(3) Location of PW collection point.
(4) Instructions for interaction with local civil populace (ROE).
(5) Number of expected replacements.
(6) Cross-leveling procedures.
e. Miscellaneous.

5. Command and Signal.

a. Command.

(1) Location of troop commander/troop TOC/squadron command post.
(2) Succession of command.

b. Signal.

(1) SOI index and edition in effect.

- Key frequencies.
- Key call signs.
- Current item number identifier.

(2) KY-57 fill and change over.

(3) Current sign and countersign.
(4) Code words.
(5) Action if jamming or "hot mike" occurs.
(6) Pyrotechnics use.
(7) Periods and conditions of listening silence.

Appendix D

Operations Other Than War

Operations other than war are military activities during peacetime and conflict that do not necessarily involve armed clashes between two organized forces. A cavalry troop will conduct reconnaissance, security, offensive, defensive, and other tactical operations in support of operations other than war. Some examples of operations other than war that a cavalry troop may be involved in are as follows:

PEACETIME CONFLICT
Counterdrug Strikes and raids

Disaster relief Peace enforcement

Civil support Support to insurgency

Peace-building Anti-terrorism

Nation assistance Peacekeeping

Noncombatant evacuation

operations (NEO)

The primary operational constraints during operations other than war are defined by the rules of engagement (ROE). These rules range from the use of force for self-defense only to total impartiality when applying force. The rules of engagement for peacetime will be more restrictive than those for conflict. The troop commander should consider the following in relation to the rules of engagement:

- The degree of force used must only be sufficient to achieve the task at hand and prevent, as far as possible, loss of human life or serious injury.
- Leaders at all levels must ensure that the ROE do not limit the soldiers' ability to defend themselves.
- Leaders must never tape over magazines to keep soldiers from chambering rounds.
- Leaders must ensure all soldiers understand what conditions must be met before chambering rounds.
- The ROE must be realistic, simple, and easy to understand. Develop a single card that clearly outlines the ROE and issue it to all soldiers for reference, keeping in mind that the card in itself is not the answer. Soldiers must know and understand the ROE.
- During peacekeeping, forces have no mandate to prevent violations of peace agreements by the active use of force. (Observe and report only.) To maintain the peace, units may be positioned

between belligerents. Troop commanders must realize that soldiers are being placed at risk. The protection of the troop must be emphasized.

- Peace enforcement missions allow the active use of force. The peacetime ROE resemble the ROE for hostilities (wartime).
- The formulation of ROE should consider the cultural differences of multinational forces.
- Soldiers should be trained in the ROE, using tactical vignettes or simulated events. Train soldiers to avoid unnecessary collateral damage to property.

Actions on contact and the troop's ability to develop the situation will be affected by the rules of engagement for the theater in which the troop is operating. The troop commander should rehearse battle drills based on those rules.

In preparing for operations other than war, a troop may require specialized training. They could benefit from training in the following areas:

- Nature of peacekeeping.
- Regional orientation/culture of belligerents.
- Negotiating skills.
- Mine/booby trap/unexploded ordnance training.
- Checkpoint operations.
- Investigating and reporting.
- Information collection.
- Patrolling.
- Media inter-relationships.
- Establishing a buffer zone.
- Supervising a truce or cease-fire.
- Contributing to maintenance of law and order.
- Demilitarizing cities or geographical areas.
- Border surveillance/security.

Glossary

AAR	after-action review
A&O	assault and obstacle
ACE	armored combat earthmover
ACR	armored cavalry regiment
ACT	air cavalry troop
ADA	air defense artillery
A/L	administrative/logistics
ALO	air liaison officer
ALOC	administrative/logistics operations center
AO	area of operations
APERS	antipersonnel (ammunition)
AT	antitank
APC	armored personnel carrier
arty	artillery
AVLB	armored vehicle launched bridge
BHL	battle handover line
BP	battle position
BUA	built-up area
C2	command and control
cal	caliber
CAS	close air support
cbt	combat
CBU	cluster bomb unit
CEOI	communications-electronics operation instructions
CEV	combat engineer vehicle
CDR	commander
CFV	cavalry fighting vehicle
cGy/hr	centigray(s) per hour
CHEMWARN	chemical warning (message sent in the event of a friendly strike)
COA	course of action
COMSEC	communications security
COT	counter obstacle team
CP	command post
CS	combat support
CSS	combat service support

CTCP	combat trains command post
DA	Department of the Army
DS	direct support
DS/R	direct support/reinforcing
EA	engagement area
engr	engineer
EPW	enemy prisoner of war
evac	evacuate; evacuation
EW	electronic warfare
1SG	first sergeant
FA	field artillery
FASCAM	family of scatterable mines
FAC	forward air controller
FEBA	forward edge of the battle area
FIST	fire support team
FLOT	forward line of own troops
FM	frequency modulation (radio); field manual
FPF	final protective fires
FRAGO	fragmentary order
FS	fire support
FSCL	fire support coordination line
FSE	fire support element
FSO	fire support officer
FSN	fire support net
GLD	ground laser designator
GPS	global positioning system
GRREG	graves registration
GSR	ground surveillance radar
HC	heavy concentration (smoke)
HE	high explosive
HEMTT	heavy expanded mobile tactical truck
HHT	headquarters and headquarters troop
HMMWV	high mobility multipurpose wheeled vehicle
HQ	headquarters
hvy	heavy
ICM	improved conventional munitions

ID	identification
IPB	intelligence preparation of the battlefield
IR	intelligence requirements
KIA	killed in action
km	kilometers
LACR	light armored cavalry regiment
LC	line of contact
LD	line of departure
ldr	leader
LID	light infantry division
LO	liaison officer
LOA	limit of advance
LOGPAC	logistics package
LRP	logistics release point
lt	light
maint	maintenance
MBA	main battle area
MCOO	modified combined obstacle overlay
mech	mechanized
METT-T	mission, enemy, terrain, troops and time available
MFP	mortar firing point
MICLIC	mine clearing line charge
MOPP	mission-oriented protection posture
mort	mortar
MRB	motorized rifle battalion
MRC	motorized rifle company
MRE	meals ready to eat
MRP	motorized rifle platoon
MSR	main supply route
MTOE	modified table(s) of organization and equipment

NAI	named area(s) of interest
NBC	nuclear, biological, chemical
NCO	noncommissioned officer
NCOIC	noncommissioned officer in charge
NEO	noncombatant evacuation operations
NET	new equipment training
NPL	no penetration line
NUCWARN	nuclear warning (message sent in the event of a friendly strike)
obj	objective
OCOKA	observation and fields of fire, cover and concealment, obstacles, key terrain, avenues of approach (considerations in evaluating terrain as part of METT-T)
OEG	operational exposure guidance
OI	operations and intelligence
OIC	officer in charge
OP	observation post
OPCON	operational control
OPLAN	operation plan
OPORD	operations order
OPSEC	operations security
PAC	personnel and administration center
PD	point of departure
PEWS	platoon early warning system
PIR	priority intelligence requirements
PL	phase line
PLL	prescribed load list
plt	platoon
PMCS	preventive maintenance checks and services
POL	petroleum, oils, and lubricants
POSNAV	position navigation
PSG	platoon sergeant
PW	prisoner of war

recon	reconnaissance; reconnoiter
REDCON	readiness condition
RFA	restricted fire area
RFL	restricted fire line
R/GS	reinforcing/general support
ROE	rules of engagement
ROM	refuel on the move
RP	release point
rte	route
S1	adjutant (US Army)
S2	intelligence officer (US Army)
S3	operations and training officer (US Army)
S4	supply officer (US Army)
SCO	squadron commander
sec ldr	section leader
SEE	small emplacement excavator
SGLI	Serviceman's Group Life Insurance
sgt	sergeant
SIDPERS	Standardization Installation/Division Personnel System
SIGSEC	signal security
SITREP	situation report
SOI	signal operation instructions
SOP	standing operating procedure
SP	start point
sqd ldr	squad leader
TAA	tactical assembly area
TAC CP	tactical command post
TACFIRE	tactical fire direction system
TC	tank commander
TIRS	terrain index reference system
TOC	tactical operations center
TOE	table(s) of organization and equipment
TOW	tube-launched, optically-tracked, wire-guided (missile)
TRADOC	Training and Doctrine Command
TRP/trp	target reference point; troop
TTP	tactics, techniques, and procedures
UMCP	unit maintenance collection point

UMT	unit ministry team
US	United States
veh	vehicle
WIA	wounded in action
wrkr	wrecker
XO	executive officer

References

SOURCES USED

These are the sources quoted or paraphrased in this publication.

Army Publications

FM 17-15	Tank Platoon. 7 October 1987.
FM 17-95	Cavalry Operations. 19 September 1991.
FM 17-95-10	Armored Cavalry Regiment and Squadron. 22 September 1993.
FM 17-98	Scout Platoon. 9 September 1994.
FM 71-1	Tank and Mechanized Infantry Company. 22 November 1988.

Command Publications

Command publications cannot be obtained through Armywide resupply channels. Determine availability by contacting the address below.

Commander
USAARMC
ATTN: ATZK-IMO-RS
Fort Knox, KY 40121-5000

FKSM 17-97-3	Cavalry Troop Common SOP. 15 July 1994.

READINGS RECOMMENDED

These readings contain relevant supplemental information.

Army Publications

ARTEP 17-57-10-MTP	Mission Training Plan for the Scout Platoon. 27 December 1988.
ARTEP 17-237-10-MTP	Mission Training Plan for the Tank Platoon. 3 October 1988.
ARTEP 17-487-30-MTP	Mission Training Plan for the Regimental Armored Cavalry Troop. 3 September 1991.

DA Form 1155	Witness Statement on Individual. June 1966.
DA Form 1156	Casualty Feeder Report. June 1966.
DA Form 1971-R	Radiological Data Sheet Monitoring and Point Technique. September 1986.
DA Form 2028	Recommended Changes to Publications and Blank Forms. February 1974.
DD Form 551	Record of Interment. August 1984.
DD Form 1077	Collecting Point Register of Deceased Personnel. July 1984.
FM 1-114	Tactics, Techniques, and Procedures for the Regimental Aviation Squadron. 20 February 1991.
FM 1-116	Tactics, Techniques, and Procedures for the Air Cavalry/Reconnaissance Troop. 20 February 1991.
FM 3-3	Chemical and Biological Contamination Avoidance. 16 November 1992.
FM 3-4	NBC Protection. 20 May 1992.
FM 3-5	NBC Decontamination. 23 July 1992.
FM 3-50	Smoke Operations. 4 December 1990.
FM 3-100	NBC Defense, Chemical Warfare, Smoke, and Flame Operations.

	23 May 1991.
FM 3-101	Chemical Staffs and Units. 22 April 1987.
FM 5-36	Route Reconnaissance and Classification. 10 May 1985.
FM 5-100	Engineer Combat Operations. 22 November 1988.
FM 5-250	Explosives and Demolitions. 15 June 1992.
FM 6-20	Fire Support in the AirLand Battle. 17 May 1988.
FM 7-91	Tactical Employment of Antiarmor Platoons, Companies, and Battalions. 30 September 1987.
FM 7-98	Operations in a Low-Intensity Conflict. 19 October 1992.
FM 8-55	Planning for Health Service Support. 15 February 1985.
FM 10-63-1	Graves Registration Handbook. 17 July 1986.
FM 17-12-1-1	Tank Gunnery (Abrams), Volume I. 19 March 1993.
FM 17-12-1-2	Tank Gunnery (Abrams), Volume II. 19 March 1993.
FM 17-12-8	Light Cavalry Gunnery. 28 October 1994.
FM 17-18	Light Armor Operations. 8 March 1994.
FM 20-32	Mine/Countermine Operations. 30 September 1992.
FM 21-10	Field Hygiene and Sanitation. 22 November 1988.
FM 21-10-1	Unit Field Sanitation Team. 11 October 1989.
FM 21-11	First Aid for Soldiers. 27 October 1988.
FM 22-100	Military Leadership. 31 July 1990.

FM 23-1	Bradley Fighting Vehicle Gunnery. 1 March 1991.
FM 34-3	Intelligence Analysis. 15 March 1990.
FM 44-1	US Army Air Defense Artillery Employment. 9 May 1983.
FM 44-3	Air Defense Artillery Employment: Chaparral/Vulcan/Stinger. 15 June 1984.
FM 44-8	Small Unit Self-Defense Against Air Attack. 30 December 1981.
FM 44-18-1	Stinger Team Operations. 31 December 1984.
FM 71-2	The Tank and Mechanized Infantry Battalion Task Force. 27 September 1988.
FM 100-2-1	Soviet Army Operations and Tactics. 16 July 1984.
FM 100-2-3	The Soviet Army: Troops, Organization, and Equipment. 6 June 1991.
FM 100-5	Operations. 14 June 1993.
FM 101-5-1	Operational Terms and Symbols. 21 October 1985.
TC 6-40A	Field Artillery Automated Cannon Gunnery. 21 April 1989.

Command Publications

Contact the address shown on page References-1 for these command publications.

FKSM 17-15-3	Tank Platoon SOP. February 1989.

FKSM 17-98-3	Scout Platoon SOP. November 1994.
FKSM 17-98-4	Scout Platoon Leader's Notebook.

Standardization Agreements (STANAGs)

To requisition these documents, complete DD Form 1425 and send it to—

Naval Publications and Forms Center
5801 Tabor Avenue
Philadelphia, PA 19120

STANAG 2002	Warning Signs for the Marking of Contaminated or Dangerous Land Areas, Complete Equipment, Supplies and Stores. 2 December 1989.
STANAG 2014	Operation Orders, Warning Orders and Administrative/Logistics Orders. 14 August 1991.
STANAG 2029	Method of Describing Ground Locations, Areas and Boundaries. 10 June 1991.
STANAG 2036	Land Mine Laying, Marking, Recording and Reporting Procedures. 29 April 1988.
STANAG 2044	Procedures for Dealing with Prisoners of War (PW). 12 July 1988.
STANAG 2047	Emergency Alarms of Hazard or Attack (NBC and Air Attack Only). 10 October 1990.
STANAG 2082	Relief of Combat Troops. 24 July 1991.
STANAG 2083	Commanders' Guide on Nuclear Radiation Exposure of Groups.

	10 September 1990.
STANAG 2084	Handling and Reporting of Captured Enemy Equipment and Documents. 26 June 1986.
STANAG 2103	Reporting Nuclear Detonations, Biological and Chemical Attacks, and Predicting and Warning of Associated Hazards and Hazard Areas. 10 June 1991.
STANAG 2104	Friendly Nuclear Strike Warning. 12 December 1989.
STANAG 2112	NBC Reconnaissance. 5 August 1986.
STANAG 2113	Denial of a Unit's Military Equipment and Supplies to an Enemy. 11 July 1991.
STANAG 2129	Identification of Land Forces on the Battlefield. 16 May 1989.
STANAG 2889	Marking of Hazardous Areas and Routes Through Them. 8 February 1990.

*** C1, FM 23-65**

FIELD MANUAL
No. 23-65

HEADQUARTERS
DEPARTMENT OF THE ARMY
Washington, DC, 19 June 1991

BROWNING MACHINE GUN CALIBER .50 HB, M2

CONTENTS

		Page
PREFACE		iv
* **CHAPTER 1. INTRODUCTION**		
1-1.	Training Strategy	1-1
1-2.	Description	1-5
1-3.	Components	1-8
1-4.	Ground Mounts	1-9
1-5.	Accessories for Ground Mounts	1-12
1-6.	Vehicular Mounts	1-14
1-7.	Ammunition	1-17
* **CHAPTER 2. MAINTENANCE**		
2-1.	Safety	2-1
2-2.	General Disassembly	2-2
2-3.	Cleaning, Inspection, and Lubrication	2-9
2-4.	Maintenance Procedures	2-12
2-5.	Maintenance Under NBC Conditions	2-12
2-6.	General Assembly	2-12
2-7.	Function Check	2-17
* **CHAPTER 3. OPERATION AND FUNCTIONING**		
3-1.	Operation	3-1
3-2.	Loading Procedures	3-1
3-3.	Unloading Procedures	3-3
3-4.	Cycle of Functioning	3-4
3-5.	Left-hand Feed	3-11
3-6.	Headspace and Timing	3-12

DISTRIBUTION RESTRICTION: Approved for public release; distribution is unlimited.

---

*This publication supersedes FM 23-65, 19 May 1972, and TC 23-65-1, 19 September 1984.

Page

CHAPTER 4. PERFORMANCE PROBLEMS AND DESTRUCTION

4-1. Malfunctions 4-1
4-2. Stoppages 4-1
4-3. Immediate Action 4-2
4-4. Remedial Action 4-3
4-5. Destruction Procedures 4-5

CHAPTER 5. MARKSMANSHIP TRAINING

Section I. Planning 5-1
5-1. Objectives 5-1
5-2. Commander's Responsibilities 5-1
5-3. Phases of Training 5-2
5-4. Sustainment Training 5-2
5-5. Remedial Training 5-2

Section II. Fundamentals 5-3
5-6. Firing Positions 5-3
5-7. Dry Fire Training 5-6
5-8. Range Determination 5-12
5-9. Observation and Adjustment of Fire 5-16
5-10. Fire Commands 5-18
5-11. Crew Exercises 5-22
5-12. Machine Gun Fundamental Skills Test 5-36

Section III. Basic Marksmanship 5-36
5-13. Concept of Zeroing/Targeting 5-36
5-14. 10-Meter Firing Exercise 5-41
5-15. Transition Day Firing Exercise 5-41
5-16. NBC Firing 5-41
5-17. Night Fire Exercise 5-41

Section IV. Advanced Gunnery 5-41
5-18. Objectives 5-42
5-19. Tracking and Leading Exercises 5-42
5-20. Mounted Firing Exercise 5-45
5-21. Mounted NBC Firing Exercise 5-45
5-22. Predetermined Firing Exercise 5-45

CHAPTER 6. COMBAT TECHNIQUES OF FIRE

Section I. Fundamentals 6-1
6-1. Characteristics of Fire 6-1
6-2. Classes of Fire 6-3

Section II. Fire Control 6-7
6-3. Methods of Fire Control 6-8
6-4. Targets and Their Engagement 6-8
6-5. Overhead Fire 6-14
6-6. Defilade Positions 6-18
6-7. Methods of Laying the Gun for Defilade Firing 6-20
6-8. Final Protective Fires 6-21
6-9. Application of Fire 6-22
6-10. Fire Adjustment 6-23
6-11. Antiaircraft Gunnery 6-25

Page

Section III. Limited Visibility Conditions 6-26
6-12. Difficulties.. 6-26
6-13. Terminology 6-26
6-14. Target Engagement 6-27
6-15. Fire Control 6-28
6-16. Preplanned Fires... 6-28
6-17. NBC Considerations.................................. 6-29

CHAPTER 7. TRAIN THE TRAINER PROGRAM
7-1. Concept.. 7-1
7-2. Trainer Certification Program 7-2
7-3. Responsibilities and Duties of the Trainer.. 7-3

APPENDIX A. SAFETY..A-1
APPENDIX B. TRAINING AIDS AND DEVICES B-1
* **APPENDIX C. RANGES AND TRAINING EXERCISES**........................C-1
APPENDIX D. FIGHTING POSITIONS..D-1
APPENDIX E. RANGE CARDS...E-1
* **APPENDIX F. AERIAL DEFENSE** F-1
* **APPENDIX G NIGHT OPTICS** G-1

* **GLOSSARY** Glossary-1
* **REFERENCES**..References-1
* **INDEX** Index-1

* **DA Forms**

* PREFACE

This manual provides technical information, training techniques, and guidance on the caliber .50 HB machine gun, M2. Unit leaders and designated gunners will use this information to successfully integrate the weapon into combat operations. They can instruct on the range or at concurrent training stations.

The material applies as is to both nuclear and conventional warfare.

Trainers must ensure that everyone observes safety procedures at all times. Commanders, trainers, and individual students must remember that safety is everyone's responsibility. Leaders will conduct all training as though each weapon were fully loaded. At no time during training will anyone allow the desire for speed or accuracy to override the requirement to follow safety procedures. Safe training is good training.

The proponent of this publication is US Army Infantry School. Send comments and recommendations on DA Form 2028 (Recommended Changes to Publications and Blank Forms) directly to the Commandant, US Army Infantry School, ATTN: ATSH-INB-O, Fort Benning, GA 31905-5594 or lusanoh@benning.army.mil.

Unless otherwise stated, the masculine gender refers to both men and women.

CHAPTER 1

INTRODUCTION

The procedures and methods used in Army machine gun marksmanship are based on the concept that soldiers must be skilled gunners who can effectively apply their firing skills in combat. The basic firing skills and exercises outlined in this manual must be a part of every unit's machine gun training program. The soldiers' proficiency depends on proper training and application of basic gunnery fundamental, which are taught in a progressive program to prepare gunners for combat.

1-1. TRAINING STRATEGY

Training strategy is the overall concept for integrating resources into a program to train individual and collective skills needed to perform a unit's wartime mission.

a. Training strategies for marksmanship are implemented in TRADOC institutions (NCOES, basic and advanced officer's courses) and in units. The overall training strategy is multifaceted and is inclusive of the specific strategies used in institution and unit programs. Also included are the supporting strategies that use resources such as publications, ranges, ammunition, training aids, devices, simulators, and simulations. These strategies focus on developing critical soldier skills, and on leader skills that are required for the intended outcome.

b. Two primary components compose the training strategies: initial training and sustainment training. Both may include individual and collective skills. Initial training is critical because a task that is taught correctly and learned well is retained longer. Well-trained skills can be more quickly regained and sustained if an interim of nonuse occurs. The more difficult and complex the task, the harder it is to sustain the skill. Personnel turnover is a main factor in decay of collective skills, since the loss of critical team members requires retraining to regain proficiency. If a long period elapses between initial and sustainment training sessions or training doctrine is altered, retraining maybe required.

c. The training strategy for caliber .50 MG marksmanship begins in selected resident training and continues in the unit. An example of this overall process is illustrated in Figure 1-1 and provides a concept of the flow of unit sustainment training. The soldiers graduating from selected resident training courses have been trained to maintain their MGs and to hit a variety of targets. They have learned range determination, target detection, application of marksmanship fundamentals, and other skills needed to engage a target. Task training during these courses may lead to qualification.

d. Training continues in units on the basic skills taught in combat arms. Additional skills, such as suppressive fire and supporting fire, are trained and then integrated into collective training exercises, which include platoon and squad live-fire STXs. (A unit-marksmanship training program is explained in Chapter 5.) The strategy for sustaining the basic marksmanship skills taught in combat arms is periodic preliminary instruction, followed by qualification range firing. However, a unit must set up a year-round program to sustain skills. Key elements include training of trainers and refresher training of nonfiring skills.

e. Additional skills trained in the unit include techniques for employment, suppressive fires, night fire, MOPP firing, and moving targets. Related soldier skills of camouflage, cover and concealment, maneuver, and preparation and selection of a fighting position are addressed in STP 21-24-SMCT, which must be integrated into tactical training.

f. In the unit, individual and leader proficiency of marksmanship tasks are integrated into collective training to include squad, section, and platoon drills and STXs. The collective tasks in these exercises, and how they are planned and conducted, are in the MTP and battle drill books for each organization. Based on the type organization, collective tasks are evaluated to standard and discussed during leader and trainer after-action reviews. Objective evaluations of both individual and unit proficiency provide readiness indicators and future training requirements.

g. A critical step in the Army's overall marksmanship training strategy is to train the trainers and leaders first. Leader courses and unit publications develop officer and NCO proficiencies necessary to plan and conduct marksmanship training and to evaluate the effectiveness of unit marksmanship programs. Training support materials are provided by the proponent schools to include field manuals, training aids, devices, simulators, and programs that are doctrinal foundations and guidance for training the force.

h. Once the soldier understands the weapon and has demonstrated skill in zeroing, additional live-fire training and a target acquisition exercise at various ranges are conducted. Target types and scenarios of increasing difficulty must be mastered to develop proficiency.

i. Initial individual training culminates in the soldier's proficiency assessment, which is conducted on a transition/record fire range. This evaluation also provides an overview of unit proficiency and training effectiveness.

j. Unit training programs maintain the soldiers' proficiency level. The ultimate goal of a unit marksmanship program is to maintain well-trained gunners so a unit can survive and win on the battlefield. The trainer must realize that qualification is not an end, but a step toward reaching this combat requirement. (See Figure 1-1.)

(1) To reach this goal, the gunner must be able to position and use his weapon under the following combat conditions:

- Enemy personnel are seldom visible except when assaulting.
- Most combat fire must be directed at an area where the enemy has been detected or where he is suspected of being located but cannot be seen. Area targets consist of objects or outlines of men irregularly spaced along covered and concealed areas (ground folds, hedges, or borders of woods).
- Most combat targets can be detected by smoke, flash, dust, noise, or movement and are visible only for a moment.
- Some combat targets can be engaged by using nearby objects as reference points.
- The nature of the target and irregularities of terrain and vegetation may require a firer to use a variety of positions in addition to the prone or supported position to fire effectively on the target. In a defensive situation, the firer usually fires from a supported position.
- Most combat targets have a low contrast outline and are obscure. Therefore, choosing an aiming point in elevation is difficult.
- Time-stressed fire in combat can be divided into three types: a single, fleeing target that must be engaged quickly; distributed targets engaged within the time they remain available; and a surprise target that must be engaged at once with accurate, instinctive fire.

(2) The unit's program must provide fundamental training to sustain and improve the skills and proficiency the soldier has attained during his basic marksmanship training. Once basic skills have been mastered, these must be improved by conducting new or advanced individual and collective training. The program must develop collective firing skills by incorporating marksmanship into tactical exercises. This training must maintain the soldier's confidence in the weapon and his skills. A soldier's

survival may depend on his ability to defend himself or other members of the unit. Therefore, individual and collective firing skills must support the expected battlefield conditions and the unit's combat mission.

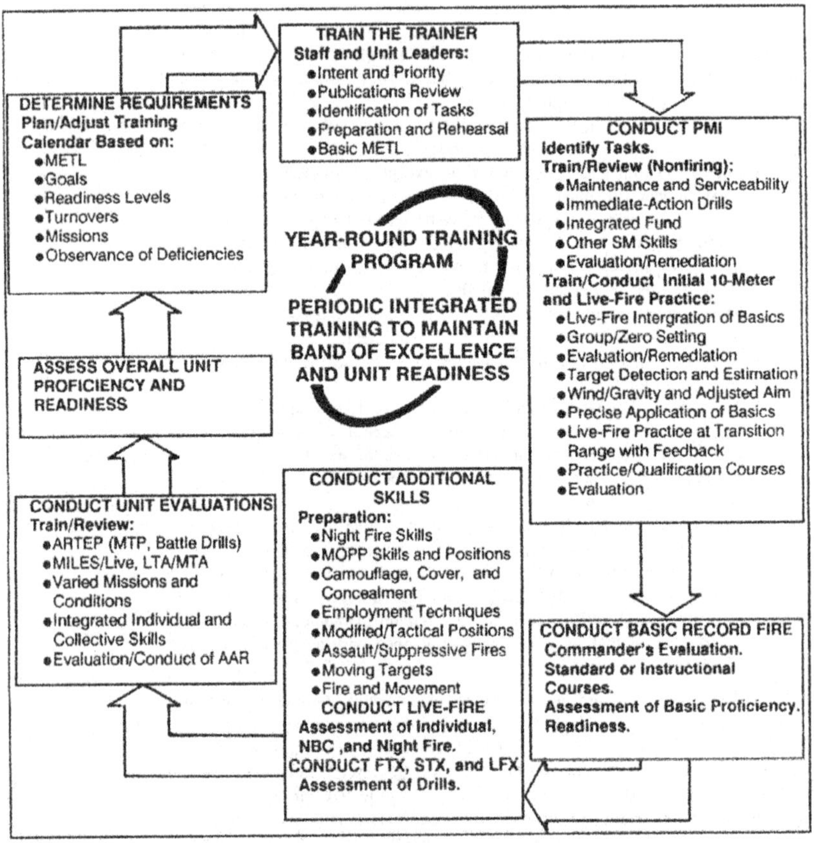

Figure 1-1. Unit marksmanship sustainment strategy.

(3) A unit's marksmanship program must be battlefield oriented. It must be based upon several individual combat tasks as well as organizational, operational, or contingency missions. It must have available resources such as ammunition, time, ranges, and qualified trainers. This manual provides the information a unit commander needs to develop an effective marksmanship program for his unit requirements.

(4) General marksmanship, training knowledge, and accurate firing are acquired skills that perish easily. Skill practice should be conducted for short periods throughout the year. Most units have a readiness requirement that all soldiers must zero their MGs within a certain time after unit assignment. Also, soldiers must confirm the zero of their assigned MGs before conducting a qualification firing.

1-2. DESCRIPTION

The Browning machine gun caliber .50 HB, M2 (Figure 1-2) is a belt-fed, recoil-operated, air-cooled, crew-served machine gun. The gun is capable of single shot, as well as automatic fire, and operates on the short recoil principle.

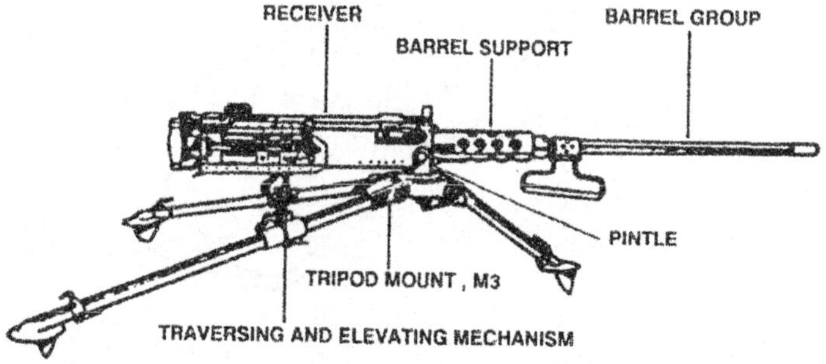

Figure 1-2. Browning machine gun.

a. The machine gun is capable of being fed from either the right or left by repositioning certain parts. The weapon has nonfixed headspace that must be set. Timing must also be adjusted to cause the gun to fire slightly out of battery to prevent damage to moving parts. The force for recoil operation is furnished by expanding powder gases, which are controlled by various springs, cams, and levers. Maximum surface of the barrel and receiver are exposed to permit air cooling. Perforations in the barrel support allow air to circulate around the breach end of the barrel and help in cooling the parts. A heavy barrel is used to retard early overheating.

b. The gun has a leaf-type rear sight (Figure 1-3), graduated in both yards and roils. The scale ranges from 100 to 2,600 in yards, and from 0 to 62 in mils. The windage knob-permits deflection changes to right or left of center. The front sight is a fixed blade type with cover (Figure 1-4).

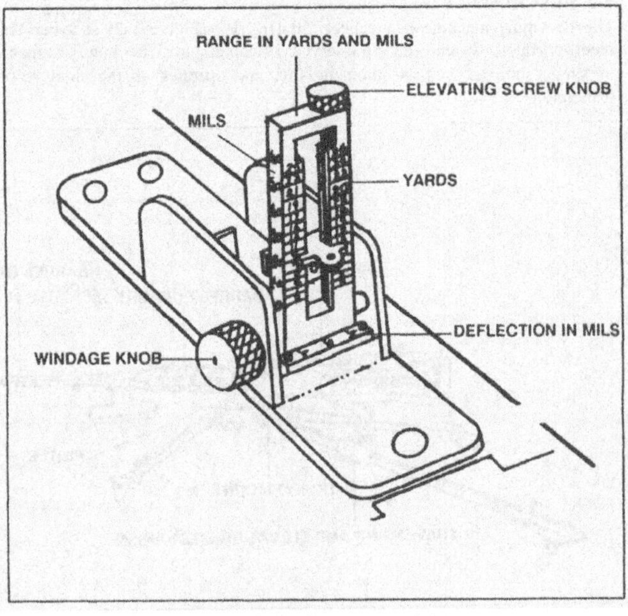

Figure 1-3. Leaf type rear sight.

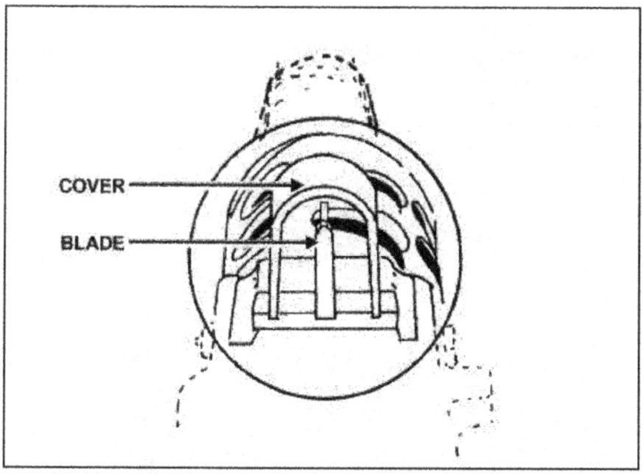

Figure 1-4. Front sight, cover, and blade.

c. Table 1-1 provides the general data on the caliber .50 MG.

Weight (approx)	84 pounds
Weight of barrel	24 pounds
Length of gun	65.13 inches
Length of barrel	45 inches
Length of rifling (approx)	41.88 inches
Number of lands and grooves	8
Twist, right-hand	one turn in 15 inches
Feed	link-belt
Operation	recoil
Cooling	air
Muzzle velocity (approx)	3,050 feet per second
Rate of fire (cyclic)	450 to 550 rounds per minute
Maximum range (approx)	7,440 yards or 6,764 meters
Maximum effective range (approx)	2,000 yards or 1,830 meters
• Area targets	1,830 meters
• Point targets, single shot	1,500 meters

Table 1-1. General data.

1-3. COMPONENTS

The major components of the caliber .50 MG and their purposes are shown in Table 1-2 and Figure 1-5.

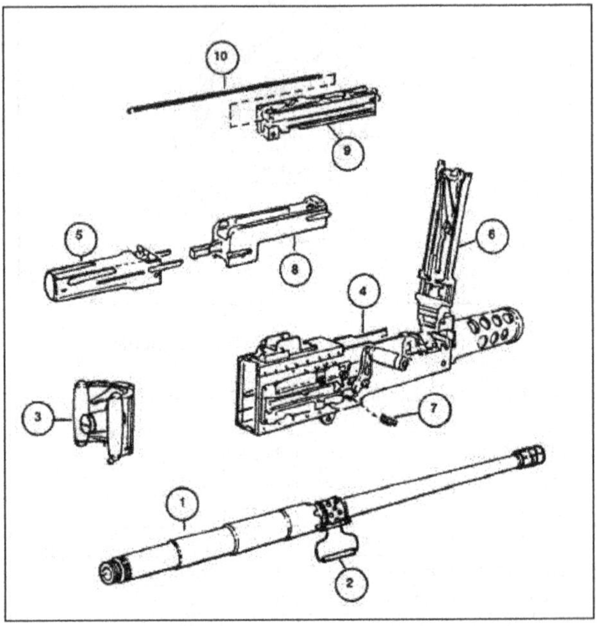

Figure 1-5. Components of the caliber .50 MG.

COMPONENTS	PURPOSES
l. Barrel Group	Houses cartridges for firing; directs projectile.
2. Carrier Assembly	Provides handle to carry barrel and to remove the barrel from the receiver.
3. Backplate Group	Houses the trigger, bolt latch release, buffer tube sleeve, and the left and right spade grips.
4. Receiver Group	Serves as a support for all major components; houses action of weapon, which controls functioning of weapon.
5. Bolt Group	Provides feeding, chambering, firing, and extracting, using the propellant gases and recoil spring for power.
6. Cover Group	Feeds linked belt ammunition; positions and holds cartridges in position for extracting, feeding, and chambering.
7. Bolt Stud	Provides a means to move the bolt to the rear with the retracting slide handle.
8. Barrel Extension Group	Secures the barrel to the recoiling parts.
9. Barrel Buffer Body	Assists in recoil and counterrecoil of the bolt group.
10. Driving Spring Rod Assembly	Drives the bolt forward when the bolt latch release is depressed.

Table 1-2. Components and their purposes.

1-4. GROUND MOUNTS

The two principal ground mounts used with the caliber .50 machine gun are the tripod mount, M3, and the antiaircraft mount, M63. The tripod mount, M3, is a ground mount designed for use against ground targets. The antiaircraft mount,

M63, is a ground mount principally designed for use against aerial targets. Its use against ground targets is limited because the mount tends to be unstable when the gun is fired at low angles.

a. **Tripod Mount, M3**. The M3 mount is the standard ground mount of the caliber .50 machine gun (Figure 1-6). It is a folding tripod with three, telescopic, tubular legs connected at the tripod head. Each leg ends in a metal shoe that can be stamped into the ground for greater stability. The two trail legs are joined together by the traversing bar. The traversing bar serves as a support for the traversing and elevating mechanism, which in turn supports the rear of the gun. The tripod head furnishes a front support for the mounted gun that is further supported by the short front leg. When the tripod is emplaced on flat terrain with all extensions closed, the adjustable front leg should form an angle of about 60 degrees with the ground. This places the gun on a low mount about 12 inches above the ground. To raise the tripod farther off the ground, extend the telescopic front and trail legs enough to keep the tripod level and maintain the stability of the mount.

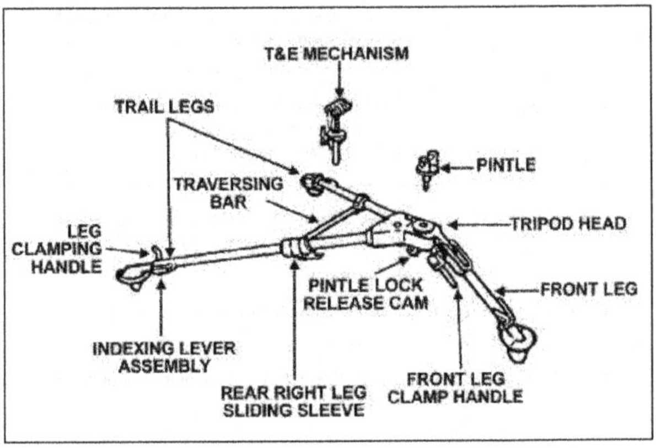

Figure 1-6. M3 tripod mount.

(1) To set the tripod trail legs--

(a) Unscrew the leg-clamping handle, press down on the indexing lever, and extend the leg to the desired length.

(b) Align the indexing lever stud with one of the holes in the tripod leg extension.

(c) Release the pressure on the indexing lever, allowing the stud to fit the desired hole. Tighten the leg-clamping handle.

(2) *To set the front leg of the tripod-*

(a) Turn the front leg clamp handle counterclockwise to loosen the front leg.

(b) Adjust the leg to the desired angle and tighten the front leg clamp.

(3) To secure the tripod legs, stamp the metal shoe on each tripod leg into the ground. Sandbag each leg to stabilize the M2 for firing.

b. **Antiaircraft Mount, M63.** The antiaircraft mount (Figure 1-7) is a four-legged, low silhouette, portable mount used for antiaircraft fire. Table 1-3 lists the general data pertaining to the M63.

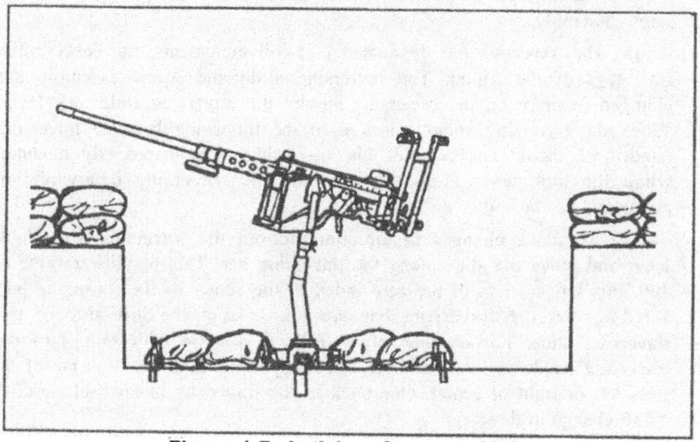

Figure 1-7. Antiaircraft mount, M63.

WEIGHTS:	
Overall	144 pounds
Four legs	24 pounds
Base assembly	54 pounds
Elevator assembly	12 pounds
Cradle assembly	44 pounds
Ammunition box-tray assembly	10 pounds
HEIGHT OVERALL	42 inches
LENGTH OF LEG	24 inches
DIAMETER OF BASE (WITH LEG ASSEMBLED)	52 inches
MAXIMUM ELEVATION	85 degrees
MAXIMUM DEPRESSION	29 degrees
MAXIMUM TRAVERSE	360 degrees

Table 1-3. M63 general data.

1-5. ACCESSORIES FOR GROUND MOUNTS

The following paragraph explains the functions of the traversing and elevating mechanism and pintle used in the mounting of the machine gun when used in the ground configuration.

a. **Traversing and Elevating Mechanism.** The T&E mechanism (Figure 1-8) is used to engage preselected target areas at night or during limited visibility conditions. Record direction and elevation readings from the traversing bar and T&E mechanism. Record all readings in mils.

(1) The traversing mechanism consists of a traversing bar, slide, and screw assembly.

(a) The traversing bar, graduated in 5-roil increments, fits between the trail legs of the tripod. The traversing slide and screw assembly are clamped in place on the traversing bar by the traversing slide lock lever. When the traversing slide is locked to the traversing bar, the traversing handwheel should be centered. The traversing slide is properly mounted when the lock lever is to the rear and the traversing handwheel is positioned to the left.

(b) To make changes in direction, loosen the traversing slide lock lever and move the slide along the traversing bar. This permits traverse of 400 mils left or right of the zero index in the center of the traversing bar. Readings on the traversing bar are taken from the left side of the traversing slide. For changes of 50 mils or less in deflection, turn the traversing handwheel of the screw assembly. This allows a traverse of 50 mils left or right of center. One click in the traversing handwheel signifies 1 mil change in direction.

(2) The elevating mechanism consists of an upper and lower elevating screw.

(a) It is connected to the gun by inserting the quick release pin assembly through the holes in the upper elevating screw yoke and the rear mounting lugs of the receiver. A scale, graduated in mils, is fitted to the upper screw to indicate elevation. This scale is marked to show 250 mils in depression and 100 mils in elevation from the zero setting.

(b) The elevating handwheel is graduated in l-mil increments up to 50 mils, and is fastened to the elevating screw by a screw lock. This synchronizes the handwheel graduations with those on the upper elevating screw. A spring-actuated index device produces a clicking sound when the handwheel is turned. Each click equals 1 mil change in elevation. The handwheel is turned clockwise to depress the barrel and counterclockwise to elevate.

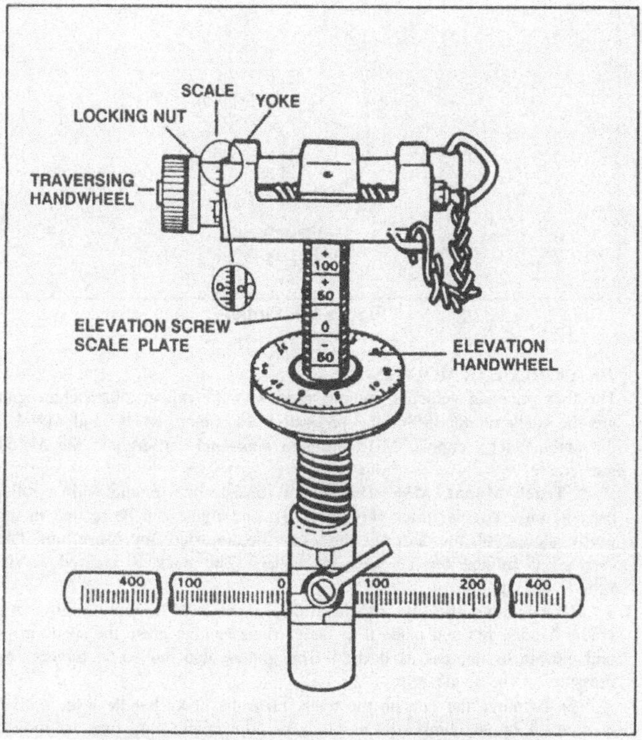

Figure 1-8. Traversing and elevating mechanism.

b. **Pintle.** The gun is connected to the tripod mount, M3, by a pintle (Figure 1-9, page 1-14). This pintle is semipermanently attached to the machine gun by a pintle bolt through the front mounting hole in the receiver. The tapered stem of the pintle seats in the tripod head. It is held secure by a pintle lock and spring. To release the pintle, raise the pintle lock, releasing the cam. The weight of the pintle and traversing and elevating mechanism are considered as part of the total weight of the tripod mount, M3 (44 pounds).

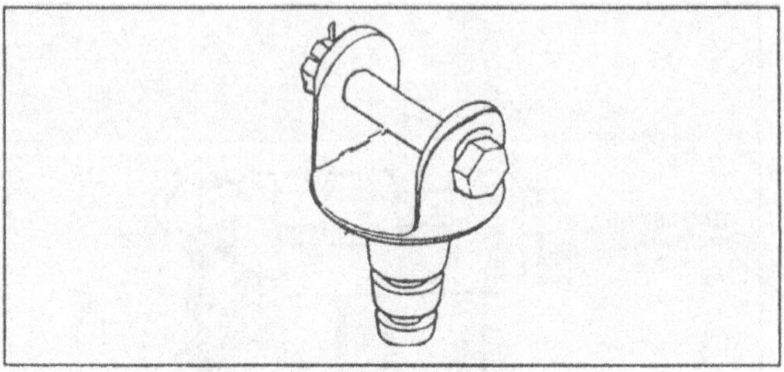

Figure 1-9. Pintle.

1-6. VEHICULAR MOUNTS

The four principal vehicular mounts used with the caliber .50 machine gun are the truck mount, M36; the pedestal truck mount, M31C and M24A2; the commander's cupola, Ml13 armored personnel carrier; and the MK64 gun cradle.

a. **Truck Mount, M36.** This mount consists of a cradle with a roller carriage on a circular track (Figure 1-10). The cradle can be rotated in the pintle sleeve of the carriage and can be adjusted for elevation. The carriage is guided on the track by rollers. The track is secured to the vehicle by supports.

(1) To move the gun in elevation on the M36 mount, remove the cradle locking pin and place it in the carriage handle; grasp the spade grips and elevate or depress as desired. The gun is also moved in traverse by pressure on the spade grips.

(2) To move the gun on the track, raise the brake handle lever until it is retained by the brake detent plungers. The cradle may then be moved on the track by applying pressure on the carriage handle.

b. **Pedestal Truck Mount, M31C.** Pedestal mounts are component assemblies designed for installation on the 1/4-ton vehicles to support a machine gun mount. They are composed of a pintle socket, pintle clamping screw column, and braces (Figure 1-11).

c. **Armored Vehicle Cupola Mount.** A caliber .50 machine gun and mount are installed in the gun support on the commander's cupola of an Ml 13 armored personnel carrier. The machine gun can be traversed 360 degrees, elevated 53 degrees, and depressed 21 degrees maximum (Figure 1-12).

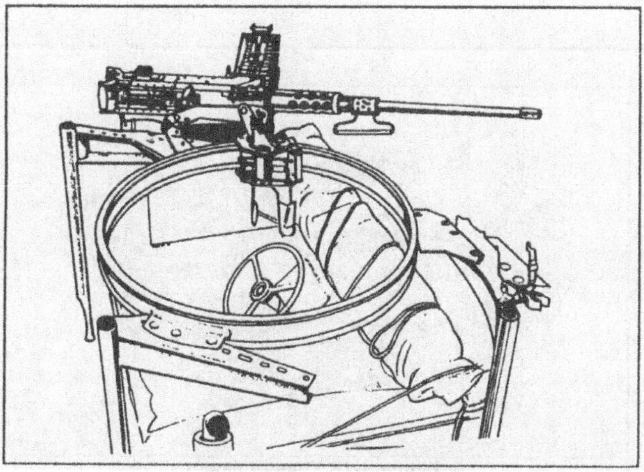

Figure 1-10. Truck mount, M36.

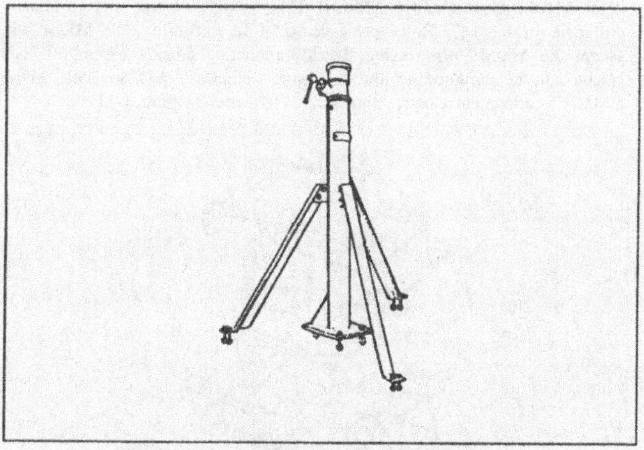

Figure 1-11. Pedestal truck mount, M31 C.

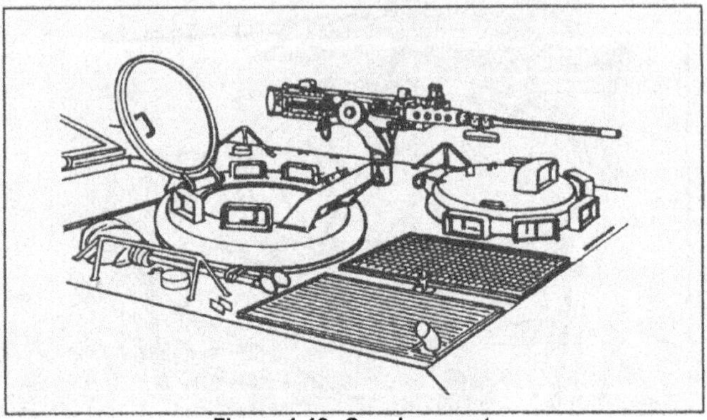

Figure 1-12. Cupola mount.

d. **MK64 Gun Cradle Mount.** This vehicle mount was primarily designed for the M2. However, because of its versatility, the MK64 will accept the MK 19 also (using the M2 mounting adapter assembly). The MK64 can be mounted on the following vehicles – M151 series, M966 HMMWV armament carrier, and the M113 series (Figure 1-13).

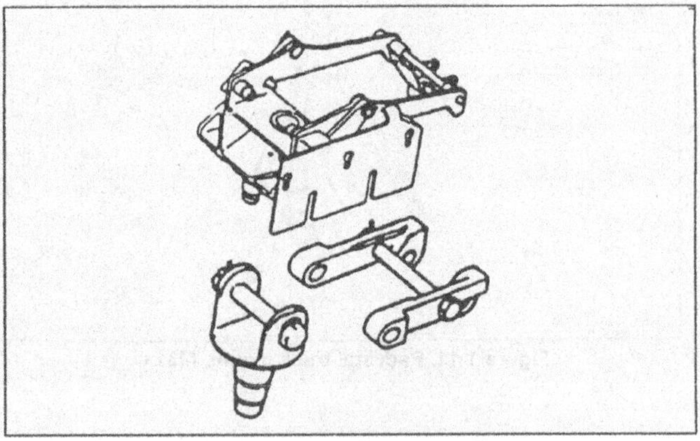

Figure 1-13. MK64 gun cradle.

1-7. AMMUNITION

This paragraph describes the ammunition used in caliber .50 machine guns (Figure 1-14). Soldiers should be able to recognize the types of ammunition available and know how to care for it. The caliber .50 cartridge consists of a cartridge case, primer, propelling charge, and the bullet. See TM 9-1300-200. The term bullet refers only to the small-arms projectile. There are eight types of ammunition issued for use in the caliber .50 machine gun. The tips of the various rounds are color-coded to indicate their type. The ammunition is linked with the M2 or M9 metallic links for use in the machine gun (Figure 1-15, page 1-18).

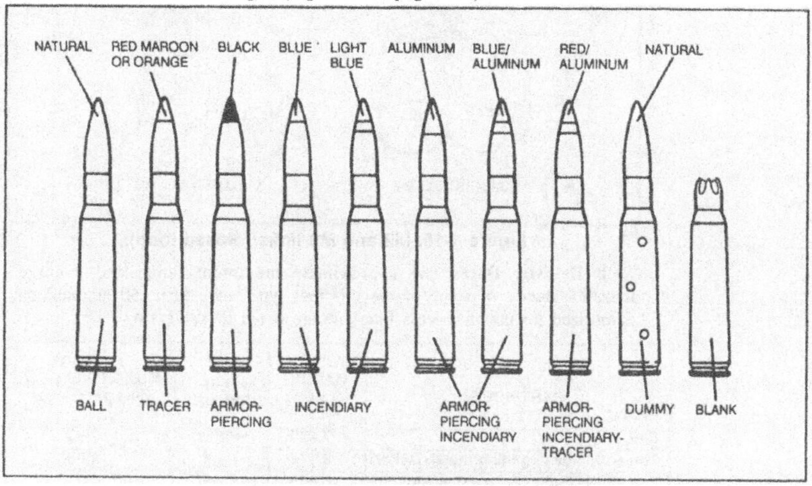

Figure 1-14. Ammunition for the M2.

a. **Classification.** The eight types of ammunition are used for the following purposes.

(1) *Ball.* For use in marksmanship training, and against personnel and light material targets.

(2) *Tracer.* To aid in observing fire. Secondary purposes are for incendiary effect and for signaling.

(3) *Armor-piercing.* For use against armored aircraft and lightly armored vehicles, concrete shelters, and other bullet-resisting targets.

(4) *Incendiary.* For incendiary effect, especially against aircraft.

(5) *Armor-piercing-incendiary.* For combined armor-piercing and incendiary effect.

(6) *Armor-piercing-incendiary-tracer.* For combined armor-piercing and incendiary effect, with the additional tracer feature.

(7) *Blank.* For simulated fire (contains no bullet).

(8) *High-pressure test.* For use only in proof firing of weapons and barrels.

(9) *Dummy.* For training (completely inert).

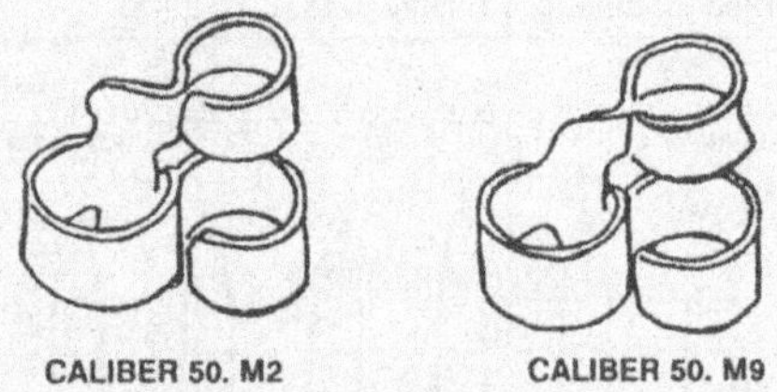

Figure 1-15. M2 and M9 links (closed loop).

b. **Ballistic Data.** The approximate maximum range and average muzzle velocity of some of the different types of caliber .50 ammunition authorized for use in the machine gun are noted in Table 1-4.

CARTRIDGE	MAXIMUM RANGE (meters)	METER TRACE	AVERAGE MUZZLE VELOCITY (feet per second)
Ball, M2	7,400	—	2,930
Tracer, M1 (with gilding metal jacket)	5,575	1,800	2,860
Tracer, M1 (with clad steel jacket)	5,450	1,800	3,030
Tracer, M17	5,450	2,450	3,030
Incendiary, M1	6,050	—	3,090
Armor-piercing, M2	7,400	—	2,930
Armor-piercing-incendiary,M8	6,470	—	3,050
Armor-piercing-incendiary-tracer, M20	6,470	*300-1,750	3,050

* This tracer is dim at near ranges but increases to bright as it moves farther from the gun.

Table 1-4. Ballistic data.

c. **Care, Handling, and Preservation**. Exercise care to prevent ammunition boxes from becoming broken or damaged. If they do, repair them immediately. Transfer all original markings to the new parts of the box. Do not open ammunition boxes until the ammunition is to be used. Ammunition removed from the airtight container, particularly in damp climates, is likely to corrode. Protect the ammunition from mud, sand, and water. If the ammunition gets wet or dirty, wipe it off at once with a clean, dry cloth. Wipe off light corrosion as soon as it is discovered. Turn in heavily corroded cartridges. Do not expose ammunition to the direct rays of the sun. If the powder is hot, excessive pressure may be developed when the weapon is fired. Do not oil or grease ammunition. Dust and other abrasives that collect on greasy ammunition are injurious to the operating parts of the gun. Moreover, oiled cartridges produce excessive chamber pressure. Do not fire dented cartridges, cartridges with loose bullets, or otherwise defective rounds.

d. **Storage**. Small-arms ammunition is not an explosive hazard, but under poor storage conditions it may become a fire hazard. Store ammunition of all classes away from radiators, hot water pipes, and other sources of heat. Whenever possible, store ammunition under cover. If it is necessary to leave ammunition in the open, keep it at least 6 inches off the ground and covered with a double thickness of tarpaulin. Place the tarpaulin so that it gives maximum protection and allows free circulation of air. Dig suitable trenches to prevent water from flowing under the ammunition pile.

e. **Miscellaneous Data**. Table 1-5 lists the maximum penetration in inches for an armor-piercing cartridge fired from the 45-inch barrel (muzzle velocity, 2,935 feet per second), which in some cases may enhance the leader's selection of targets to engage.

	INCHES AT:		
MATERIAL	**200 METERS**	**600 METERS**	**1,500 METERS**
Armor plate (homogeneous)	1.0	0.7	0.3
Armor plate (face-hardened)	0.9	0.5	0.2
Sand (100 pounds dry weight/cubic feet)	14.0	12.0	16.0
Clay (100 pounds dry weight/cubic feet)	28.0	27.0	21.0

Table 1-5. Maximum penetration for armor-piercing cartridge.

Table 1-6 lists the maximum penetration in inches for a ball cartridge fired from the 45-inch barrel (muzzle velocity, 2,935 feet per second):

	INCHES AT:		
MATERIAL	**200 METERS**	**600 METERS**	**1,500 METERS**
Sand (100 pounds dry weight/cubic feet)	14.0	12.0	6.0
Clay (100 pounds dry weight/cubic feet)	28.0	27.0	21.0
Concrete	2.0	1.0	1.0

Table 1-6. Maximum penetration for ball cartridge.

CHAPTER 2

MAINTENANCE

This chapter addresses the proper care of the machine gun to ensure its overall effectiveness and efficient functioning. The information includes the gunner's knowledge in disassembly and assembly, inspection, and maintenance procedures of the gun, its mount, the T&E, and its ammunition.

2-1. SAFETY

The paramount consideration while training with the machine gun is safety. It is imperative that the weapon be cleared properly before disassembly and inspection. (See Figure 2-1 for step-by-step procedures.)

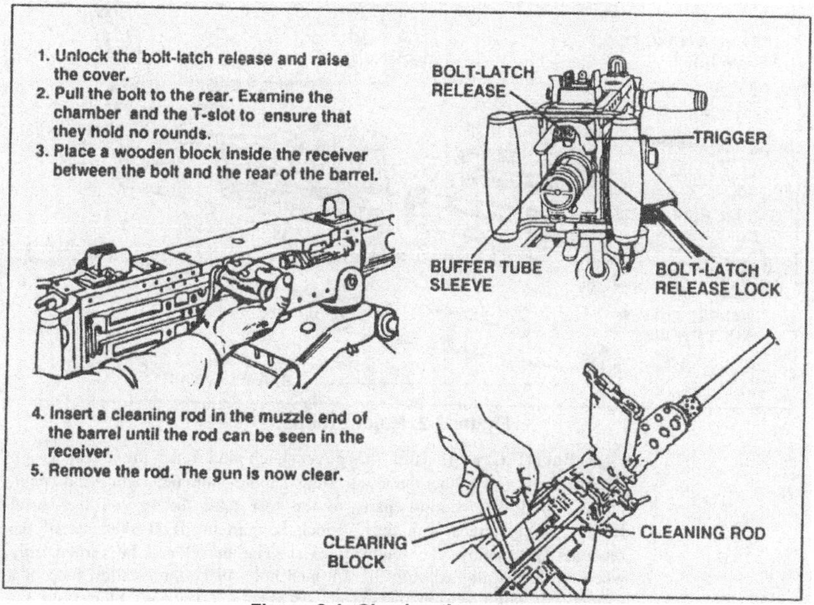

Figure 2-1. Clearing the weapon.

2-2. GENERAL DISASSEMBLY

The crew operating the MG must be fully familiar with its general disassembly, which consists of removing the major groups and assemblies for inspection and cleaning. The eight major groups that must be disassembled in the following order are barrel group, backplate group, driving spring rod assembly, bolt group, barrel extension group, barrel buffer body group, barrel buffer assembly, and receiver group (Figure 2-2).

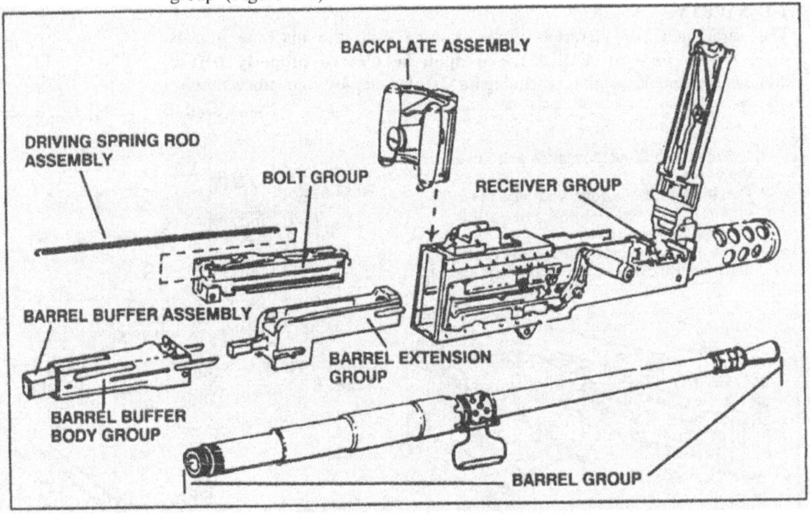

Figure 2-2. Major groups.

a. **Barrel Group.** Turn the cover latch and raise the cover group (Figure 2-3). Grasp the retracting slide handle with the right hand, palm up, and pull the recoiling parts to the rear until the lug on the barrel locking spring aligns with the 3/8-inch hole in the right sideplate of the receiver (just below the feedway exit). The barrel can be turned only when the lug is aligned with the 3/8-inch hole. Place the smallest loop of a caliber .50 link, or suitable spacer, between the trunnion block and the barrel extension (Figure 2-4). This holds the barrel locking spring lug aligned with the 3/8-inch hole in the right sideplate. Unscrew the barrel from the receiver (Figure 2-5, page 2-4). Be careful not to damage the threads or barrel locking notches when setting the barrel down. Pull back slightly on the retracting slide handle and remove the link or spacer from the receiver.

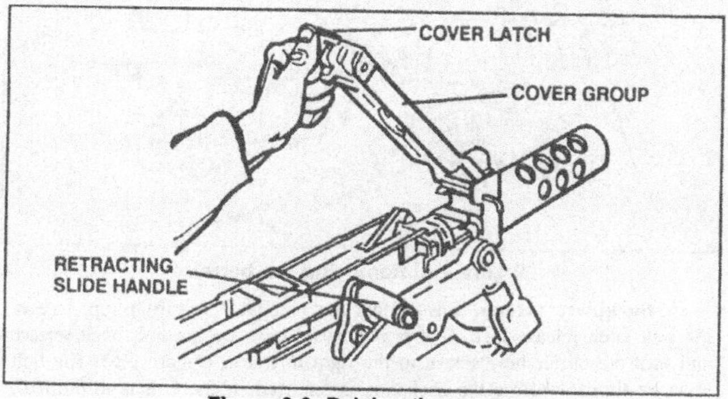

Figure 2-3. Raising the cover.

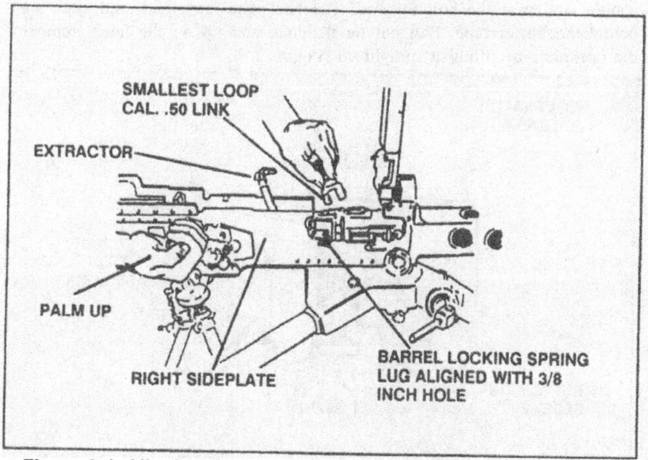

Figure 2-4. Aligning the lug on the barrel locking spring with the 3/8-inch hole in the right sideplate.

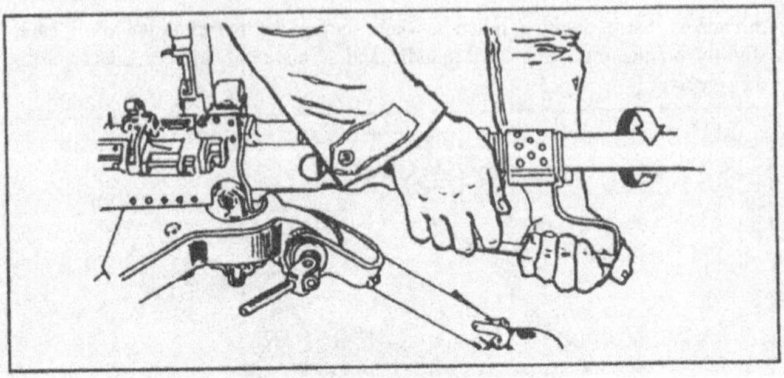

Figure 2-5. Removing the barrel.

b. **Backplate Group.** Ensure that the bolt latch release is up, free of the bolt latch release lock. If it is not, push down on the bolt latch release and turn the buffer tube sleeve to the right to free it (Figure 2-6). The bolt must be forward before the backplate is removed. If the bolt is to the rear, push down on the bolt latch release, place palm up on the retracting slide handle, and ease the bolt forward. The backplate latch lock and latch are below the buffer tube. Pull out on the lock and up on the latch; remove the backplate by lifting it straight up (Figure 2-7).

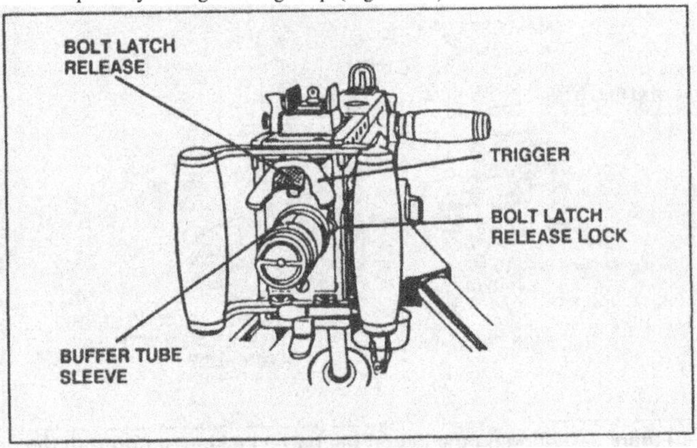

Figure 2-6. Releasing the bolt latch.

CAUTION

Care must be taken to prevent the bolt from slamming forward with the barrel removed.

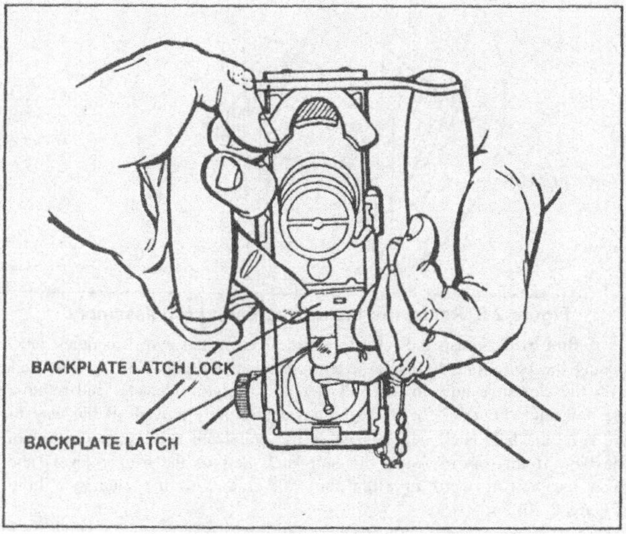

Figure 2-7. Removing the backplate.

c. **Driving Spring Rod Assembly.** The inner and outer driving springs and driving spring rod are located inside the receiver next to the right sideplate (Figure 2-8, page 2-6). Push in on the head of the driving spring rod and push to the left to remove the driving spring rod retaining pin from its seat in the right sideplate. Pull the driving spring assembly to the rear and out of the receiver.

WARNING

Never attempt to cock the gun while the backplate is off and the driving spring assembly is in place. If the backplate is off and the driving spring assembly is compressed, the retaining pin on the driving spring rod can slip from its seat in the sideplate and could cause serious injury to anyone behind the gun.

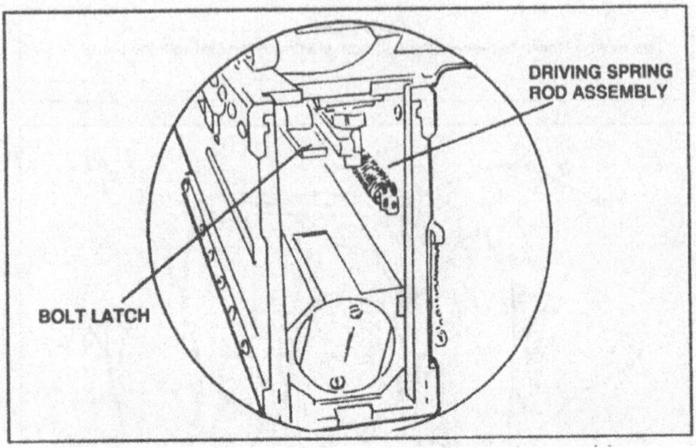

Figure 2-8. Removing the driving spring rod assembly.

d. **Bolt Stud.** Grasp the retracting slide handle and give it a quick jerk, freeing the bolt from the barrel extension. Align the collar of the bolt stud with the clearance hole in the bolt slot on the right sideplate, and remove the bolt stud (Figure 2-9). If the bolt is accidentally moved all the way to the rear, the bolt latch will engage in the bolt latch notches in the top of the bolt. If this occurs, raise the bolt latch (left of the trigger lever) and push the bolt forward to align the bolt stud with the clearance hole (Figure 2-10).

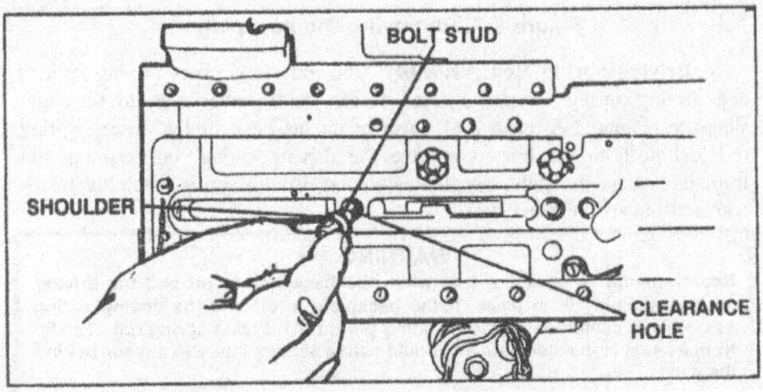

Figure 2-9. Removing the bolt stud.

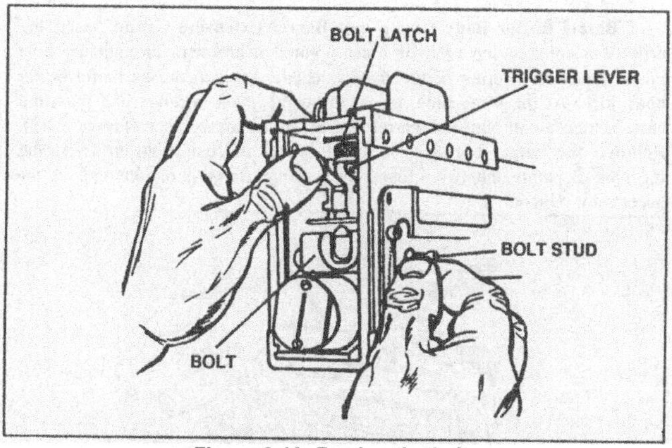

Figure 2-10. Freeing the bolt.

e. **Bolt Group.** After freeing the bolt, slide it to the rear and out of receiver (Figure 2-11). Place the bolt down on its right side (with the extractor arm up), so that the extractor will not fall from the bolt.

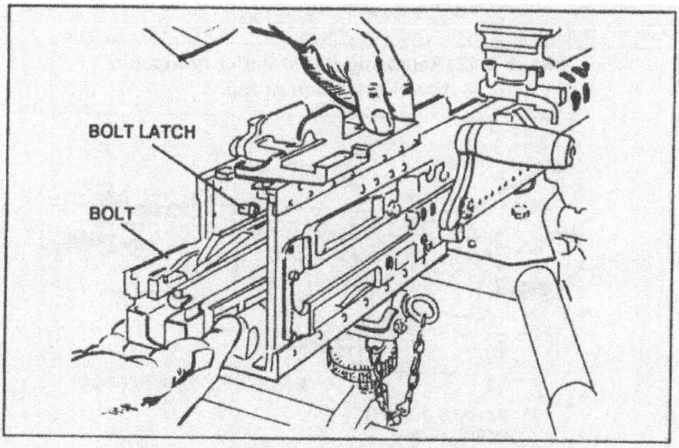

Figure 2-11. Removing the bolt from the receiver.

f. **Barrel Buffer Body Group and Barrel Extension Group.** Insert the drift of a combination tool, or other pointed instrument, through the hole in the lower rear corner of the right sideplate. Push in on the barrel buffer body lock. At the same time, place one hand in the receiver and push the barrel extension group and barrel buffer group to the rear (Figure 2-12). Remove the barrel buffer group and barrel extension group from the receiver. Separate the two groups by pushing forward on the tips of the accelerator (Figure 2-13).

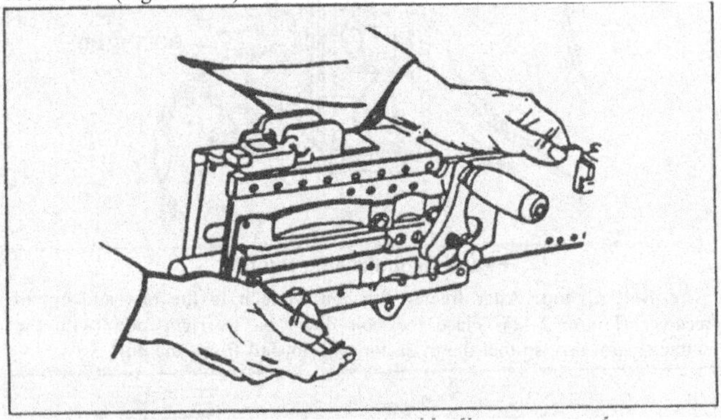

Figure 2-12. Removing barrel buffer group and barrel extension group.

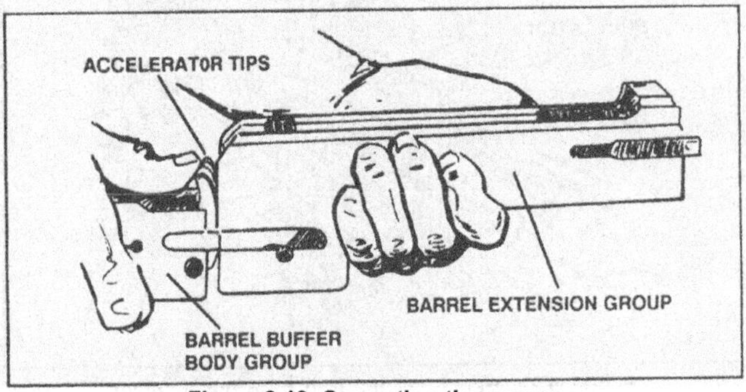

Figure 2-13. Separating the groups.

g. **Barrel Buffer Assembly.** Pull the barrel buffer assembly from the rear of the barrel buffer body group. The barrel buffer assembly will not be disassembled (Figure 2-14). This completes general disassembly.

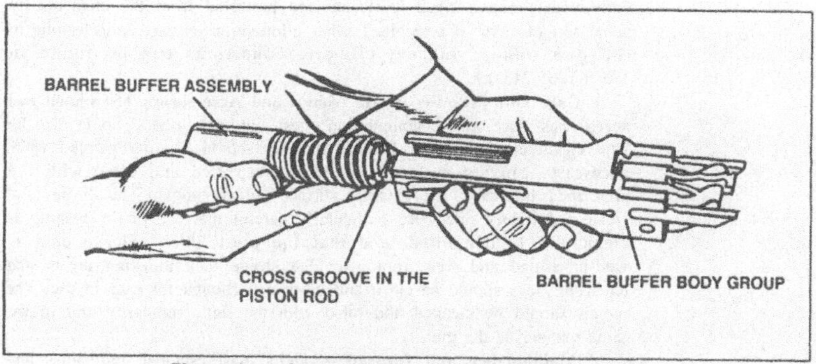

Figure 2-14. Separating the barrel buffer assembly from the barrel buffer body group.

2-3. CLEANING, INSPECTION, AND LUBRICATION

To ensure proper care of the MG, it is necessary to have a system of maintenance or an SOP for the frequency of cleaning. Each gun should be cleaned as soon after firing as possible and each time it is exposed to field conditions. Under combat conditions, the gun should be cleaned and oiled daily. Under extreme climatic and combat conditions, it maybe necessary to clean and lubricate more frequently. Under ideal conditions, where the gun is not used, and is stored in a clean, dry place, it may only be necessary to inspect, clean, and lubricate every 5 days. The gun should be disassembled, cleaned, and oiled in a clean, dry location. If possible, keep the gun covered with a gun cover, canvas, tarpaulin, or poncho when not in use.

a. **Routine Care and Cleaning.** Before firing (when the situation permits), take the following steps to ensure efficient functioning of the machine gun:

- Disassemble the gun into its major groups or assemblies.
- Clean the bore and chamber, and lightly oil them.
- Clean all metal parts thoroughly with CLP. See paragraph 2-3f for lubrication procedures.

b. **Care and Cleaning Under Unusual Conditions.** Extreme cold, hot, dry, and tropical climates affect the gun and its functioning. Care should be taken under these climatic conditions to ensure that the gun is cleaned daily with the prescribed lubricants and protected from the elements by some sort of cover if possible. Further information on care and cleaning of the gun under unusual climatic conditions can be found in TM 9-1005-213-10.

c. **Care and Cleaning of M3 Mount and Accessories.** The mount and accessories, such as the ammunition chest and spare parts, should also be kept clean and lubricated. Painted surfaces should be spot painted when necessary. Moving surfaces should be inspected and oiled with the prescribed lubricant. All external surfaces of the mount should be kept clean and lightly oiled. Be particularly careful that the pintle bushing is clean and lightly oiled, and that the pintle lock release cam is well-lubricated and free from grit. The sleeve lock indexing levers and telescopic legs should be clean and lubricated enough for ease in use. The mount should be cleaned and oiled with the same regularity and in the same manner as the gun.

d. **Maintenance and Inspection.** Units must establish guidelines and conduct regular maintenance and inspection to keep the machine gun and its mounts in operational conditions.

(1) *Gun maintenance.* The importance of a thorough knowledge of care, cleaning, and maintenance of the machine gun cannot be overemphasized because these actions determine whether or not the gun will function properly when needed. The bore and chamber must be properly maintained to preserve accuracy. Because of the close fit of working surfaces and the high speed at which the gun operates, the receiver and moving parts must be kept clean, correctly lubricated, and free from burrs, rust, dirt, or grease to ensure proper, efficient functioning.

(2) *Mount maintenance.* The care, cleaning, lubrication, and adjustment of the mounts used with the gun are no less important. The functioning of the gun and mount together determine overall effectiveness. All accessories and equipment used with the gun and mount, including ammunition, must also be properly maintained.

(3) *Inspection.* When inspected, the machine gun should be completely disassembled. Inspecting personnel should look for dirt, cracks, burrs, and rust.

e. **Inspection Checklist.** Table 2-1 is an inspection checklist to be used as a guide for crewmembers or inspecting personnel to ensure that the gun and equipment are properly maintained.

UNIT	INSPECTION
1. GUN.	
a. Barrel.	Inspect the bore and chamber for rust. See that they are clean and lightly oiled.
b. Moving parts.	See that they are clean and lightly oiled. Operate the retracting slide handle and bolt latch release several times to see that the parts function without excessive friction.
c. Headspace and timing.	Check with the gauges to ensure that headspace and timing are correct.
d. Rear sight and windage knob.	Ensure that the sight is in good condition, clean, free of grease or dirt, and lightly oiled. Elevation should be set at 1,000, windage zero, and the sight should be down.
2. MOUNT (M3, MK64, M36, or M4).	See that it is clean, lightly lubricated, and that all clamps are securely tightened. It should function properly and be complete.
3. SPARE PARTS AND TOOLS.	Inspect to see that they are clean and lightly oiled. See that spare parts kits are complete and in good condition. Replacement parts should be requisitioned and newly drawn parts examined.
4. T&E.	See that it is clean, lightly lubricated, and that both hand wheels work properly.
5. AMMUNITION.	See that ammunition is properly stored, and that boxes and ammunition are in good condition and not oiled.

Table 2-1. Inspection checklist.

f. **Lubrications.** Use cleaner, lubricant, preservative to clean the machine gun. As its name implies, it cleans, lubricates, and preserves all in one application.

(1) After cleaning the gun with CLP, wipe it dry and reapply a thin coating. Allow this thin coat to dry on the parts for a short time before reassembly. CLP deposits a thin coating on the metal which minimizes carbon buildup and prevents foreign material from sticking. It is this coating that provides the frictionless operation of the weapon parts, not liquid oil deposited on them. A gun treated with CLP will operate better and remain clean longer than one treated with any other cleaning material. Use of CLP will reduce maintenance costs and extend the life of the weapon.

(2) Rifle bore cleaner is a cleaning solvent which can be used to clean powder residue, carbon, and dirt from weapons. RBC does not preserve or lubricate a weapon. If you clean a weapon with RBC, dry the weapon and lubricate it with lubricating oil, semifluid (LSA); lubricating oil, special purpose (PL-S); or lubricating oil, general purpose (PL-M). The use of these oils will cause sand or grit to stick to the weapon. RBC and oil should be used only when CLP is not available.

2-4. MAINTENANCE PROCEDURES

There are certain actions that must be taken before, during, and after firing to properly maintain the gun. See Table 2-2 for a preoperation checklist.

PART	BEFORE	DURING FIRING OR TEMPORARY CESSATION	AFTER
Bore	Make sure it is clear and clean.		Clean and oil lightly.
Moving parts	Oil lightly and test for worn or broken parts. They should function without excessive friction.	Lubricate working parts. Observe the functioning of the gun to anticipate failures.	Inspect, clean, and oil lightly.
Headspace and timing	Check adjustment by gauge and correct if necessary.	Watch for bulged cases to prevent a ruptured (separated) case. If a separated case occurs, remove it and readjust headspace.	Check adjustment.
Rear sight and windage knob	See that the sight is clean and functions properly. Set sight at 1,000, windage zero.	Keep properly set.	Clean and oil; set at 1,000, windage zero.
Spare parts and tools	Clean and oil spare parts and tools. Check kits for completeness. Examine newly drawn parts.	Keep available.	Clean and oil. Check and replace damaged or missing parts.
Ammunition	Have an adequate supply; clean, correctly loaded, and in good condition. DO NOT OIL.	Keep correctly aligned with the feedway; check resupply. Protect from sun, moisture, and dirt. Watch for link stoppage.	Clean, store carefully, and replenish supply.

Table 2-2. Operational checklist.

2-5. MAINTENANCE UNDER NBC CONDITIONS

If contamination is anticipated, apply oil to all outer metal surfaces of the weapon. DO NOT OIL AMMUNITION. Keep the weapon covered as much as possible. If the weapon is contaminated, decontaminate it as prescribed by FM 3-5 and then clean and lubricate it.

2-6. GENERAL ASSEMBLY

To assemble the gun, replace the groups and assemblies in reverse order of their removal in disassembly.

a. **Barrel Buffer Assembly and Barrel Buffer Body Group.** Replace the barrel buffer assembly in the barrel buffer body group, with the key on the spring guide to the right. This key must fit in its slot in the right side of the barrel buffer body. Turn the barrel buffer tube until the screwdriver slot (in the rear of the tube) is vertical and the arrow is pointing to the right. The stud on the tube lock will now engage the serrations in the barrel buffer tube to keep the tube from turning. Push the barrel buffer assembly fully forward (Figures 2- 15 and 2-16).

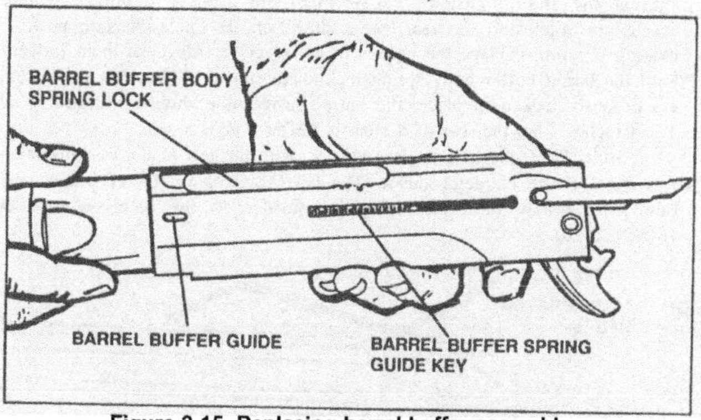

Figure 2-15. Replacing barrel buffer assembly.

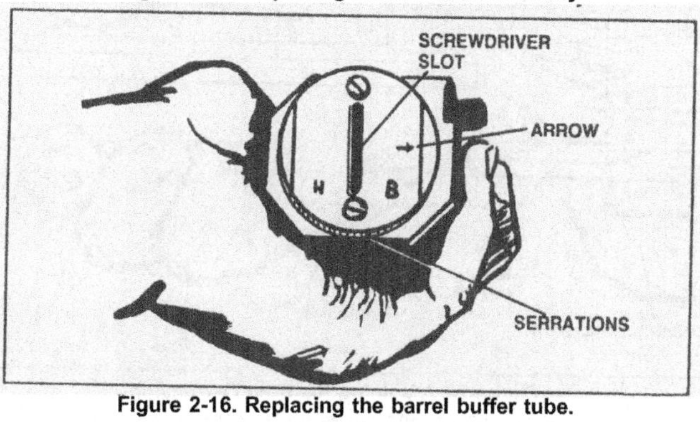

Figure 2-16. Replacing the barrel buffer tube.

b. **Barrel Buffer Group and Barrel Extension Group.** To join the two groups together, hold the barrel buffer group in the right hand, with the index finger supporting the accelerator. Join the notch on the shank of the barrel extension group with the cross-groove in the pistol rod of the barrel buffer assembly. At the same time, align the breech lock depressors with their guideways in the sides of the barrel extension, ensuring that the tips of the accelerator are against the rear end of the barrel extension (claws against the shank) (Figure 2-17). Push the groups together. As the accelerator rotates to the rear, press down on its tips to ensure positive locking of groups. Place the groups in the receiver, and push them forward until the barrel buffer body spring lock snaps into position. When the parts are properly locked in place, the barrel buffer tube should protrude about 1 1/8 inches from the rear of the barrel buffer body group.

c. **Bolt.** Place the bolt in the receiver, with the top of the cocking lever forward and the extractor down. The barrel extension, barrel buffer, and bolt groups may be assembled and returned to the receiver together (Figure 2-18).

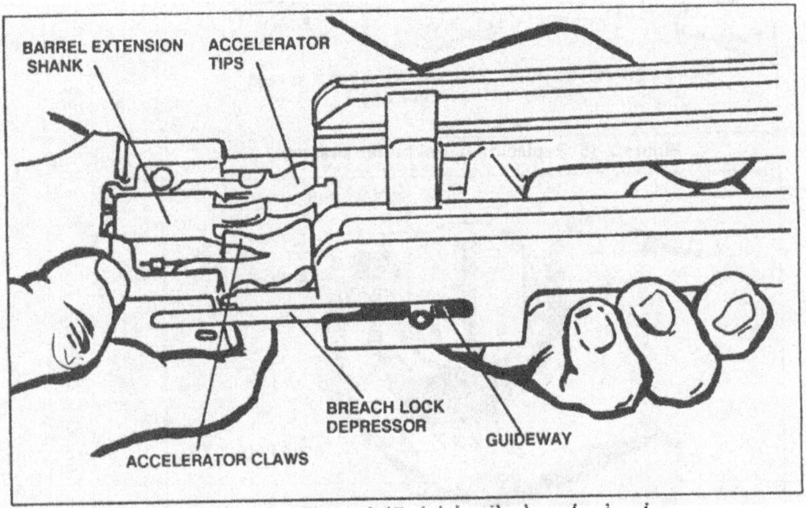

Figure 2-17. Joining the barrel extension group and barrel buffer group.

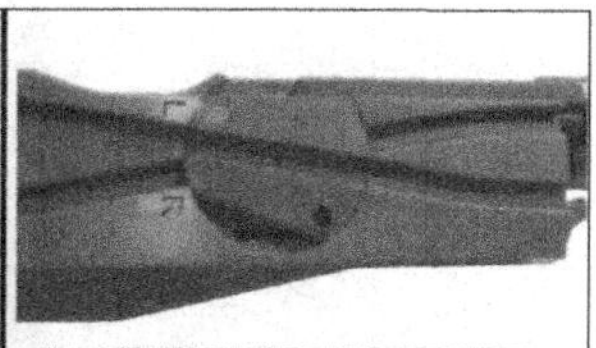

* Figure 2-18. Returning the barrel extension, barrel buffer, and bolt groups together.

*** CAUTION**

Before inserting the bolt group, ensure bolt switch mechanism is positioned on **L** (left-hand) feed for the weapon.

* d. **Bolt Stud**. Align the stud hole in the bolt with the clearance hole and replace the bolt stud, ensuring that the collar of the stud is inside the sideplate (Figure 2-19).

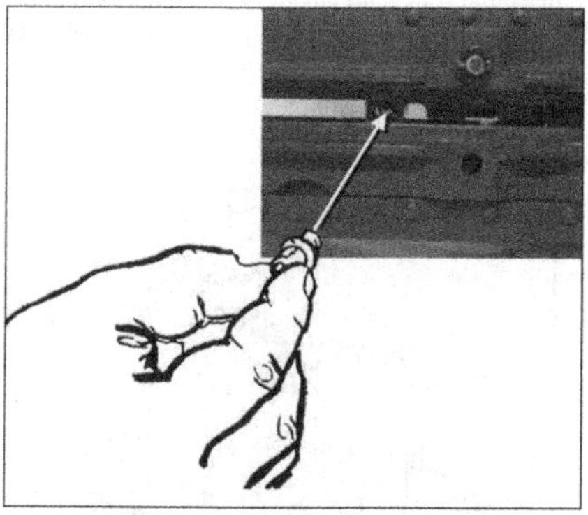

* Figure 2-19. Replacing the bolt stud.

* e. **Drive Spring Assembly**. Press up on the bolt latch and push the bolt all the way forward by pushing on the bolt stud only. Place the end of the drive spring rod in its hole in the rear of the bolt, and push forward on the drive spring assembly and the barrel buffer tube. Press in and to the right on the head of the drive spring rod and place the retaining pin in its seat in the right sideplate (Figure 2-20).

* **Figure 2-20. Inserting drive spring and drive spring rod (with drive spring rod inside drive spring).**

* NOTE: At this time, the barrel buffer tube should be completely inside the receiver. If not, the barrel buffer body spring is not properly seated.

* f. **Backplate Group**. Hold the backplate with the latch down and the trigger up; place the backplate guides in their guideways. Hold out on the latch lock and tap the backplate into position until the latch snaps into place (Figure 2-21). Release the latch lock and pull up on the backplate group to ensure it is firmly seated.

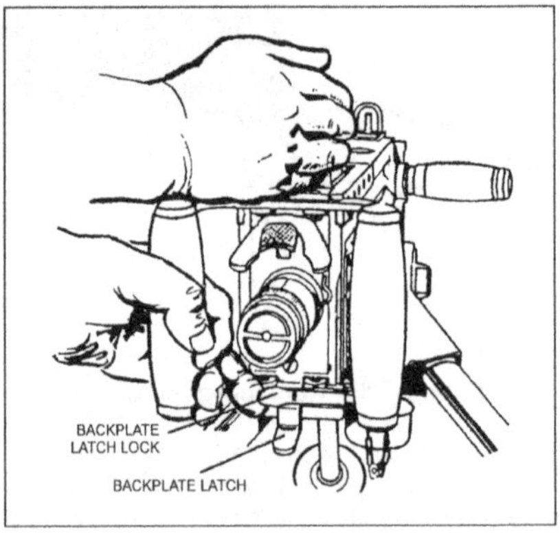

* **Figure 2-21. Replacing the backplate group.**

*** CAUTION**

Do not use the driving rod to drive the bolt forward from the rear position. This may damage the drive spring group and cause a stoppage.

* g. **Barrel**. Pull the retracting slide handle to the rear until the lug on the barrel locking spring is visible through the 3/8-inch hole in the right sideplate. Place the smallest loop of a caliber .50 link, or suitable spacer, between the trunnion block and the barrel extension. Screw the barrel all the way into the barrel extension; then unscrew the barrel two notches. Remove the link and close the cover. This completes general assembly.

2-7. FUNCTION CHECK

A function check must be performed as soon as the weapon is assembled to ensure that the weapon has been assembled correctly. The following procedures should be taken to check the function of the weapon.

a. Place the weapon in the single-shot mode.

b. Open the cover and lock the bolt to the rear (bolt should stay to rear while in the single-shot mode).

c. Hold the retractor handles, press the bolt latch release, and ride the bolt forward.

d. Press down on the trigger; weapon should fire. (Check T-slot to ensure that firing pin does protrude.)

e. Place the weapon in the automatic-fire mode

f. Pull the retractor slide handle to the rear and hold. (Bolt should not lock to rear.)

g. Release the pressure on the slide handles and ride the bolt forward

h. Make sure firing pin does not protrude

i. Press trigger; weapon should fire

j. Make sure firing pin does protrude.

NOTE: Before firing a newly assembled weapon, first set the headspace and timing. Chapter 3 describes these procedures.

* CHAPTER 3

OPERATION AND FUNCTIONING

This chapter explains the operation of the MG. It discusses the loading, unloading, and clearing procedures, and the cycle of functioning of the weapon. When training the cycle of functioning using dummy ammunition, it is imperative that all safety procedures be followed.

* 3-1. OPERATION

The overall operation of the MG includes how to load, unload, and clear the weapon. During the weapon's operation, it is mandatory that all ammunition be free of dirt and corrosion, that the ammunition be properly linked, and that the double-linked end be at the top of the ammunition can.

* 3-2. LOADING PROCEDURES

Trainers must ensure that the weapon functions correctly and that proper headspace and timing have been set before loading. When loading in either mode, the ammunition is fed into the MG in the same manner (Figure 3-1). Ensure that the bolt is forward and the cover is closed. Insert the double-loop end of the ammunition belt into the feedway until the first round is engaged by the belt-holding pawl. Figure 3-2 shows the correct position of the bolt latch in the single-shot or automatic mode.

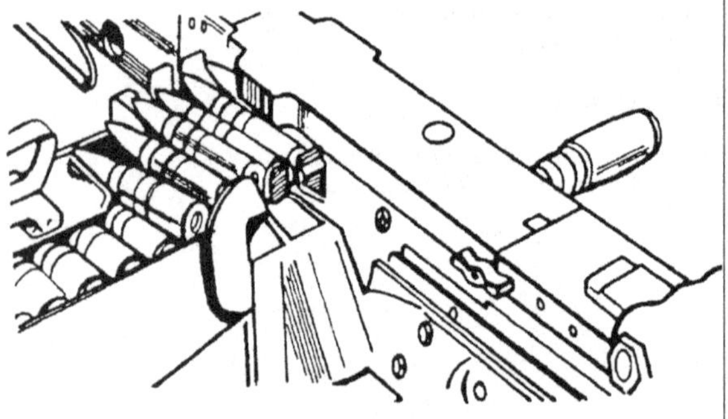

Figure 3-1. Inserting ammunition.

a. **Single-Shot Mode**. When engaging targets at ranges greater than 1,100 meters, using the single-shot mode (firing one round at a time) allows the gunner to deliver well-aimed fire on the target. To load in the single-shot mode--

(1) Keep the bolt-latch release unlocked in the up position and release it manually for each round.

(2) Jerk the retracting slide handle to the rear and lock it in position. Return the retracting slide handle to the forward position and then release the bolt by pressing the bolt latch release. The gun is now half-loaded.

(3) To complete loading, jerk the retracting slide handle to the rear and lock it in position. Return the retracting slide handle to the forward position. Press the bolt latch release. When the bolt goes forward for the second time, the gun is loaded.

b. **Automatic Mode**. To load in the automatic mode--

(1) Lock the bolt-latch release down with the bolt- latch release lock.

(2) Jerk the retracting slide handle to the rear and release it. The gun is now half-loaded.

(3) To complete loading, jerk the retracting slide handle to the rear a second time and release it. When the bolt goes forward for the second time, the gun is loaded.

Figure 3-2. Firing modes.

* 3-3. UNLOADING PROCEDURES

To unload the MG, the gunner must first ensure that the weapon is in the single-shot mode. The cover is then lifted and the assistant gunner removes the ammunition belt from the feedway. The bolt is then locked to the rear. If a round is chambered, it will release, unfired, when the bolt locks to the rear. Once the bolt is locked to the rear, the chamber and

T-slot are examined to ensure that they are not holding rounds. In darkness, this must be done by feeling the areas. After the examination has been done (during training), a wooden block is inserted in the receiver between the bolt and the rear of the barrel, extending above and below the receiver about one inch. Then a cleaning rod is inserted in the muzzle end of the barrel and pushed through the bore until it can be seen in the receiver. Remove the rod, the gun is now clear (Figure 3-3).

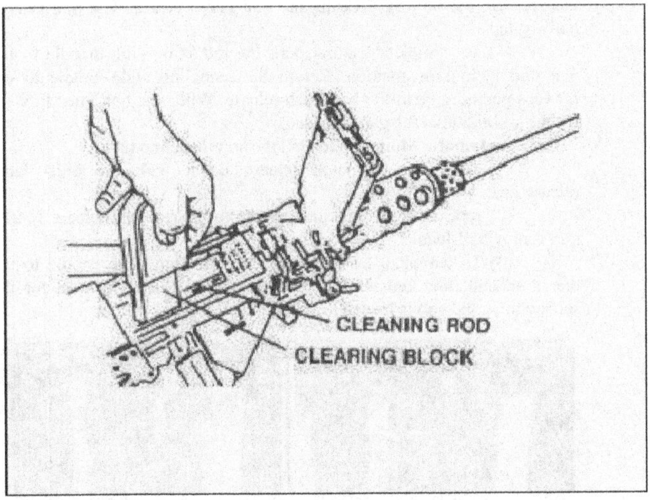

Figure 3-3. The clearing block.

*** 3-4. CYCLE OF FUNCTIONING**

The cycle of functioning is broken down into basic steps: feeding, chambering, locking, firing, unlocking, extracting, ejecting, and cocking. Some of these steps may occur at the same time.

a. **Feeding**. Feeding is the act of placing a cartridge in the receiver, approximately in back of the barrel, ready for chambering. When the bolt is fully forward and the top is closed, the ammunition belt is held in the feedway by the belt-holding pawl (Figure 3-4).

(1) As the bolt is moved to the rear, the belted ammunition is moved over and then held in a stationary position by the belt-holding pawl. At the same time, the belt-feed pawl rides up and over the link, holding the first round in place. When the bolt is all the way to the rear, the belt-feed slide moves out far enough to allow the belt-feed pawl spring to force the pawl up between the first and second rounds (Figure 3-5).

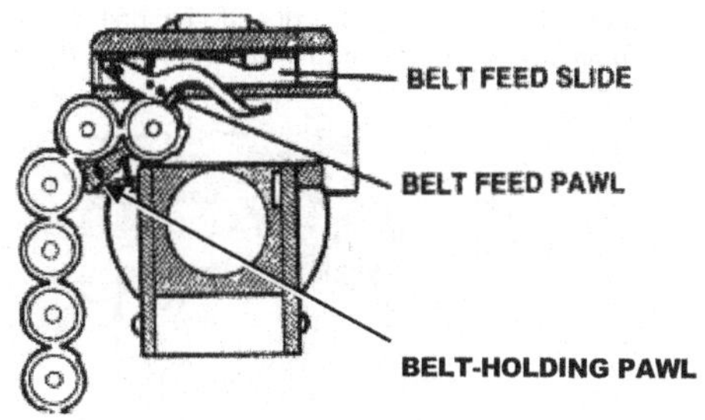

Figure 3-4. Feeding--step 1.

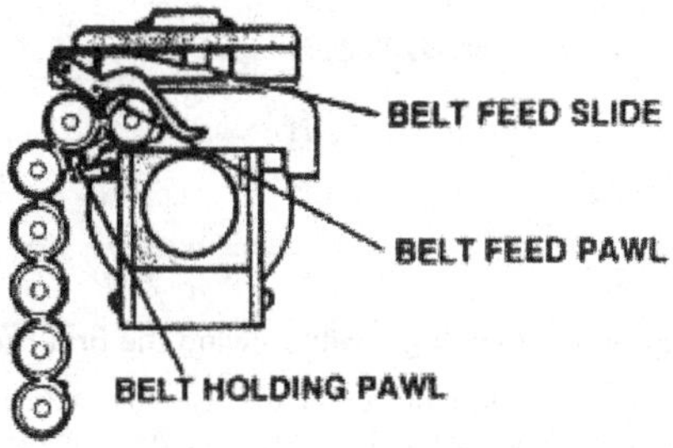

Figure 3-5. Feeding--step 2.

(2) As the bolt moves forward, the belt-feed slide is moved back into the receiver, pulling with it the next linked cartridge. When the bolt reaches the fully forward position, the belt-holding pawl will snap into place behind the second linked cartridge (Figure 3-7), holding it in place. The extractor will then grasp the rim of the first cartridge, preparing to release it from the belt on the next rearward motion (Figure 3-8).

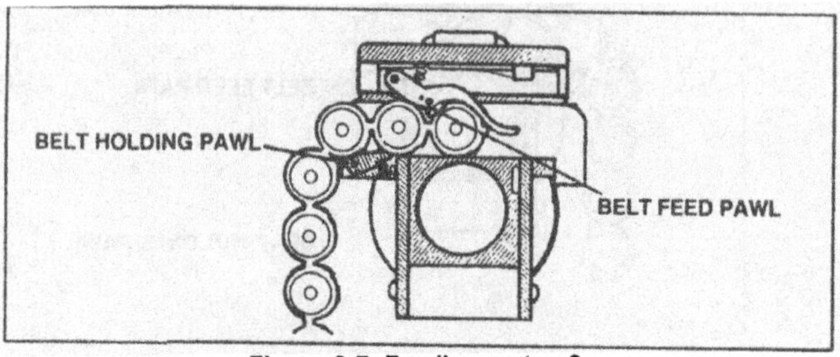

Figure 3-7. Feeding – step 3.

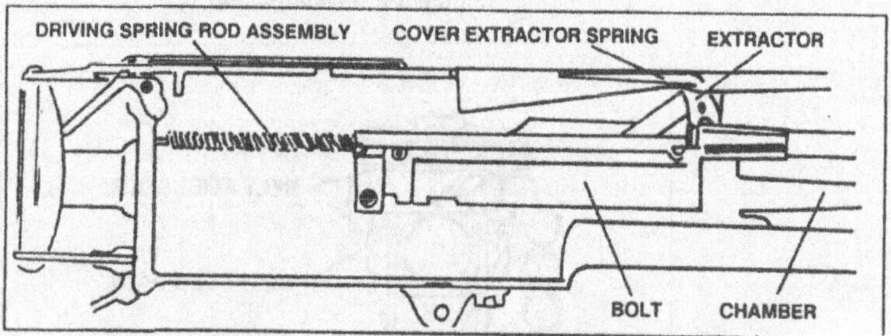

Figure 3-8. Feeding – withdrawing the first round from the feedway.

(3) As the bolt then moves to the rear, the extractor will pull the cartridge with it, releasing it from the belt. As it moves to the rear, the extractor is forced down by the extractor cam, causing the cartridge to be moved into the T-slot in the bolt face, preparing the cartridge to be chambered (Figure 3-9, page 3-6). It is connected under the extractor switch on the side of the receiver until it is repositioned by the forward movement of the bolt, and pressure of the cover extractor spring forces it over the next round.

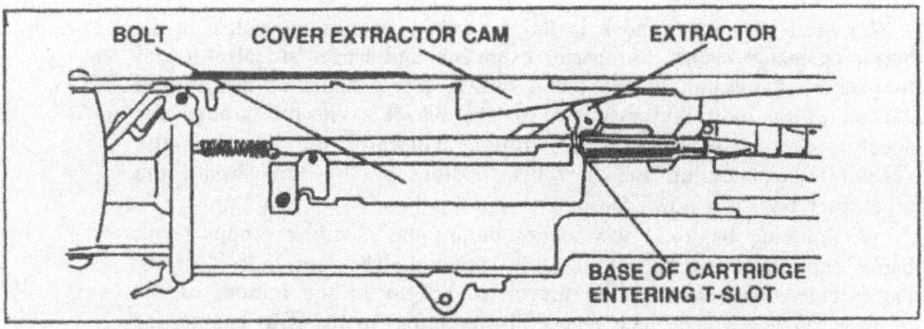

Figure 3-9. Feeding - cartridge entering the T-slot in the bolt.

b. **Cambering.** Cambering is placing the cartridge into the chamber of the weapon. During this cycle, the bolt moves forward, carrying the cartridge in the T-slot in a direct route to the chamber of the weapon. At the same time, the extractor rides up the extractor cam and when the bolt is fully forward, the extractor grasps the next linked cartridge (Figure 3-10).

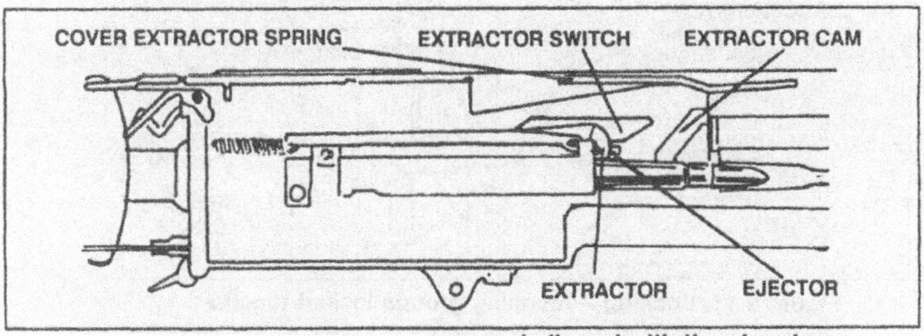

Figure 3-10. Cambering – new round aligned with the chamber.

c. **Locking.** The bolt is locked to the barrel and barrel extension.

(1) Initially, the bolt is forced forward in counter-recoil by the energy stored in the driving spring assembly and the compressed buffer disks. At the start of counter-recoil, the barrel buffer body tube lock keeps the accelerator tips from bounding up too soon and catching in the breech lock recess in the bolt. After the bolt travels forward about 5 inches, the lower rear projection of the bolt strikes the tips of the accelerator, turning the accelerator forward. This unlocks the barrel extension from the barrel buffer body group and releases the barrel buffer spring. The barrel buffer spring expands, forcing the piston rod forward.

(2) Since the cross groove in the piston rod engages the notch on the barrel extension shank, the barrel extension and barrel are also forced forward by the action of the barrel buffer spring. Some of the forward motion of the bolt is transmitted to the barrel extension through the accelerator. As the accelerator rotates forward, the front of the accelerator speeds up the barrel extension; at the same time, the accelerator tips slow down the bolt.

(3) Locking begins 1 1/8 inches before the recoiling groups (bolt, barrel extension, and barrel) are fully forward. The breech lock in the barrel extension rides up the breech lock cam in the bottom of the receiver into the breech lock recess in the bottom of the bolt, locking the recoiling groups together. The recoiling groups are completely locked together three-fourths of an inch before the groups are fully forward (Figure 3-11).

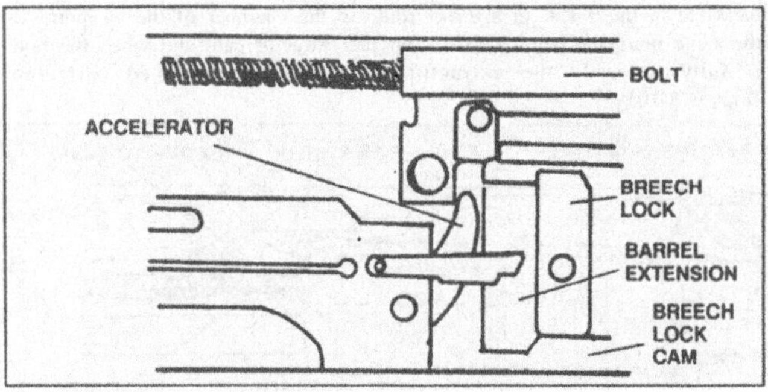

Figure 3-11. Locking – recoiling groups locked together.

d. **Firing.** The firing pin is released, igniting the primer of the cartridge.

(1) As the trigger impressed down, it pivots on the trigger pin, so that the trigger cam on the inside of the backplate engages and raises the rear end of the trigger lever. This in turn pivots on the trigger lever pin assembly, causing the front end of the trigger lever to press down on the top of the sear stud. The sear is forced down until the hooked notch of the firing pin extension is disengaged from the sear notch. The firing pin and firing pin extension are driven forward by the firing pin spring; the striker of the firing pin hits the primer of the cartridge, firing the round (Figures 3-12 and 3-13, page 3-8).

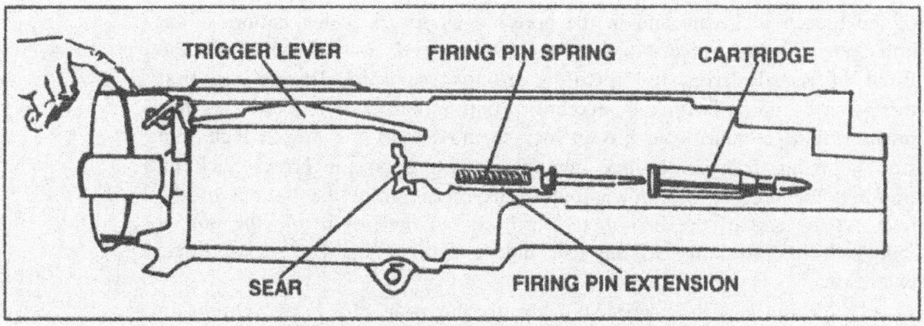

Figure 3-12. Firing – ready to fire.

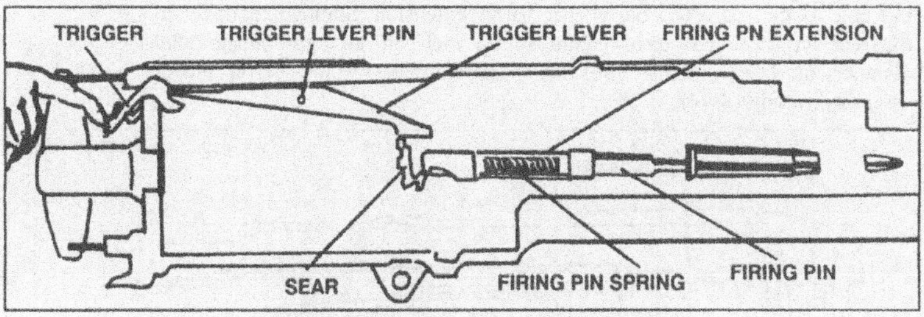

Figure 3-13. Firing – round ignited.

(2) For automatic firing, the bolt-latch release must be locked or held depressed, so that the bolt latch will not engage the notches in top of the bolt, holding the bolt to the rear as in single-shot firing. The trigger is pressed and held down. Each time the bolt travels forward in counter-recoil, the trigger lever depresses the sear, releasing the firing pin extension assembly and the firing pin. This automatically fires the next round when the forward movement of the recoiling groups is nearly completed. The gun should fire about one-sixteenth of an inch before the recoiling groups are fully forward. Only the first round should be fired with the parts fully forward. The gun fires automatically as long as the trigger and bolt latch are held down and ammunition is fed into the gun.

e. **Unlocking.** The bolt is unlocked from the barrel and barrel extension.

(1) At the instant of firing, the bolt is locked to the barrel extension and against the rear end of the barrel by the breech lock, which is on top

of the breech lock cam and in the breech lock recess in the bottom of the bolt. When the cartridge explodes, the bullet travels out of the barrel; the force of recoil drives the recoiling groups rearward. During the first three-fourths of an inch, the recoiling groups are locked together. As this movement takes place, the breech lock is moved off the breech lock cam stop, allowing the breech lock depressors (acting on the breech lock pin) to force the breech lock down, out of its recess from the bottom of the bolt. At the end of the first three-fourths of an inch of recoil, the bolt is unlocked, free to move to the rear independent of the barrel and barrel extension.

(2) As the recoiling groups move to the rear, the barrel extension causes the tips of the accelerator to rotate rearward. The accelerator tips strike the lower rear projection of the bolt, accelerating the movement of the bolt to the rear. The barrel and barrel extension continue to travel to the rear an additional three-eighths of an inch, or an approximate total distance of 1 1/8 inches until they are stopped by the barrel buffer assembly (Figure 3-14).

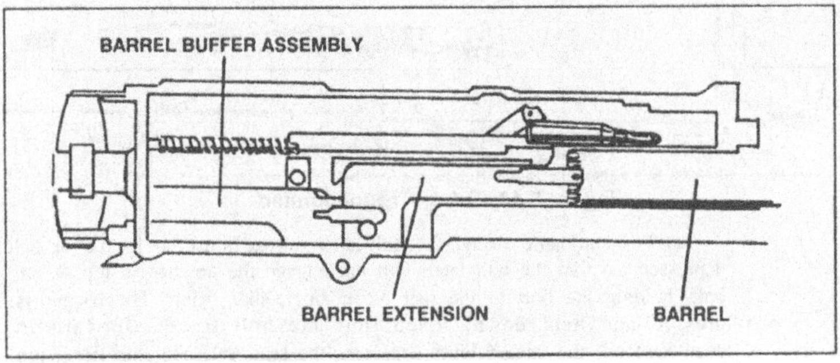

Figure 3-14. Unlocking – barrel and barrel extension stopped by the barrel buffer assembly.

(3) During the recoil of 1 1/8 inches, the barrel buffer spring is compressed by the barrel extension shank, since the notch on the shank is engaged in the cross groove in the piston rod head. The spring is locked in the compressed position by the claws of the accelerator, which engage the shoulders of the barrel extension shank. After its initial travel of three-fourths of an inch, the bolt travels an additional 6 3/8 inches to the rear, after it is unlocked from the barrel and barrel extension, for a total of

7 1/8 inches. During this movement, the driving springs are compressed. The rearward movement of the bolt is stopped as the bolt strikes the buffer plate. Part of the recoil energy of the bolt is stored by the driving spring rod assembly, and part is absorbed by the buffer disks in the backplate (Figure 3-15).

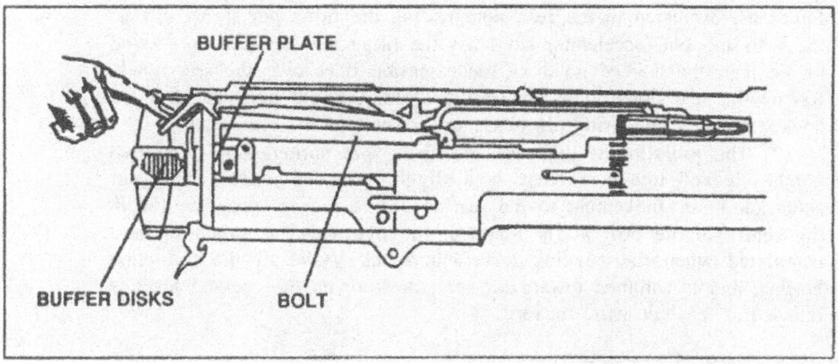

Figure 3-15. Unlocking – recoil movement completed.

f. **Extracting.** The empty cartridge case is pulled from the chamber.

(1) The empty case, held by the T-slot, has been expanded by the force of the explosion; therefore, it fits snugly in the chamber. If the case is withdrawn from the chamber too rapidly, it may be torn. To prevent this, and to ensure slow initial extraction of the case, the top forward edge of the breech lock and the forward edge of the lock recess in the bolt are beveled. As the breech lock is unlocked, the initial movement of the bolt away from the barrel and barrel extension is gradual.

(2) The slope of the locking faces facilitates locking and unlocking and prevents sticking. The leverage of the accelerator tips on the bolt speeds extraction after it is started by kicking the bolt to the rear to extract the empty case from the chamber.

g. **Ejecting.** The empty cartridge case is expelled from the receiver.

(1) As the bolt starts its forward movement (counter-recoil), the extractor lug rides below the extractor switch, forcing the extractor assembly farther down until the round is in the center of the T-slot of the bolt.

(2) The round, still gripped by the extractor, ejects the empty case from the T-slot. The last empty case of an ammunition belt is pushed out by the ejector.

h. **Cocking.** The firing pin is withdrawn into the cocked position.

(1) When the recoiling groups are fully forward, the top of the cocking lever rests on the rear half of the V-slot in the top plate bracket. As the bolt moves to the rear, the top of the cocking lever is forced forward. The lower end pivots to the rear on the cocking lever pin. The rounded nose of the cocking lever, which fits through the slot in the firing pin extension, forces the extension to the rear, compressing the firing pin spring against the sear stop pin (accelerator stop). As the firing pin extension is pressed to the rear, the hooked notch of the extension rides over the sear notch, forcing the sear down. The sear spring forces the sear back up after the hooked notch of the firing pin extension has entered the sear notch.

(2) The pressure of the sear and firing pin springs holds the two notches locked together. There is a slight overtravel of the firing pin extension in its movement to the rear to ensure proper engagement with the sear. As the bolt starts forward, the overtravel is taken up and completed when the cocking lever enters the V-slot of the top plate bracket, and is caromed toward the rear; pressure on the cocking lever is relieved as the bolt starts forward.

3-5. LEFT-HAND FEED

By repositioning some of the components, the MG is capable of alternate feed. Ammunition can be fed into the weapon from the right or left side of the receiver; however the Army uses only left-hand feed. (See Table 3-1).

PART	POSITION
Belt-feed lever, shoulder headless pin, and spring.	Upper (rear) hole; lug of feed lever is on left side of cover.
Belt-feed slide.	Feed pawl is on left side of cover.
Belt-feed pawl arm.	Arm toward latch end of cover, pointing right.
Cover latch shaft lever	Left side of cover.
Cartridge stops and link stripper.	Right-hand rear cartridge stop assembly and front cartridge stop on right side of feedway.
Retracting slide with handle.	Right sideplate.
Belt-holding pawl.	Left side of feedway.
Bolt switch.	Cam grove in line with "L" on bolt.

Table 3-1. Position of parts for left-hand feed.

DANGER

Clear the gun of ammunition before setting headspace.

3-6. HEADSPACE AND TIMING

Headspace is the distance between the face of the bolt and the base of the cartridge case, fully seated in the chamber. Timing is the adjustment of the gun so that firing takes place when the recoiling parts are in the correct position for firing. Because the cartridge is held by the T-slot of the bolt, headspace with the MG is measured as the distance between the rear of the barrel and the face of the bolt. This occurs when the recoiling parts are forward and there is positive contact between the breech lock recess in the bolt and the lock in the barrel extensions. Periodic calibration checks should be made of the gauge by direct support personnel at least annually.

WARNING

Firing a weapon that has improperly set headspace and timing could result in damage to the machine gun, or injury to the gunner. Damage may also occur in the trunnion block, base of the barrel, or face of the bolt. This warning applies whether the gun is firing service ammunition or M1E1 blanks. (The weapon has improper early timing when two rounds are fired – and firing stops.)

a. **Gauges.** The headspace and timing gauge consists of a headspace gauge and two timing gauges (Figure 3-16). These gauges provide an accurate means of checking the adjustment of headspace and timing.

NOTE: The headspace and timing gauge should be kept with the gun at all times.

Figure 3-16. Headspace and timing gauge.

b. **Headspace.** Check and set headspace before firing, after assembling the gun, and after replacing the barrel or receiver group. Use the following procedures to set headspace.

(1) Raise the cover all the way up. Grasp the retracting slide handle (Figure 3-17). Using the retracting slide handle, retract the bolt until the barrel-locking-spring lug is centered in the 3/8-inch hole on the right side of the receiver (Figure 3-18).

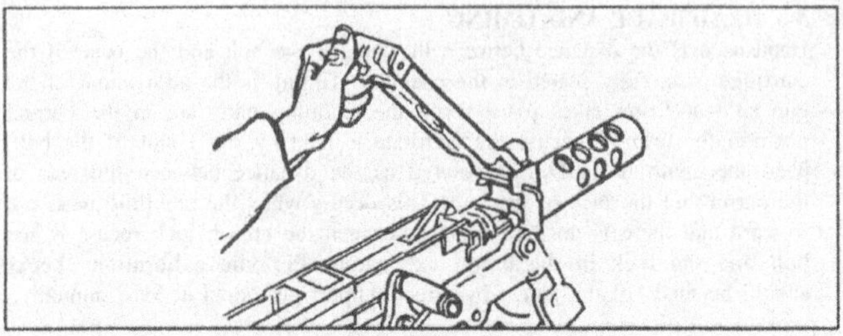

Figure 3-17. Raising the cover.

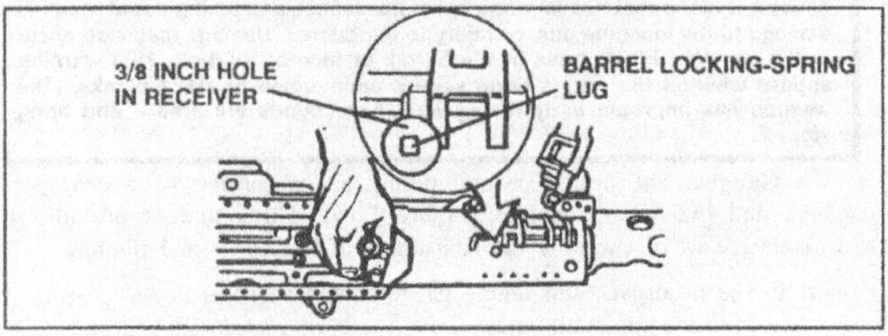

Figure 3-18. Retracting the bolt.

(2) Hold the bolt in this position and screw the barrel fully into the barrel extension (Figure 3-19).

Figure 3-19. Screwing in the barrel.

WARNING
When resetting the headspace and timing of a gun that has been fired, use an asbestos mitt to avoid burns.

(3) With the bolt still retracted, unscrew the barrel two notches (clicks). Release the retracting slide handle and allow the bolt to go forward.

NOTE: At this point, check the barrel for rotation. Attempt to turn the barrel in either direction. The barrel should not turn. If the barrel does turn, stop here and check barrel notches and the barrel-locking spring for damage.

(4) Pull the bolt to the rear with the retracting slide handle and hold. This cocks the weapon. Otherwise, the headspace gauge will not fit.

(5) Holding the retracting slide handle, release the bolt, and slowly return the bolt fully forward (Figure 3-20). Do not press the trigger or let the bolt slam forward.

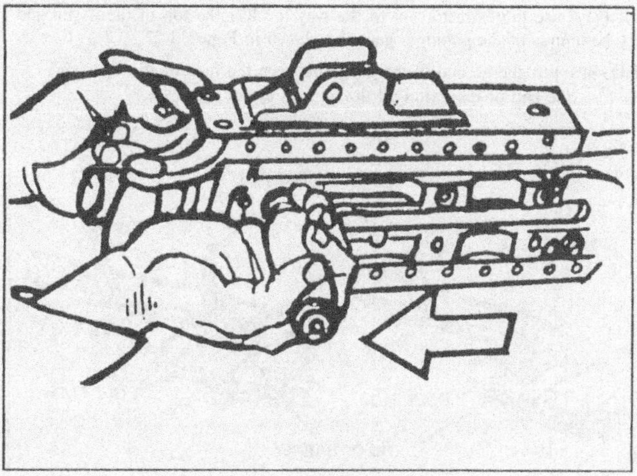

Figure 3-20. Releasing the bolt.

(6) Retract the retracting slide handle and separate the barrel extension from the trunnion block by a 1/16-inch gap (Figure 3-21).

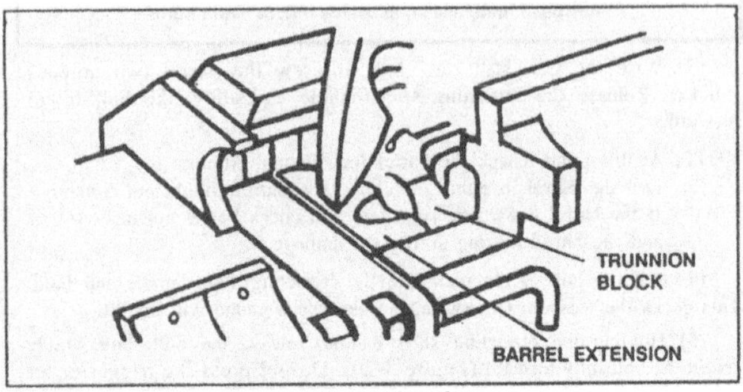

Figure 3-21. Setting the gap.

(7) Raise the extractor out of the way to clear the top of the T-slot and try both ends of the go/no-go gauge as shown in Figure 3-22.

NOTE: Insert the go end of the gauge between the face of the bolt and the end of the barrel all the way up to the ring.

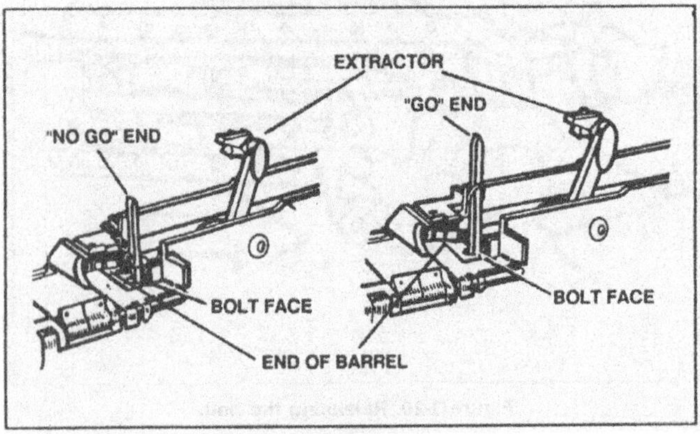

Figure 3-22. Using the gauge.

(8) If the go end of the gauge enters the T-slot freely to the center ring of the gauge, and the no-go end will not enter, headspace is correct. Remove gauge. Headspace setting is now complete. If the go end of the gauge will not enter the T-slot freely, headspace is too tight. Continue as follows.

(9) Retract the bolt so you can see the barrel-locking lug in the center of the 3/8th-inch alignment hole on the right side of the receiver.

(10) Unscrew the barrel one notch (click).

(11) Return the bolt fully forward.

(12) Recheck headspace (step 9).

(13) Repeat steps 10 through 13 until the go gauge fits but the no-go gauge does not fit.

NOTE: You should not have to unscrew the barrel more than five notches (clicks) beyond the first setting of two clicks. If this condition does occur, turn in the machine gun to your unit armorer for inspection.

(14) If the no-go end of the gauge enters the T-slot, headspace is too loose. Adjust it using the same procedures as above, screwing the barrel into the barrel extension rather than out.

(15) Repeat steps 10 through 13, one click at a time, until the no-go gauge does not fit but the go gauge does fit.

c. **Timing.** Timing is the adjustment of the weapon so that firing takes place when the recoiling parts are between .020 and .116 inch out of battery to prevent contact between the front end of the barrel extension and the trunnion block. Use the following procedures to set timing.

WARNING

Make sure the gun is clear of ammunition before starting.

(1) Check headspace first as previously described.

(2) Pull the bolt to the rear with the retracting slide handle and then ease bolt fully forward to cock the machine gun.

(3) Grasp the retracting slide handle and retract the bolt just enough (1/16 inch) to insert the no-fire gauge between the barrel extension and the trunnion block. Release the retracting slide handle (Figure 3-23).

(4) Depress the trigger. Gun should not fire.

Figure 3-23. Inserting the no-fire gauge.

NOTE: Insert timing gauge with bevel against barrel notches.

NOTE: If the gun does not fire, go to step 5. If the gun does fire, you have early timing. Go on to steps 7 through 14.

(5) Grasp the retracting slide handle and retract the bolt just enough to remove the no-fire gauge and insert the fire gauge in the same place (Figure 3-24). Release the retracting slide handle.

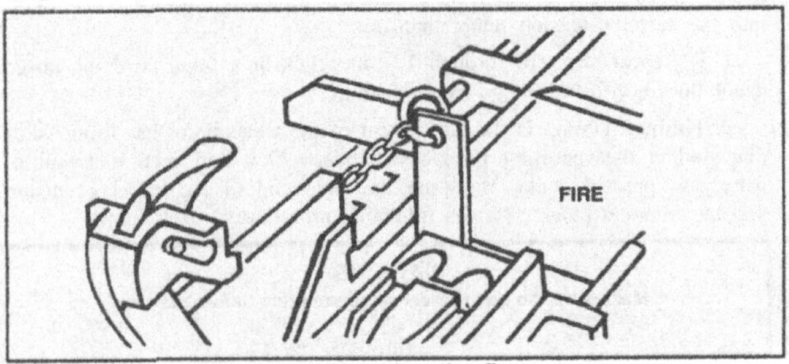

Figure 3-24. Inserting the fire gauge.

(6) Depress the trigger. Gun should fire. If it does, timing adjustment is now complete.

NOTE: If the gun does not fire, you have late timing. Go to steps 7 through 14.

(7) Remove the gauge, cock the gun, and return the bolt forward

(8) Insert the fire gauge.

(9) Remove the backplate.

(10) Screw the timing adjustment nut all the way down until it touches the trigger lever (Figure 3-25).

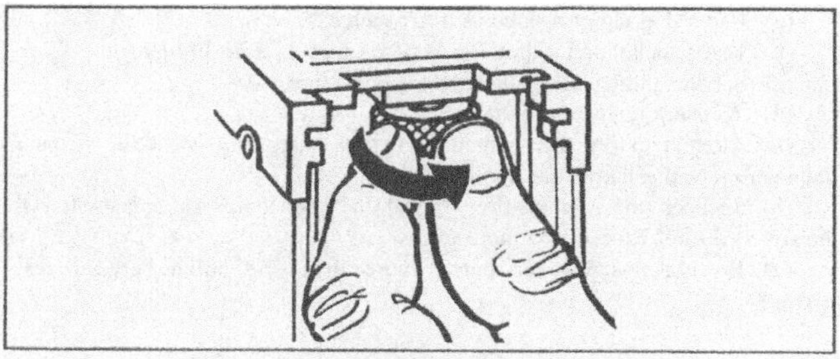

Figure 3-25. Adjusting the timing nut.

WARNING

Never cock the gun with the backplate off.

(11) Try to fire the MG by pushing up on the rear of the trigger lever. Gun should not fire.

(12) Screw the timing adjustment nut up (to the right) one click at a time. Push up on the trigger lever after each click. Keep doing this until the gun fires.

(13) Turn the timing adjustment nut no more than two more clicks up (to the right).

(14) Remove the gauge, replace the backplate, and pull the bolt to the rear to cock the machine gun. Ease the bolt forward with the charging handle. Do not allow the bolt to slam forward.

(15) Recheck the timing with the fire/no-fire gauge twice after the backplate is installed to ensure that the setting is correct.

d. **Field Expedient Methods.** When a go/no-go gauge is not available, you can still set the headspace and timing using field expedient methods. However, this method should be used only in combat.

(1) *To set headspace:*

(a) Raise the cover and retract the bolt in the normal manner until the barrel-locking-spring lug is centered in the 3/8-inch hole on the right side of the receiver.

(b) Hold the bolt in this position and screw the barrel fully into the barrel extension; then unscrew the barrel two clicks or notches.

(2) *To set timing:*

(a) Use a dog tag or a dime as a fire gauge.

(b) Use a nickel and a dime or four dog tags as a no-fire gauge.

(c) Set the timing using the normal procedure.

(3) *To check for correct settings:*

(a) Attempt to fire the weapon. If it fires sluggishly, clear the weapon then unscrew the barrel one more notch.

(b) Recheck the rate of fire. Repeat the procedures in paragraph (1); however, do not exceed two more clicks.

(c) Do not unscrew the barrel more than one notch between test firings.

CHAPTER 4

PERFORMANCE PROBLEMS AND DESTRUCTION

This chapter discusses the malfunctions, stoppages, immediate actions, remedial actions, and destruction procedures of the MG. All corrective actions must comply with safety procedures.

4-1. MALFUNCTIONS

A malfunction is any failure of the gun to function satisfactorily. Examples of malfunctions are:

a. **Failure to Function Freely.** Sluggish operation is usually due to human failure to eliminate excessive friction caused by dirt, lack of proper lubrication, burred parts, incorrect headspace adjustment, or incorrect timing.

b. **Uncontrolled Automatic Fire.** Uncontrolled automatic fire (runaway gun) is when fire continues even when the trigger or trigger control mechanism is released. If the cause is present before the gun is fired, the gun will start to fire when the recoiling groups move into battery the second time. If the defect occurs during firing, the gun will continue firing when the trigger control mechanism is released. A runaway gun may be caused –

- By a bent trigger lever, forward end of the trigger lever sprung downward.
- By burred beveled contacting surfaces of the trigger lever and sear.
- By a jammed or broken sideplate trigger.

To stop the uncontrolled automatic fire:

(1) Keep the gun laid on target.

(2) Twist the belt, causing the gun to jam.

(3) Caution, do not unlatch the cover.

(4) Wait 5 minutes to guard against cook off.

(5) Clear weapon, replace broken, worn, or burred parts. Check the sideplate trigger and trigger control mechanism, when applicable.

4-2. STOPPAGES

A stoppage is any interruption in the cycle of operation caused by the faulty action of the gun or ammunition. Stoppages are classified as follows:

a. **Failure to Feed.** Prevents the round from being properly positioned in the receiver group.

b. **Failure to Chamber.** Prevents the complete chambering of the round.

c. **Failure to Lock.** Prevents the breech lock from correctly entering its recess in the bolt.

d. **Failure to Fire.** Prevents the ignition of the round.

e. **Failure to Unlock.** Prevents the breech lock from moving out of its recess in the bolt.

f. **Failure to Extract.** Prevents the extraction of the expended cartridge from the chamber.

g. **Failure to Eject.** Prevents the ejection of the expended cartridge from the receiver.

h. **Failure to Cock.** Prevents the firing pin extension from being engaged with the sear.

Table 4-1 lists the causes of most stoppages.

NATURE OF STOPPAGE	USUAL CAUSES	OTHER CAUSES
Failure to feed.	Defective ammunition belt. Defective feed mechanism. Defective extractor.	Improperly loaded belt. Short round
Failure to chamber.	Broken part or obstruction in T-slot or chamber. Separated (ruptured) case.	Thick or thin rim, bulged round, protruding primer.
Failure to lock	Incorrect headspace.	Broken parts. Battered breach lock. Battered breach lock cam. Faulty breech lock cam adjustment.
Failure to fire.	Defective parts in firing mechanism. Defective ammunition. Incorrect timing.	
Failure to unlock.	Broken parts in receiver.	Worn or faulty breech lock cam, or faulty adjustment.
Failure to extract.	Dirty chamber.	Defective cartridge case.
Failure to eject.	Defective ejector.	Burred T-slot.
Failure to cock.	Broken sear. Worn sear notch. Weak sear spring. Worn hooked notch on firing pin extension.	Broken cocking lever.

Table 4-1. Stoppages and their causes.

4-3. IMMEDIATE ACTION

The first thing to do when a stoppage occurs is to apply immediate action, which is the prompt action taken by the firer to reduce a stoppage.

Immediate action is performed by the gunner; however, every crewmember must be trained to apply immediate action. The following procedures will assist in reducing most stoppages without analyzing their cause in detail.

WARNING

Failure to follow procedures may result in damage to weapon and injury to personnel.

a. If gun fails to fire, take the following action:

(1) Wait 5 seconds; a hangfire maybe causing the misfire. In the next 5 seconds, pull the bolt to the rear (check for ejection and feeding of belt), release it, re-lay on the target, and attempt to fire. If the bolt-latch release and trigger are depressed at the same time, the bolt goes forward and the weapon should fire automatically.

(2) If the gun again fails to fire, wait 5 seconds, pull the bolt to the rear (engage with bolt latch if applicable), and return the retracting slide handle to its forward position. Open the cover and remove the belted ammunition. Inspect the gun to determine the cause of stoppage.

b. A hangfire or cookoff can cause injury to personnel or damage to the weapon. To avoid these, the gunner must take the following precautions:

(1) Always keep the round locked in the chamber the first 5 seconds after a misfire occurs. This prevents an explosion outside of the gun in event of a hangfire.

(2) If the barrel is hot, the round must be extracted within the next 5 seconds to prevent a cookoff. When more than 150 rounds have been fired in a 2-minute period, the barrel is hot enough to produce a cookoff.

(3) If the barrel is hot and the round cannot be extracted within the 10 seconds, it must remain locked in the chamber for at least 5 minutes, to guard against a cookoff.

(4) Keep the gun cover closed during the waiting periods.

4-4. REMEDIAL ACTION

When immediate action does not correct the malfunction, the quickest way to resume firing is to replace the defective part.

a. **Removal of a Cartridge from the T-Slot.** If the cartridge does not fall out, hold the bolt to the rear, and with the extractor raised, use a screwdriver to push the cartridge out the bottom of the receiver.

b. **Removal of a Ruptured Cartridge.** A ruptured (separated) cartridge case may be removed with a cleaning rod or ruptured cartridge extractor. When using the ruptured cartridge extractor, raise the cover and pull the bolt to the rear. Place the extractor in the T-slot of the bolt in the same manners as that of a cartridge, so that it is held in line with the bore by the ejector of the extractor assembly of the gun. With the extractor aligned with the bore and held firmly in the T-slot, let the bolt go forward into the ruptured case, and the shoulders will spring out in front of the case. Pull the bolt to the rear and remove the ruptured case and extractor (Figures 4-1 and 4-2).

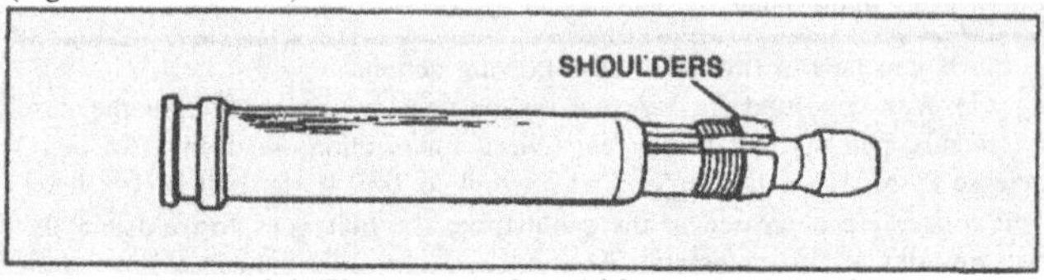

Figure 4-1. The ruptured cartridge case extractor.

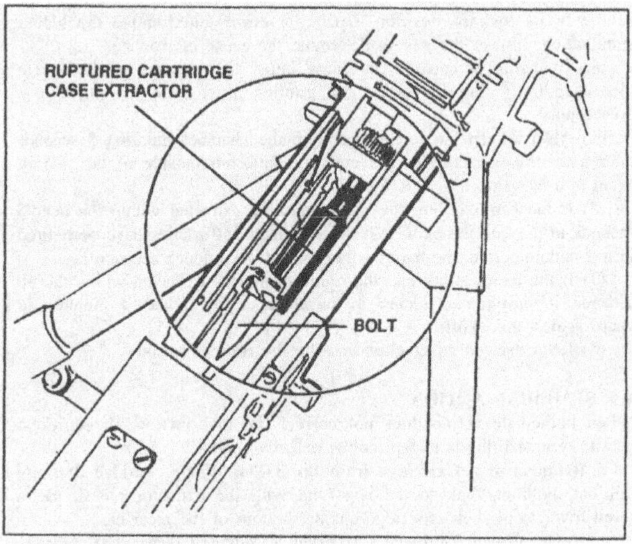

Figure 4-2. Ruptured cartridge case extractor aligned with the chamber.

4-5. DESTRUCTION PROCEDURES

The decision to destroy the gun to prevent its capture and use by the enemy is a command decision. It will be ordered and carried out only on authority delegated by the major unit commander.

- Destroy the machine gun and mount only when they are subject to capture or abandonment. Destruction must be as complete as circumstances permit.
- Lacking time for complete destruction, destroy the parts essential to operation of the gun, beginning with those parts most difficult for the enemy to duplicate.
- Destroy the same parts of each gun to prevent the reconstruction of a complete gun from several damaged guns.

The following methods may be used to destroy specific components of the MG.

a. **Gun.** Field strip the gun. Use the barrel as a sledge. Raise the cover and smash the cover forward and down toward the barrel support. Smash the backplate group. Remove the firing pin from the bolt; place the striker in the hole in the face of the bolt and bend it until broken. Remove the barrel buffer tube lock assembly from the barrel buffer body group and bend and deform it. Smash and bend the breech lock depressors. Place the barrel extension in the rear of the receiver with the barrel extension shank protruding; knock off the shank by striking it with the barrel from the side. Deform and crack the receiver by striking it with the barrel at the sideplate corners nearest the feedway. Smash the extractor.

b. **Tripod Mount, M3.** Leave the pintle on the tripod by removing the pintle bolt from the gun. Use the barrel as a sledge. Strike the sides of the pintle and deform it. Fold the trail legs and turn the mount over. Stand on the folded trail legs and knock off the pintle latch (pintle lock release cam). Smash the elevating mechanism with the barrel. If possible, smash the rear legs to prevent unfolding.

c. **Antiaircraft Mount, M63.** Remove the sideplate trigger control mechanism from its container or the gun, and deform it by using the barrel. Lock the cradle and yoke assembly in the horizontal position and beat the trigger frame assembly and cradle until they are bent down along the elevator assembly. Strike the elevator from the side with the barrel until it is bent so that the elevator will not rotate in the base.

d. **Spare Parts.** Destroy the bolt, barrel extension, firing pins, and barrel buffer groups. Break or deform all other parts.

e. **Ammunition.** When time permits and material is available, destroy ammunition by burning. Unpack all ammunition from boxes or cartons, stack the ammunition in a heap and, using flammable material available, ignite and take cover immediately.

f. **Burning.** To destroy the gun by burning, place a thermite grenade in the receiver on the bolt (with the cover resting on the grenade) and fire the grenade. (This method may require the use of more than one grenade.) Remove the backplate group, place a thermite grenade in the rear of the receiver and fire the grenade.

g. **Disposal.** Bury in suitable holes, or dump parts into streams, mud, snow, sumps, latrines, or scatter the parts over a wide area.

CHAPTER 5

MARKSMANSHIP TRAINING

The procedures, exercises, and techniques for implementing a proper marksmanship program are based on the concept that all machine gunners must understand common firing principles, be proficient marksmen, and be confident in applying these skills in combat. This depends on their understanding of the machine gun and their application of marksmanship fundamentals. Proficiency is accomplished through practice that is supervised by qualified trainers and through objective performance amassments by the unit leaders. The structure of this chapter is in four sections: planning, fundamentals, basic marksmanship, and advanced gunnery. All advanced exercises are conducted under conditions that are as much like tactical conditions as possible.
(See Appendix B for training aids and devices.)

Section I. PLANNING

The planning of MG training is no different from other marksmanship training. Guidelines are provided to assist the trainers in understanding, preparing, and ensuring that all training is conducted to standard. This section addresses the objectives, the responsibilities of the commanders, and the phases of training. It also introduces the training devices that assist in training, and designates when remedial and sustainment training should be conducted.

5-1. OBJECTIVES

The objectives of the MG training program are to guide the trainers, leaders, and gunners through a sequence of training to standard that produces a gunner who is able to maintain the gun and effectively engage targets in combat.

5-2. COMMANDER'S RESPONSIBILITIES

The responsibilities of the commander are to ensure the instructors and their assistants are thoroughly trained and rehearsed in the planning, knowledge, and presentation of all MG training. He will ensure that safety is emphasized and never overlooked during training. Serviceable weapons are a must for good training and commanders must make sure they are used properly. He also ensures sufficient time is scheduled.

5-3. PHASES OF TRAINING

The initial training strategy is divided into three phases of marksmanship instruction — the fundamental phase, the basic marksmanship phase, and the advanced gunner phase. Gunner safety is continuously stressed during all phases of training.

a. In the fundamental phase, the gunner learns the necessary common skills, such as dry-fire exercises, preparation of positions, manipulation of the T&E, range determination, and sighting and aiming. The gunner must master these skills before he is allowed to progress to the basic marksmanship phase.

b. During the basic marksmanship phase, the gunner learns the basics of loading, zeroing, laying, and engaging of single and multiple targets from the tripod and vehicle mode.

c. The advanced gunnery phase trains the gunner in engaging moving targets, night firing, NBC firing, mounted firing, and firing using fire commands. The gunner will be placed under the stress and strain of simulated combat conditions.

5-4. SUSTAINMENT TRAINING

Once individuals and units have trained to a required level of proficiency, leaders must structure collective and individual training plans to conduct critical task training at the frequency necessary for the sustainment training strategy. Mission training plans and individual training evaluation programs help achieve and sustain collective/individual proficiency. Sustainment training prevents skill degradation. Army units must be prepared to accomplish their wartime missions at any time – they cannot rely on infrequent peaking to the appropriate level.

5-5. REMEDIAL TRAINING

Remedial training will be conducted in any of the phases of training where the gunner does not meet the standard. The trainer must be instantly aware of any gunner that seems to be having trouble. Once the problem has been identified, the gunner should be retrained as soon as possible so that he will maintain the same level of proficiency as the other gunners.

Section II. FUNDAMENTALS

The fundamentals are necessary basic skills that a gunner must be trained in before he can be expected to effectively engage targets. Personnel conducting marksmanship training must fully understand the fundamentals and be well rehearsed in applying them. The basics in MG training are assuming a proper firing position, sighting, aiming, determining range, and manipulating the T&E mechanism.

5-6. FIRING POSITIONS

Before a gunner can hit targets, he must learn to get behind the weapon in a position that allows him to be comfortable, affords him protection, and enhances mission accomplishment.

a. The tripod firing positions are prone, sitting, and standing. They are assumed in the following manner.

(1) The prone position is used when firing from the tripod that is set in a low position. It is assumed by lying on the ground directly behind the gun. The gunner then spreads his legs a comfortable distance apart with his toes turned outward. His left elbow rests on the ground, and his left hand grasps the elevating handwheel of the T&E. His right hand lightly grasps the right spade grip with his right thumb in a position to press the trigger. The position of his body can then be adjusted to position his firing eye in alignment with the sights of the weapon (Figure 5-1).

Figure 5-1. Prone position with tripod mount.

(2) The sitting position can be used when the tripod is set in a high or low position. The gunner sits directly behind the gun between the legs of the tripod. He may extend his legs under the tripod or cross them, depending on his physique. The gunner then places both elbows on the inside of his thighs to get the best support. He grasps the elevating

handwheel of the T&E with the left hand, and lightly grasps the right spade grip with his right hand. He must ensure that the right thumb is in position to press the trigger (Figure 5-2).

Figure 5-2. Sitting position with tripod mount.

(3) The standing position is used when the gunner is firing from a fighting position. This position is assumed by standing directly behind the gun with the feet spread a comfortable distance apart. The gunner grasps the elevating handwheel of the T&E with the left hand. He lightly grasps the right spade grip with the right hand, ensuring that the right thumb is in a position to press the trigger. Adjustment of the body is allowed in order to align the firing eye with the sights on the weapon (Figure 5-3).

Figure 5-3. Standing position with tripod mount.

b. The vehicular firing position for the MG is standing. It is assumed by constructing a solid platform to stand on, using sandbags or ammunition boxes; or, in the case of the Ml13 APC, using the commander's seat. The gunner must then ensure that his platform is high enough to place the spade grips of the gun about chest high. He grasps the spade grips with both hands and places both thumbs in a position to press the trigger. The gunner holds the gun tightly to his chest for stabilization; his elbows should be locked tightly to his sides. He sights over the weapon and adjusts his position by flexing his knees and leaning forward to absorb any recoil (Figure 5-4).

Figure 5-4. Standing position using cupola mount.

c. The antiaircraft firing position uses a standing position when firing from the M63 mount. To assume the position, the gunner stands with his feet spread comfortably apart with his shoulders squarely behind the gun (Figure 5-5, page 5-6). When the gunner is engaging aerial targets, he grasps the upper extension handles with both hands. When engaging low-level aircraft or ground targets, he grasps the lower extension handles with both hands.

NOTE: The kneeling position may be used; it has the advantage of presenting a lower profile of the gunner and also aligns the gunner's eye closer to the axis of the barrel.

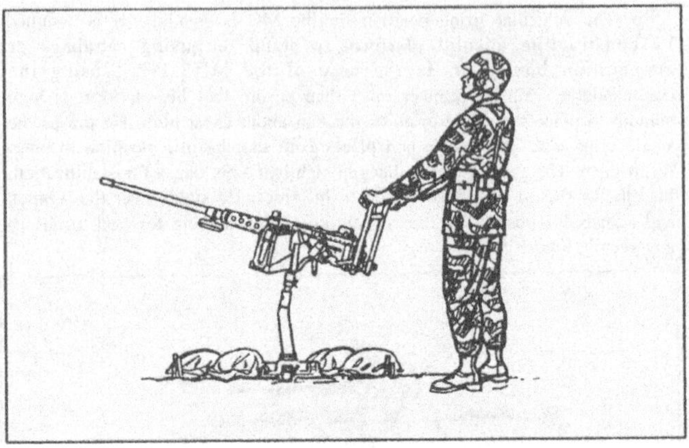

Figure 5-5. Antiaircraft firing position.

5-7. DRY-FIRE TRAINING

Dry-fire training is designed to teach the gunner the essentials of MG gunnery, including safety. Dry-fire training also includes sighting, aiming, sight setting, laying, manipulating the gun, manipulating the T&E mechanism, and determining the range. Thorough, carefully supervised training of these essentials is necessary to conserve time and ammunition during live fire. Practical exercises should be used to determine gunners' proficiency. Mastery of these skills is a must before the gunner is allowed to move on to the next phase of training. Practice is a must to achieve mastery.

a. **Sighting and Aiming.** Sighting is the ability of the gunner to use correct sight alignment and correct sight picture to engage targets.

(1) The first step in proper sighting is finding a natural, comfortable spot where the gunner is able to see the front sight blade through the rear peep sight. It is important the gunner understands that the spot he chooses to sight from must be constant throughout his firing.

(2) The second step in sighting is to move the weapon until the top center of the front sight blade is exactly in the center of the rear sight peep hole. The gunner can achieve this by drawing imaginary lines that bisect in the center of the rear peephole and then placing the top of the front ight blade center of them. This is correct sight alignment (Figure 5-6).

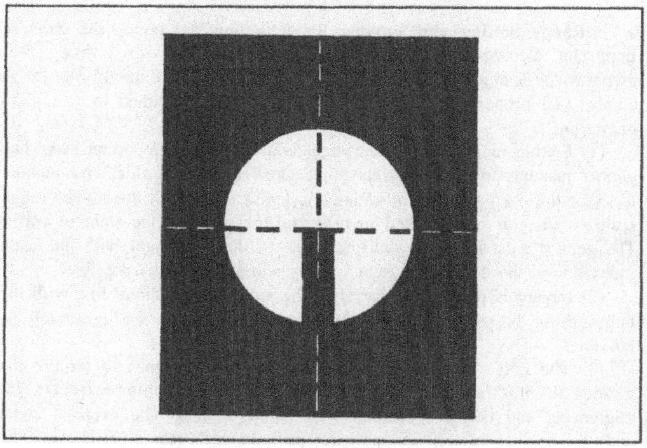

Figure 5-6. Correct sight alignment.

(3) The third step is to establish correct sight picture. To perform this task, the gunner adjusts the weapon until the top center of the front sight blade is bottom center of the intended target. Correct sight picture is a combination of sight alignment and placement of point of aim (Figure 5-7).

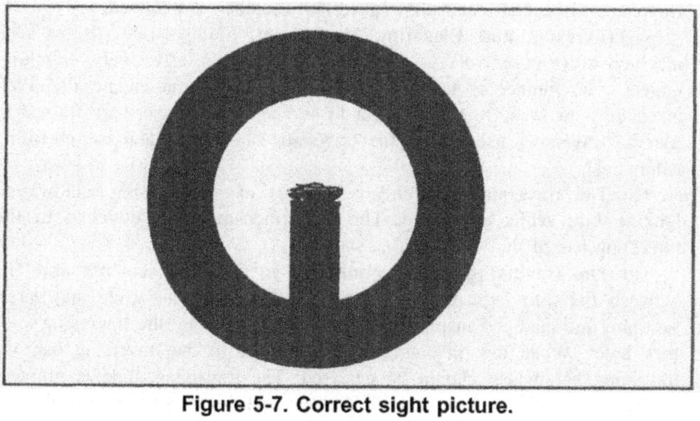

Figure 5-7. Correct sight picture.

b. **Range Setting and Laying.** Range setting and laying the gun are important elements in marksmanship training. It is this training that prepares the gunner to accurately and rapidly place fire on his target in combat. To properly set ranges, the gunner must be trained in rear sight operation.

(1) Setting ranges on the rear sight is a simple but important task. The gunner just has to rotate the elevating screw knob in a clockwise manner to move the peep sight up or counterclockwise to move it down. The range scale on the left is graduated in mils and the scale on the right in yards. The gunner must align the hairline index of the peep sight with the scale index line at the desired range as quickly and accurately as possible.

(2) Laying is placing the barrel of the weapon on a direct line with the target, using the sights. This must also be done as quickly and accurately as possible.

(3) The range setting and laying exercises are designed to require the gunner to practice and the instructor to evaluate both correct sight alignments and correct placements of aiming points. The exercise starts with the gunner in the sitting firing position with rear sight down. The coach will announce a range and general aiming point. The gunner will then repeat the range and direction of target. The coach will then announce, "Begin." The gunner will then raise and set his sights, and align the weapon on the aim point. When the gunner completes this task, he will announce "Up." The coach will then get behind the weapon and check the range setting and aim point and critique the gunner on his findings. This exercise will be continued until speed and accuracy is obtained.

c. **Traversing and Elevating Mechanism.** Manipulation of the T&E mechanism (Figure 5-8) is another key factor in effectively engaging targets. The gunner is taught how to instinctively manipulate the T&E mechanism to shift from one target to another. The gunners are trained to use the traversing handwheel, the traversing slide lock, and the elevating handwheel.

(1) The traversing mechanism consists of a traversing handwheel, locking nut, scale, and yoke. The T&E mechanism is attached to the traversing bar of the M3 bipod.

(a) The traversing bar is graduated in 5-mil increments and fits between the trail legs of the M3 tripod. The traversing slide and screw assembly are clamped in place on the traversing bar by the traversing slide lock lever. When the traversing slide is locked to the traversing bar, the traversing handwheel should be centered. The traversing slide is properly mounted when the lock lever is to the rear and the traversing handwheel is positioned to the left.

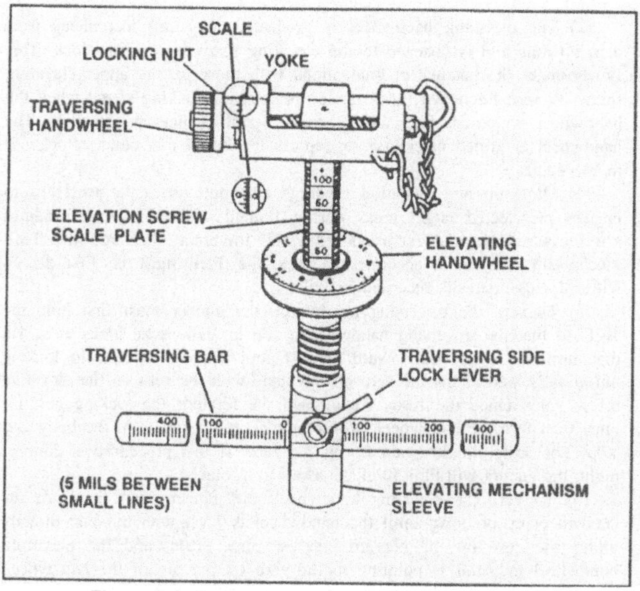

Figure 5-8. Traversing and elevating mechanism.

(b) To make changes in direction, loosen the traversing slide lock lever and move the slide along the traversing bar. This permits traverse of 400 mils left or right of the zero index in the center of the traversing bar. Readings on the traversing bar are taken from the left side of the traversing slide. For changes of 50 mils or less in deflection, turn the traversing handwheel. This allows a traverse of 50 mils left or right of center. One click in the traversing handwheel signifies 1 mil change in direction.

(2) The elevating mechanism consists of an upper and lower elevating screw, which is connected to the gun by inserting the quick release pin assembly through the holes in the upper elevating screw yoke and the rear mounting lugs of the receiver. A scale, graduated in mils, is fitted to the upper screw to indicate elevation. This scale is marked to show (-) minus 250 mils in depression and (+) plus 100 mils in elevation from the zero setting.

(3) The elevating handwheel is graduated in 1 mil increments from 0 to 50 mils and is fastened to the elevating screw by a screw lock. This synchronizes the handwheel graduations with those on the upper elevating screw. A spring-actuated index device produces a clicking sound when the handwheel is turned. Each click equals 1 mil change in elevation. The handwheel is turned clockwise to depress the barrel and counterclockwise to elevate.

(4) Direction and elevation readings constitute the data necessary to engage preselected target areas during limited visibility. These readings are measured by and recorded from the traversing bar and the T&E mechanism. To obtain accurate readings, the T&E must be first zeroed with all measurements recorded in mils.

(a) To zero the traversing handwheel, the gunner must first hold the T&E so that the traversing handwheel is on his left as he looks at it. He then turns the handwheel toward himself until it stops, loosens the locking nut slightly, and aligns the zero on the scale with the zero on the elevating screw yoke. Once the zeros are aligned, he tightens the locking nut. He must then turn the handwheel two complete turns away from the body and stop. The scale should again be on the zero. If this procedure is done at night, the gunner will turn 50 clicks away from him.

(b) To zero the elevating handwheel, the gunner must first turn the handwheel up or down until the handwheel is level with the line directly under the zero on the elevating screw plate scale, and the elevating handwheel indicator is pointing to the zero on the top of the handwheel. The elevating mechanism sleeve is then rotated up until it is stopped by the handwheel. The gunner then rotates the sleeve down until it stops, making sure he counted each complete rotation. He then divides the number of rotations by two, rotates the sleeve back up that number, and stops. The T&E mechanism is now ready to be attached to the tripod.

(c) To obtain and record direction readings, the gunner sets the sight on the proper range to hit the target, loosens the traversing slide lock lever, and slides the T&E mechanism along the traversing bar until the weapon is sighted on the aiming point of the target. The T&E mechanism is then locked down by tightening the traversing slide lock lever. All readings are taken from the left side of the sleeve mechanism. If the left side of the sleeve is not exactly on one of the 5-mil tick marks, the gunner must slide the sleeve to the next smaller tick mark to align it exactly. The traversing handwheel is then used to move the weapon back on point of aim. The direction is now ready to be recorded. The reading is taken from the number on the traversing bar and the direction from the direction of the barrel of the weapon. If the sleeve mechanism is on the right side of the zero on the traversing bar, then the reading is left; if it is on the left side of the zero, then it is a right reading. The width of a target may also be measured and recorded by first moving the traversing handwheel until the sights are aligned with the right or left side of the target. The clicks required to do this measure the width.

NOTE: Before repositioning the weapon for another target, the gunner must realign the handwheel.

(d) To obtain an elevation reading, the gunner must first ensure that the sights are aligned and at the desired aim point of the target. The elevation reading is made up of two portions, a major reading and a minor reading. The major reading is taken from the elevating screw plate scale. The scale is graduated in 50-mil increments and ranges from a minus (-) 250 mils to a plus (+) 100 mils with a zero between them. There is an index line below each number and a plus or minus sign above each number, with the exception of the zero. The zero does not have a plus or minus sign. To obtain the elevation reading, the gunner should lower his head until his eyes are level with the elevating handwheel. The major reading is the first number with a plus or minus sign, with its index line just visible above the elevating handwheel. The minor reading is taken from the top surface of the elevating handwheel. It is graduated in 1-mil increments for a total of 50 mils. The handwheel is also equipped with an indicator that points to each number on the handwheel as it is turned. Once the gunner has the major reading from the screw plate scale, he will then get the minor reading by looking at the number at which the indicator is pointing. Both portions of the elevation reading are recorded by placing a slash (/) mark between the two portions.

(e) An elevation reading is valid only on one T&E mechanism. If the same data is placed on another T&E mechanism using the same weapon, the data may be inaccurate. The number of threads exposed on the T&E must remain the same both when obtaining and recording data. If the number of exposed threads is changed in any manner, the firing will be off target. For example, when a gun is freed to engage targets in the secondary sector, the data will be correct if the gunner ensures that the same amount of threads is exposed when he returns to his primary sector of fire.

(f) To ensure that the data is correct, the gunner should fire and adjust his weapon.

(5) The T&E manipulation exercise gives the gunner practice and the instructor a tool to evaluate the gunner's progress (Figure 5-9, page 5-12). The exercise is conducted in two stages. Both stages require the coach to give directions and the gunner to respond. In the first stage, the coach

positions himself about 10 paces to the front of the gun. He then directs the gunner to manipulate the weapon in certain directions. He indicates the direction by the use of hand signals. The gunner responds by manipulating the T&E mechanism with his left hand. The gunner must keep his eyes on the coach; at no time during this exercise is he permitted to look at the T&E mechanism. The coach must be very attentive during the first stage because the gunner will be manipulating using the elevating handwheel and the traversing handwheel. The second stage is done in the same way except the gunner must make bold changes in elevation and deflection. The exercise continues until the instructor is satisfied that the gunner can manipulate the weapon by T&E without looking at the device. This exercise can also be conducted using the basic MG target. The gunner will be shifted from one selected target to another. The coach must observe all movements of the gunner during this training.

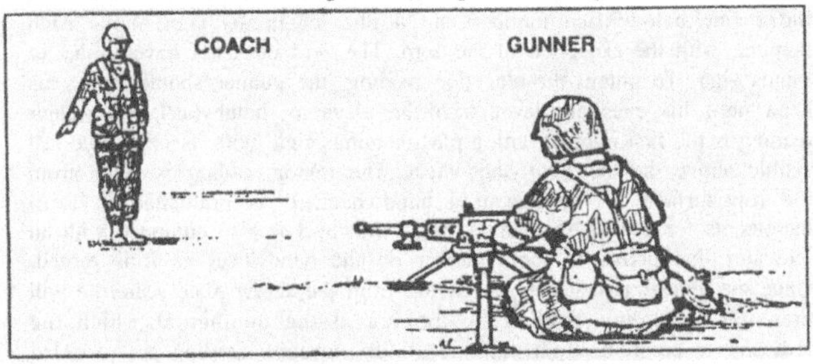

Figure 5-9. Manipulation exercise.

5-8. RANGE DETERMINATION

Range determination is the process of estimating the distance to a target from a gunner's position. The ability of the gunner to get the range to, sight on, and destroy a target is the realism of combat. Under combat conditions, ranges are seldom known in advance; therefore the effectiveness of fire depends largely upon the accuracy and speed of the gunner in determining range. Some methods of determining range are estimating by eye (Table 5-1, page 5-15), firing the gun, measuring range from a map or aerial photograph, stepping off the distance, or securing information from other units. Ranges are determined to the nearest 100 meters for machine gun firing. In combat, the most commonly used methods are estimating by eye and firing the gun. There is also a method used for measuring lateral distance.

a. The two techniques of eye estimation are the 100-meter unit of measure method and the appearance of objects method.

(1) When using the 100-meter unit of measure method, the gunner must be able to visualize what 100 meters looks like on the ground. With this distance in mind, the gunner can mentally determine the number of 100-meter units between his position and the target. The accuracy of this method is limited to 500 meters or less, and it requires constant practice (Figure 5-10).

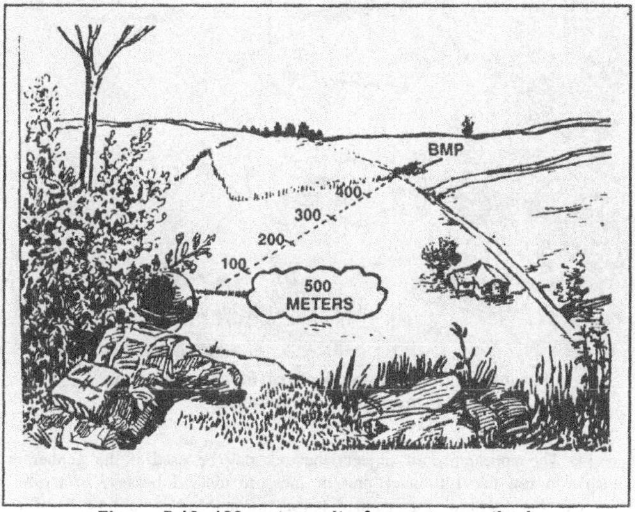

Figure 5-10. 100-meter unit of measure method, less than 500 meters.

(2) For targets that appear to be more than 500 meters, the gunner must modify this technique. The gunner selects what he thinks is the halfway point between the target and his position. He then mentally counts the number of 100-meter units to the halfway point and doubles it. This method of range determination is not accurate beyond 1,000 meters (Figure 5-11, page 5-14).

(3) Some terrain affects the appearance of 100-meter units of measure. When the terrain slopes upward toward the target, 100 meters appears longer than on level terrain. It appears shorter on downward

sloping terrain. The gunner must consider these two factors when using the 100-meter unit of measure method.

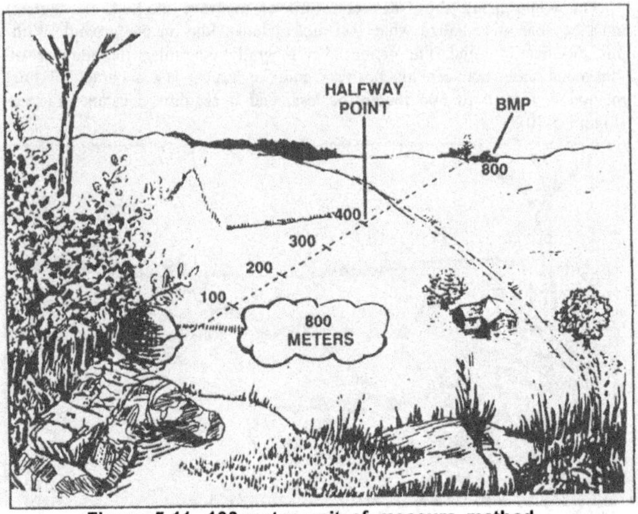

Figure 5-11. 100-meter unit of measure method, more than 500 meters.

(4) The appearance of objects method may be used if the gunner is unable to use the 100-meter unit of measure method because of terrain. To use this method, the gunner learns through practice how familiar objects look at various known ranges. This can be achieved by studying the appearance of a man standing 100 meters away. The gunner must then fix the appearance of the man firmly in his mind to include the size and details of his uniform and equipment. Next, he studies the same man at the same distance in the kneeling and prone positions. This procedure is used at 200, 300, 400, and 500 meters. By comparing the appearance of the man at these known ranges, he can establish a series of mental images that will help him determine range on unfamiliar terrain out to 500 meters. This training could also be conducted to familiarize the gunner with other objects, such as weapons and vehicles, at various ranges.

FACTOR	APPEARS NEARER (Range is underestimated when)—	APPEARS MORE DISTANT (Range is overestimated when)—
Target visibility.	Most of the target is visible and offers a clear outline.	Only a small part of the target is seen or target is small in relation to its surroundings.
Terrain, or position of the observer.	Looking across a depression, most of which is hidden from view. Looking down from high ground. Looking down a straight, open road or along a railroad track. Looking over uniform surfaces such as water, snow, desert, or grain fields.	Looking across a depression, all of which is visible. Looking from low toward high ground. When vision is narrowly confined as in streets, draws, or forest trails.
Light and atmosphere	In bright light or the sun is shining from behind the observer. The target is in sharp contrast with the background, or is silhouetted by reason of size,shape, or color,or is seen in the clear atmosphere of high altitudes.	In poor light such as dawn and dusk, in rain, snow, fog, or when the sun is in the observer's eyes. The target blends into the background or terrain.

Table 5-1. Factors affecting range estimation by eye.

b. Firing the gun is another method of determining range. In this method, the gunner opens fire on the target at the estimated range and moves the center of the beaten zone into the center base of the target by means of the T&E handwheels. He resets the sight so the new line of aim is at the center base of the target and notes the range setting on the rear sight. This range setting may apply only to this gun. When the ground in the vicinity of the target does not permit observation of the strike of the rounds, or when surprise fire on the target is desired, fire is adjusted on a point that offers observation and is known to be the same range as the target. The gunner then lays his gun on the target when ordered. When moving into position occupied by other units, range cards prepared by those units can furnish valuable range information on targets, suspected targets, and various terrain features. When the tactical situation and time permit, range may be determined by pacing off the distance.

c. Lateral distance measure is a method that the gunner may use to determine the distance from one target to another from left to right or right to left. When the gun is mounted on the M3 tripod, width can be measured by aiming on a point and manipulating the traversing handwheel, counting the clicks from one point to another point of aim. Each click equals one meter at 1,000 meters or one-half meter at 500 meters. This method is accurate but time-consuming. The finger measurement method is not a method of range determination; it is a method of measuring the lateral distance (in fingers or mils) between two points. To measure the distance in fingers between a reference point and a target, extend the arm with palm outward, the fingers cupped, and elbow locked. Close one eye, raise the index finger, and sight along its edge, placing the edge of the finger along the flank of the target or reference point (Figure 5-12). The remaining space is then filled in by raising fingers until the space is covered. The measurement is then stated as being one or more fingers or so many mils, depending on the number of fingers used (Figure 5-13).

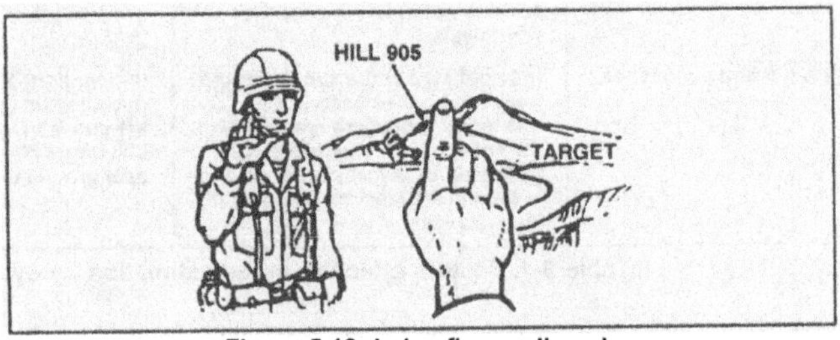

Figure 5-12. Index finger aligned.

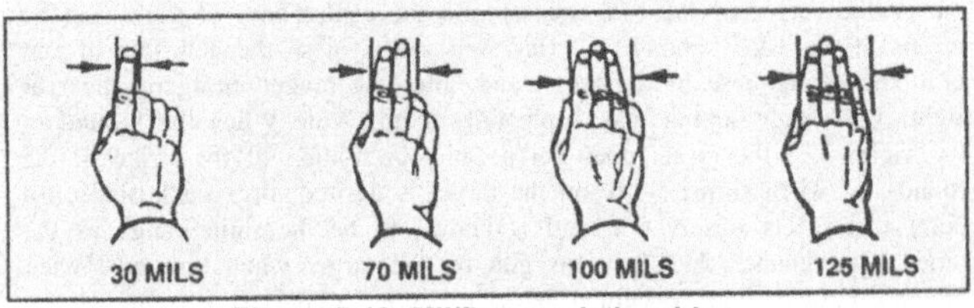

Figure 5-13. Mil/finger relationships.

5-9. OBSERVATION AND ADJUSTMENT OF FIRE

The purpose of observation and adjustment of fire practice is to teach the adjustment of fire by observing the strike of the bullets and the flight of the tracers, or by frequent re-laying on the target using sights.

a. Observation is used when firing on the 10-meter range because the impact of the round is visible on the target. When firing at greater distances, the strike of the round on the ground may cause dust to rise that is visible to the gunner; however, during wet weather, the strike cannot always be seen. In this case, use tracer ammunition that allows the gunner or crew to note the strike of the burst in relation to the target.

b. Adjustments on the target can be made using the mil relation; that is, one click of traversing or elevating handwheel moves the strike of the round one-half inch on the target at 10 meters. When firing on field targets, adjustment is made by moving the burst into the target. One click of traverse will move the strike of the round one-half meter at 500 meters, or one meter at 1,000 meters (Figure 5-14). However, the distance one click of elevation will move the strike of the round depends on the range to the target and the slope of the ground. The gunner determines the number of mils necessary to move the center of the strike into the target, and he manipulates the gun the required number of mils. This does not require the use of sights. For example, should the gunner fire on a target at 500 meters and observe the strike 10 meters to the right of the target and short about 50 meters, he would traverse the gun to the left 20 clicks (mils) and add one or more clicks (mils), depending on the slope.

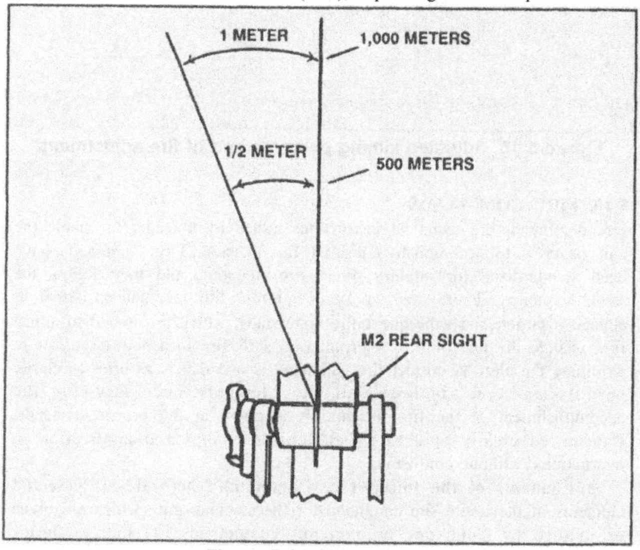

Figure 5-14. Mil relation.

c. The gunner may use the adjusted aiming point method to adjust the fire. In this method, the gunner must use his sights. He selects an aiming point that will place the next burst on target. For example, when the gunner fires on a target at 500 meters and estimates that the rounds impacted 20 meters short and 10 meters to the right, he would rapidly select an aiming point about 20 meters beyond the target and 10 meters to the left of it and lay on that aiming point and fire (Figure 5-15).

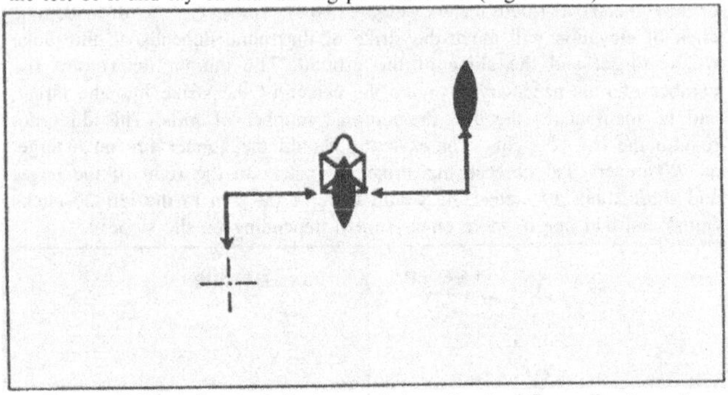

Figure 5-15. Adjusted aiming point method of fire adjustment.

5-10. FIRE COMMANDS

Fire commands are technical instructions issued by a leader to enable the unit or crew to accomplish a desired fire mission. Fire commands have been standardized for infantry direct fire weapons, and they follow the same sequence. There are two types – initial fire commands, issued to engage a target; and subsequent fire commands, which are issued to adjust fire, change the rate of fire, interrupt fire, shift fire to a new target, or to terminate the alert. A correct fire command is one that is as brief as clarity permits and yet includes all the elements necessary for the accomplishment of the fire mission. It is given in the proper sequence, transmitted clearly at a rate that permits receipt and application of instructions without confusion.

a. **Elements of the Initial Fire Command.** There are six essential elements of the initial fire command for the machine gun, which are given or implied by using one or more of the methods of control. During training, the gun crew repeats each element as it is given. This is done to avoid confusion and to train the crew to think and act in the proper sequence. The six elements are the alert, direction, description, range, method of fire, and the command to open fire.

(1) *Alert.* This element brings the crew to a state of readiness to receive further instructions. Once alerted, the gunner ensures the gun is loaded. The assistant gunner continuously checks with the leader for orders or instructions and passes them on to the gunner. The oral alert is announced as FIRE MISSION. At this command, the gunners are alerted that a target has been detected and fire may be delivered upon it. When the leader announces the alert, such as FIRE MISSION, both gun crews react to the alert. If only a certain gun is to engage, the leader announces NUMBER 1 (or 2). The other crew follows the fire mission, loads, and lays on the target to take up the fire, if required.

(2) *Direction.* This element indicates the general direction to the target and may be given in one or a combination of the following ways:

(a) The leader gives the direction orally to the target in relation to the position of the gun(s). For example, FRONT, RIGHT FRONT, LEFT FRONT.

(b) The leader can designate a small or obscure target by pointing with his arm and hand or aiming the machine gun. When pointing with his arm and hand, a man standing behind him should be able to look over his shoulder and sight along his arm and index finger to locate the target. When a gun has been aimed at a target, a soldier looking through the sights should be able to see the target.

(c) Tracer ammunition is a quick and sure method of designating a target that is not clearly visible. When using this method, the leader should first give the general direction to direct the gun crew's attention to the desired area. To minimize the loss of surprise when using tracer ammunition, the leader does not fire until he has given all the elements of the fire command except the command to fire. The leader may use his individual weapon or fire one or more bursts from the machine gun. The firing of the tracer(s) then becomes the last element of the fire command and is the signal to open fire. For example:

FIRE MISSION.
FRONT.
BUNKER.
WATCH MY TRACER(S).
SLOW (or SINGLE SHOT).

The leader fires his individual weapon or a machine gun at the enemy bunker, then his gun crew(s) opens fire.

(d) Another method of designating obscure targets is by using easily recognizable reference points. Prominent terrain features and man-made objects make good reference points. All leaders and members of the crew(s) must be familiar with the terrain features and the terminology used to describe them. The general direction to the reference point should be given. For example:

FIRE MISSION, NUMBER 2.
FRONT.
REFERENCE: LONE PINE TREE.
TARGET: TRUCK.

Sometimes a target must be designated by using successive reference points. For example:

FIRE MISSION, NUMBER 1.
RIGHT FRONT.
REFERENCE: RED-ROOFED HOUSE, LEFT TO HAYSTACK, LEFT TO BARN.
TARGET: MACHINE GUN.

Finger measurements can be used to direct the gun crew's attention to the right or left of reference points. For example:

FIRE MISSION.
LEFT FRONT.
REFERENCE: CROSSROAD. RIGHT FOUR FINGERS.
TARGET: LINE OF TROOPS.

When the guns are mounted on tripods, lateral distance from reference points can be accurately announced. When gunners are firing the tripod-mounted gun, lateral distance is assumed to be in mils unless otherwise indicated, so the word "mils" is not necessary. For example:

FIRE MISSION.
FRONT.
REFERENCE: KNOCKED-OUT TANK. LEFT FOUR ZERO.
TARGET: COLUMN OF TROOPS.

(3) *Description.* The target description is used to create a picture of the target in the minds of the gun crew. The gun crew must know the type of target they are to engage to properly apply their fire. The leader should describe it briefly but accurately. For example:

Dismounted enemy personnel .... TROOPS
Automatic weapons .... MACHINE GUN
Armored vehicles .... TANK
Artillery or antitank weapon .... ANTITANK
Airplanes or helicopters .... AIRCRAFT

If the target is obvious, no description is necessary. Finger measurements or mil measurements can be used to designate the width of a linear target when the flanks cannot be pinpointed.

(4) *Range.* The range to the target is given so the gun crew knows how far to look for the target and immediately knows what range setting to place on the rear sight. Range is determined and announced in meters. Since the meter is the standard unit of range measurement, the word "meters" is not announced. With machine guns, the range is determined and announced in even hundreds and thousands. For example: THREE HUNDRED, ONE THOUSAND, ONE ONE HUNDRED. This element may be omitted when the gunners can obviously determine the range; however, it is desirable in some situations to announce the range.

(5) *Method of fire.* This element includes manipulation and rate of fire.

(a) Manipulation is used to prescribe the class of fire with respect to the gun. It is announced FIXED, TRAVERSE, SEARCH, TRAVERSE AND SEARCH, SWINGING TRAVERSE, or FREE GUN.

(b) To control the rate of fire, the gunner may use single shot, slow, rapid, or cyclic.

- Single shot. Place the gun in the single-shot mode and engage the target with aimed shots. The MG is accurate out to 1,500 meters.
- Slow fire. Slow fire consists of less than 40 rounds per minute, in bursts of five to seven rounds, fired at 10- to 15-second intervals.
- Rapid fire. Rapid fire consists of more than 40 rounds per minute, fired in bursts of five to seven rounds, at 5- to 10-second intervals.
- Cyclic fire. Cyclic fire is when the weapon fires 450 to 550 rounds per minute.

(6) *Command to open fire.* If surprise fire is not desired, the command FIRE is given without pause. It is often important that machine gun fire be withheld for maximum effect of surprise fire. To ensure this, the leader may preface the command to commence firing with the words AT MY COMMAND or AT MY SIGNAL. When the gunner(s) is ready to engage the target, he reports UP to the assistant gunner(s) who signals READY to the leader. For example:

FIRE MISSION.
FRONT.
TROOPS.
AT MY COMMAND. (Pause until crew members are ready and fire is desired.)
FIRE (or appropriate arm-and-hand signal).

When the leader makes a mistake in the initial fire command, he corrects it by announcing CORRECTION, and then gives the corrected element(s). For example:

FIRE MISSION.
FRONT.
TROOPS.
FIVE HUNDRED.
CORRECTION.
SIX HUNDRED.
TRAVERSE.
AT MY COMMAND.

When the leader makes an error in the subsequent fire command, he may correct it by announcing CORRECTION, and then repeating the entire subsequent fire command. For example:

LEFT FIVE, DROP ONE.
CORRECTION
LEFT FIVE, DROP ONE ZERO.

b. **Subsequent Fire Commands.** If the gunner fails to adjust his fire on the target, the leader must promptly correct him by announcing or signaling the desired changes. When changes are given, the gunner makes the required corrections and continues to engage the target without further command. When firing under the control of a leader, the assistant gunner checks with the leader for instructions, which he passes on to the gunner. Changes in the rate of fire are given orally and by arm-and-hand signals. To interrupt firing, the leader announces CEASE FIRE or gives a signal to cease fire. The gun crew(s) remains on the alert and firing can be resumed on the same target by announcing FIRE. To terminate the alert, the leader announces CEASE FIRE, END OF MISSION.

5-11. CREW EXERCISES

The purpose of crew exercise is to develop precision, speed, skill, and teamwork in examining equipment, placing the gun into action, and taking it out of action. In crew exercise, precision must be stressed. Once that is attained, speed, skill, and teamwork will follow. Duties are rotated during crew exercise to allow each member of the gun crew to become familiar with all the duties. During crew exercise, all oral or visual signals are repeated. When the fire command is completed, the gunner will give the assistant gunner an UP. The assistant gunner will extend his hand and arm into the air in the direction of the leader (to indicate READY) and announce, UP. With the M3 mount, the crew must consist of at least four men, including the leader. There is no designated crew in the TOE for a dismounted caliber .50 MG. The following paragraphs are only suggestions for the breakdown of equipment and member designation that may be established by the commander.

a. **Crew Equipment.** In addition to individual arms and equipment, crew members carry the following equipment for the tripod-mounted machine gun:

CREW MEMBER	SUGGESTED MINIMUM EQUIPMENT
Squad or crew leader.	**Binoculars, compass, one box of ammunition.**
No. 1 Assistant gunner.	**Tripod.**
No. 2 Gunner.	**Receiver, T&E mechanism attached, and headspace and timing gauge.**
No. 3 Ammunition bearer.	**Barrel, barrel cover, and box of ammunition.**

b. **Form for Crew Exercise.** The crew leader commands, FORM FOR CREW DRILL.

(1) *Positions with equipment.* The crew forms in column, facing the crew leader with five paces between men (Figure 5-16, page 5-24). They are in the following order: assistant gunner, gunner, and ammunition bearer. When the crew members reach their correct positions, they assume the prone position with equipment arranged as follows:

(a) No. 1: Tripod to his left, trail legs to the rear, front leg uppermost.

(b) No. 2: Receiver across his front, backplate to the right, retracting slide handle uppermost.

(c) No. 3: Barrel to his right, muzzle to the rear, ammunition box to his left front with latch to the right (latch to the front for the new box).

(d) Other members, if present: Ammunition boxes in front, one foot apart, latches to the right (front).

(e) Crew leader: Ammunition box to his right as he faces the crew, latch to the right (front).

(2) *Rotation of duties.* Duties are rotated to ensure that each member learns and is capable of performing the duties of the other members.

(a) The command to rotate all personnel is, FALL OUT LEADER. At this command, each member of the crew rises, moves forward, and assumes a new duty. The crew leader becomes the ammunition bearer. The assistant gunner moves forward and becomes the crew leader. The gunner moves forward and becomes the assistant gunner. The ammunition bearer moves forward and becomes the gunner.

(b) If the leader is not changed, the command, FALL OUT ASSISTANT GUNNER, is given. At this command, the crew members rise, the gunner becomes the assistant gunner, the ammunition bearer becomes the gunner, and the assistant gunner becomes the ammunition bearer. When the crew members have assumed their new position, they call out their new duties in order, ASSISTANT GUNNER, GUNNER, AMMUNITION BEARER.

NOTE: An additional crew exercise, which the crew maybe required to practice, is the setting of headspace and timing. These procedures are outlined in paragraph 3-6.

Figure 5-16. Crew formed in column.

c. Inspection of Equipment Before Firing. When the crew is formed with equipment, the command is, INSPECT EQUIPMENT BEFORE FIRING. At this command, the crew proceeds as follows:

(1) The assistant gunner inspects the tripod M3 mount to ensure that:

(a) The indexing levers and clamps on the front and trail legs function properly, and the legs are in the short (low) position.

(b) The front leg and trail legs are closely folded, and the front leg clamp is hand tight.

(c) The sleeve lock latch and pintle lock release cam are in working order, and the pintle lock release cam is down.

(d) The pintle bushing is free from dirt and burrs.

(2) The gunner inspects the receiver group to ensure that:

(a) The barrel support and breech bearing are free of dirt.

(b) The gun pintle is free of dirt.

(c) The feed mechanism and bolt switch are properly assembled to feed from left (soldier raises cover for proper inspection).

(d) The striker projects through the aperture in the face of bolt (soldier closes the cover).

(e) The rear sight is set at 1,000 yards (900 meters) windage zero.

(f) The T&E mechanism is securely attached to the receiver.

(g) The traversing handwheel is centered.

(h) The elevating screws are equally exposed (about 2 inches) above and below elevating handwheel.

(i) The backplate is latched and locked in place.

(j) The bolt latch release is locked in the down position by the bolt latch release lock.

(3) The ammunition bearer inspects the barrel and ammunition box to ensure that:

(a) The barrel is clear.

(b) The barrel carrier assembly is securely attached to the barrel.

(c) The barrel threads are free of dirt.

(d) The metallic links are clean (soldier opens ammunition box).

(e) The belt is properly loaded and placed in box with the double-looped end up.

(f) Dummy ammunition is used during crew exercise, and no live ammunition is present.

(g) The box is closed and latched.

(4) When the ammunition bearer completes his inspection, he moves to the gunner's position with the barrel in his right hand and ammunition box in his left hand. With the aid of the gunner, he screws the barrel into the barrel extension (Figure 5-17, page 5-26). The headspace and timing adjustment is made. The ammunition bearer remains on the left and on line with the gunner.

(5) The crew leader examines his ammunition as described in paragraph (3).

(6) At the completion of the inspection, a report is rendered as follows:

(a) The ammunition bearer reports: AMMUNITION CORRECT (or any deficiencies).

(b) The gunner reports: GUN AND AMMUNITION CORRECT (or any deficiencies).

(c) The assistant gunner reports: ALL CORRECT (or any deficiencies).

Figure 5-17. Screwing the barrel into the barrel extension.

d. **Placement of the Gun Into Action.** To place the gun into action, the crew leader commands and signals, GUN TO BE MOUNTED HERE (pointing to the position where the gun is to be mounted), FRONT (pointing in the direction of fire), ACTION (vigorously pumping his fist in the direction of the designated gun position).

(1) At the command or signal ACTION, the assistant gunner grasps the left trail leg near the center with his left hand. Springing to his feet and grasping the tripod head with his right hand, he lifts the tripod across the front of his body with the front leg up, and carries the tripod to the desired location. Upon arrival at the position, he places the trail leg pointing upward. Steadying the tripod with his left hand on the front leg, he loosens the front leg clamp with his right hand (Figure 5-18), positions the front leg with his left hand, and tightens the front leg clamp with his right hand. With his right hand on the tripod head, he slides his left hand down on the left trail leg and with a snapping motion, pulls the left leg (to the left), engaging the sleeve latch (Figure 5-19). He then aligns the tripod for direction, drops the mount to the ground, stamps the right and left trail shoes with his right or left foot, and assumes the prone position behind the mount.

Figure 5-18. Assistant gunner-(No.1) opening the tripod.

Figure 5-19. Emplacing the mount.

(2) The gunner and ammunition bearer move together. When the tripod is nearly mounted, the gunner and ammunition bearer spring to their feet. The gunner places both hands on the spade grips, the ammunition bearer grasp the ammunition box in his left hand and the gun carrying handle with this right hand. Both men lift the gun from the ground and move rapidly to the gun position. Then the ammunition bearer places the ammunition box on line and in front of the leader's ammunition box. The gunner, assisted by the ammunition bearer, inserts the pintle into the pintle bushing. The ammunition bearer stamps the front shoe into the ground with his left foot (Figure 5-20), releases the carrying handle, lowers his right hand, and lifts the gun pintle lock release cam. When the gun pintle is fully seated, he presses down the pintle lock release cam with his right hand, turns to his left, and returns to his original position.

Figure 5-20. Mounting the gun.

(3) The assistant gunner holds the left spade grip with his left hand. With his right hand, he lowers the T&E mechanism to the traversing bar, ensuring that the traversing handwheel is to the left and the lock lever is to the rear. He then assumes a semiprone position to the left of the gun with his feet to the rear and his head on line with the feedway. He unlatches and raises the cover of the ammunition box, removes the ammunition belt,

inserts the double-looped end into the feedway, and taps the feedway with his right hand to ensure it is closed.

(4) With his right hand palm-up, the gunner grasps the retracting slide handle, vigorously jerks it to the rear, and releases it. The gun is half-loaded (Figure 5-21). He then assumes a prone position directly behind the gun with his legs spread and heels down. His right hand lightly grasps the right spade grip with the thumb in position to depress the trigger. His left hand is on the elevating handwheel (palm down) with his thumb near the traversing slide lock lever. He rests on his left elbow with his head as close as possible to the rear sight (Figure 5-22, page 5-30). He then gives the assistant gunner an UP. The assistant gunner announces UP and extends his hand and arm into the air in the direction of the leader.

Figure 5-21. Gun mounted, gunner half-loading the gun.

Figure 5-22. Crew ready for action.

e. **Removal of the Gun From Action.** To take the gun out of action, the command is OUT OF ACTION.

(1) The gunner raises the cover with his left hand, the assistant gunner lifts the ammunition out of the feedway, replaces the ammunition in the box, and closes and latches the box. The gunner closes the cover, pulls the retracting slide handle to the rear with right hand (palm up) and releases it, presses the trigger with his right thumb, and loosens the traversing slide lock with his left hand. He rises to his feet, grasping both spade grips. At this time, the ammunition bearer will arrive at the gun position. With his left hand, the ammunition bearer grasps the ammunition box and places his right foot on the front leg shoe. With his right hand, he reaches down and lifts up the pintle lock release cam and grasps the carrying handle (Figure 5-23).

(2) The gunner and ammunition bearer lift the gun from the tripod, turn right, and carry the gun back to their original position. They set the gun down with the muzzle to the left and the retracting slide handle up. The gunner pulls the retracting slide handle to the rear and aligns the lug on the barrel locking spring with the 3/8-inch hole in the right side of the receiver. The ammunition bearer unscrews the barrel from the receiver, picks up the ammunition box and barrel, moves five paces to the rear, places the barrel to his right with the muzzle to the rear and the ammunition box to his left, and assumes the prone position.

Figure 5-23. Crew taking the gun out of action.

(3) The gunner rotates the bolt latch release lock to the right and releases the bolt latch. He checks his sights to ensure they are properly set.

(4) The assistant gunner assumes a kneeling position behind the tripod and releases the sleeve lock latch with his right hand. He rises to his feet, grasps the front leg with his left hand, and rotates the tripod to a vertical position on the trail legs. With his right hand, he loosens the front leg clamp, folds down the front leg, then tightens the clamp. With his right hand on the tripod head, he rotates the tripod on the right trail leg and releases the sleeve latch. Then he folds the left trail leg against the right with his left hand (Figure 5-24, page 5-32). Holding the tripod head with his right hand, trail legs with his left, he lifts the tripod across the front of his body with the front leg up. He turns to the right and returns to his original position. At this time, the crew leader picks up his ammunition box and faces the crew. The assistant gunner places the tripod on the ground, assumes a prone position to the right of the tripod, and announces UP.

Figure 5-24. Folding the trail legs.

f. **Duties of The Crew.** To hand-carry the gun and equipment, the command is SECURE EQUIPMENT, FOLLOW ME. At this command,–

(1) The crew leader carries his ammunition box in his right hand.

(2) The assistant gunner carries the tripod over either shoulder.

(3) The gunner carries the receiver on either shoulder with the spade grips to the front.

(4) The ammunition bearer carries his ammunition box in his left hand and the barrel in his right hand with the muzzle to the rear (or on his left shoulder with the muzzle to the front).

g. Relocation of Tripod-Mounted Gun. When the gun is mounted on the tripod, it can be moved for short distances by dragging or by a two- or three-man carry. (In the latter, the men should move in step to make carrying easier.)

(1) *Dragging.* The gun is dragged when there is limited cover, or when the situation requires the gun to be moved in this manner. The gunner and assistant gunner drag the mounted gun to the desired position (Figure 5-25).

(2) *Two-man carry* With the gunner on the right and assistant gunner on the left, each grasps the front leg with his forward hand and a trail leg with the other hand, just above the traversing bar (Figure 5-26).

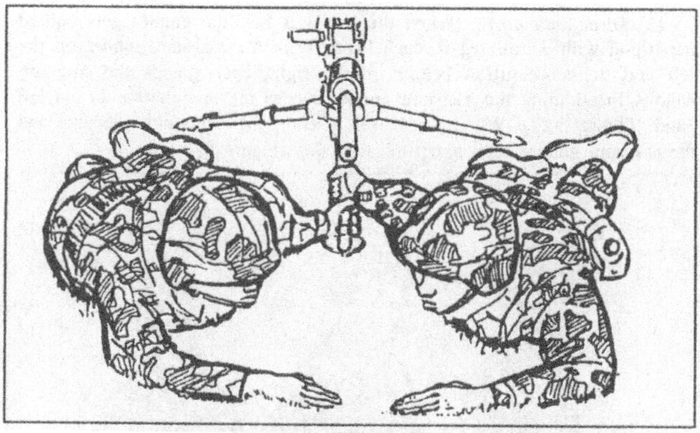

Figure 5-25. Dragging the gun into position.

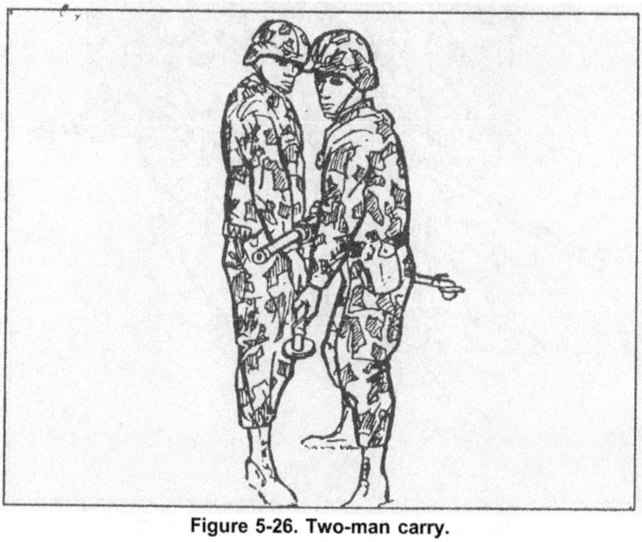

Figure 5-26. Two-man carry.

(3) *Three-man carry.* When the barrel is hot, the gunner gets behind the tripod with a trail leg in each hand. With the assistant gunner on the left and the ammunition bearer on the right, each grasps the carrying handle. In addition, the assistant gunner carries the ammunition in his left hand (Figure 5-27). When the barrel is cool, the ammunition bearer and the assistant gunner each grasp the front leg (Figure 5-28).

Figure 5-27. Three-man carry (hot barrel).

Figure 5-28. Three-man carry (cola barrel).

NOTE: Carrying the gun by the barrel may cause damage to the barrel support and the barrel extension.

h. **Movement of the Gun to Other Mounts.** With the mount prepared to receive the gun, the cradle of the mount is placed in a horizontal position. To move the gun to the mount, the gunner carries the right spade grip in his left hand and a box of ammunition in his right. The assistant gunner grasps the carrying handle with his left hand and a box of ammunition in his right hand. When they get to the mount, the gunner and assistant gunner place their ammunition boxes near the mount. The gunner removes the rear mounting (gun-locking) pin with his right hand. The assistant gunner removes the front mounting (gun-locking) pin with his right hand. They place the gun on the mount. The gunner aligns the holes in the rear mounting lugs of the receiver with the rear mounting bracket and inserts the rear mounting pin. The assistant gunner aligns the front mounting hole in the front of the receiver with the front mounting bracket and inserts the front mounting pin. (For use of the sideplate trigger with the M63 mount, see TM 9-1005-213-10.)

5-12. MACHINE GUN FUNDAMENTAL SKILLS TEST

The machine gun fundamental skills test should be held periodically to ensure that proficiency with the MG is maintained by all crewmen. It consists of 10 fundamental skills; all tasks are hands-on (Figure 5-29). The test should be given prior to range firing on a go/no-go basis.

TASK	GO/NO-GO	
1. Clear the MG.	✓	
2. Disassemble the MG.	✓	
3. Assemble the MG.	✓	
4. Do a function check.	✓	
5. Maintain the MG.	✓	
6. Set the headspace on the MG.	✓	
7. Set the timing on the MG.	✓	
8. Load the MG.	✓	
9. Perform immediate action on the MG.	✓	
10. Unload the MG.	✓	

Figure 5-29. The MG fundamental skills test.

Section III. BASIC MARKSMANSHIP

This phase of training is designed to allow the gunner to apply the fundamentals of marksmanship that he learned earlier. During this training, the gunner is introduced to the basic machine gun target (see Appendix C), procedures for both the 10-meter and field fire ranges, and how to acquire targets. He also fires practice and qualification.

5-13. CONCEPT OF ZEROING/TARGETING

The concept of zero is very simple; it is no more than the best way to adjust the sights of the weapon so the point of aim of the sights and the point of impact of the rounds are the same at any given range. A properly zeroed M2 gives the gunner the highest probability of hit for most combat targets with the least adjustment to the point of aim. There are three methods of zeroing/targeting used with the .50 caliber MG.

a. Ten-meter zero is the basic and the most common method of zeroing the M2 MG. Once zeroed on a 10-meter range using the standard machine gun target, the weapon is ready for field fire. As other weapons, the sight on the M2 must also be set at an initial start point (Figure 5-30). The initial sight setting for field zero is basically the same; except the range setting during field zero will depend on the range to the target, and it is always 1,000 yards for 10 meters.

(1) *Set elevation.* Raise the rear sight by lifting straight up until it snaps into the upright position. Adjust the range scale to 1,000 yards by rotating the elevation screw knob in the necessary direction. (Clockwise moves the scale up; counterclockwise moves the scale down.)

(2) *Set windage.* Rotate the windage knob until the zero index mark on the base rear sight is aligned with the index mark on the top of the receiver. (Clockwise moves the windage scale to the left; counterclockwise moves it to the right.)

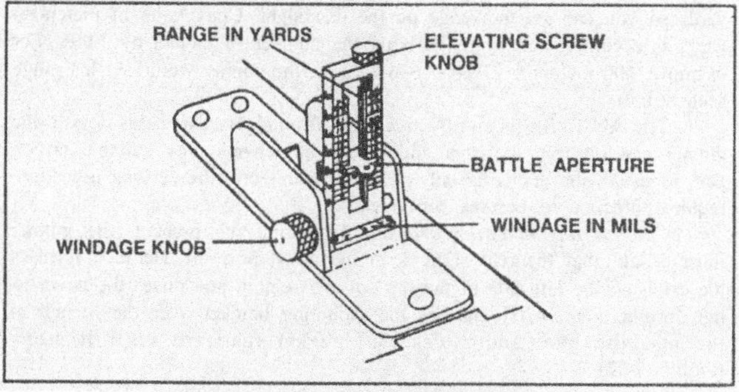

Figure 5-30. Rear sight setting.

(3) *Obtain proper sight picture.* Obtain the proper sight picture by looking through the zero aperture and centering the front sight blade in it. Once the sight alignment is obtained, place this combination on the center base of the selected target (Figure 5-31).

Figure 5-31. Proper sight picture.

b. Field zeroing/targeting is an expedient method of obtaining a zero when a 10-meter zero cannot be conducted. When preparing to field zero, make sure the M2 is mounted securely on the M3 tripod, make sure the T&E is working properly, and finally, know the distance to your zero target. The only difference in initial sight setting for field zero is range setting on the scale. The gunner must also remember that the range scale on the M2 is indicated in yards. Therefore, in order to get as close to the target as possible, you may have to convert the meters to the target into yards so you can set the range on the rear sight. Conversion of meters to yards is accomplished by multiplying the number of meters by 1.094. For example, 600 meters x 1.094 = 656.4 yards; the gunner would set his range scale at 650.

c. The AN/TVS 5 is an effective night fire aid. By using this device, the gunner can observe the area and detect and engage any suitable target. But, as usual, the device is only as good as its zero; the zeroing procedure requires practice to become proficient.

(1) *Mount the mounting bracket.* To mount the bracket, the gunner must ensure that the rear sight is in the down position. He then releases the catch on the left side of the top cover assembly and raises the cover to the upright position. He pushes the mounting bracket over the breech of the machine gun and slides the bracket rearward until it stops (Figure 5-32).

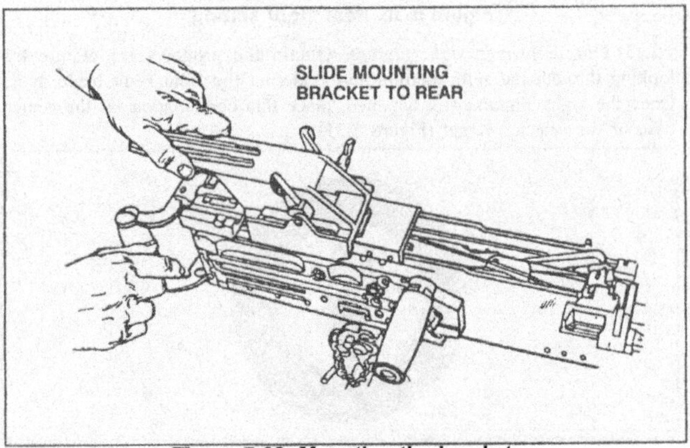

Figure 5-32. Mounting the bracket.

He pushes the three locking cams to secure the bracket (Figure 5-33) and closes the top cover assembly.

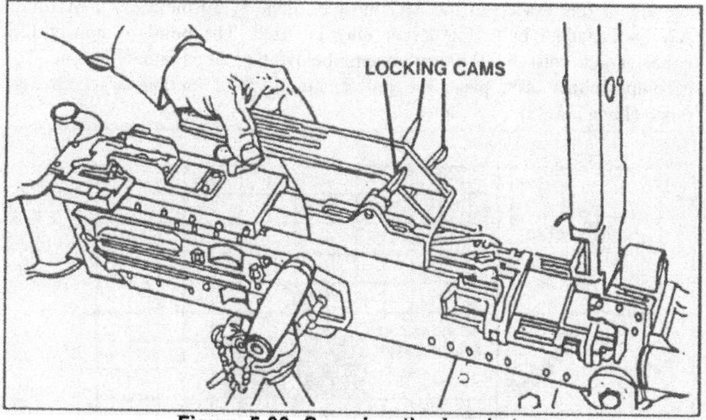

Figure 5-33. Securing the bracket.

(2) *Install the sight on the mount.* To install the sight, the gunner must align the scribe lines on the sight with the scribe lines on the bracket. He places the sight in the groove at the top of the bracket and tightens the lever screw to secure the sight to the bracket (Figure 5-34).

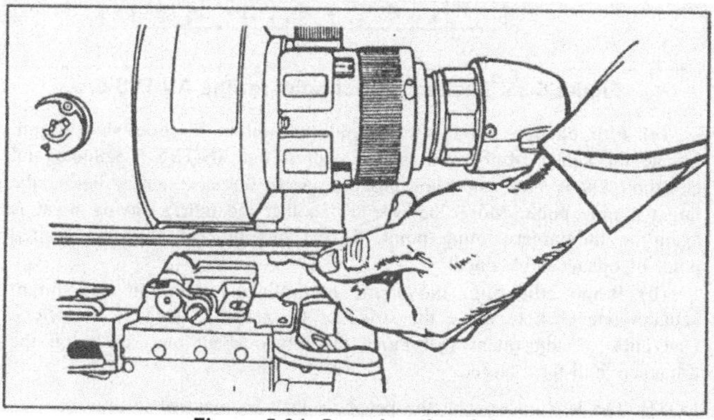

Figure 5-34. Securing the sight.

(3) *Zero the AN/TVS-5.* A gunner must zero the MG with its NVD on it before he can effectively engage targets with the AN/TVS-5. The zeroing of this device to the MG must be done at 50 meters. An M16A1, A2, or Canadian bull zero target may be used. The point of impact for either target must be 9.8 centimeters below the target aiming point. To accomplish this task, place the reticle aiming point on the target aiming point (Figure 5-35).

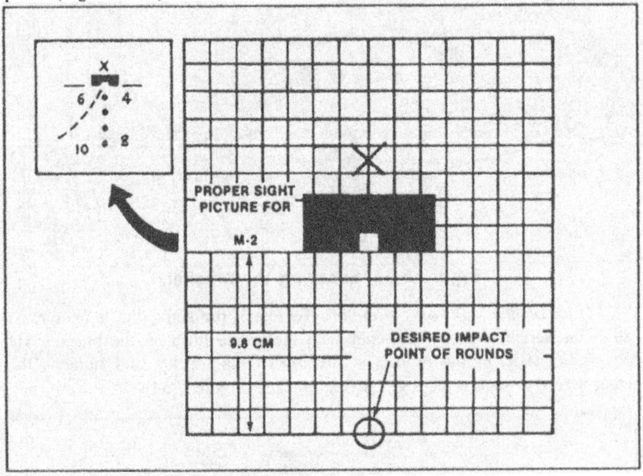

Figure 5-35. Desired impact point for the AN/TVS-5.

(a) Fire enough single-shot rounds to obtain a good shot group. Locate the center of the shot group. Adjust the AN/TVS-5 azimuth and elevation knobs until the shot group impacts 9.8 centimeters below the target aiming point. Move the weapon so that the reticle aiming point is again on the target aiming point. Repeat this process until the desired point of impact is obtained.

(b) When adjusting, move the azimuth or elevation adjustment actuator one click to move the strike of the round .5 inch at 50 meters. One click of adjustment will move the reticle about one square on the Canadian bull-type target.

NOTE: The lens cover with the peephole may be required to prevent scope washout from the muzzle flash.

(4) *Dismount the sight.* The gunner dismounts the sight from the bracket first by loosening the lever screws. Then he lifts the sight from the bracket. He must then dismount the bracket in the reverse mounting procedure.

5-14. TEN-METER FIRING EXERCISE

The purpose of 10-meter firing is to develop skills in the delivery of initial burst on target. When conducted properly, it will train gunners in the basic skills, such as 10-meter zero, controlled bursts, traverse, and traverse and search firing techniques. (See Appendix C for details on setup and conduct of firing.)

5-15. TRANSITION DAY FIRING EXERCISE

Transition day firing of the M2 machine gun will teach the gunner some techniques of fire that he may encounter in combat situations. The gunner will field zero his weapon and engage point and area targets from the tripod-mounted firing position. Within this training, the gunner will be required to apply all the fundamentals of gunnery learned in preparatory gunnery training and 10-meter firing. (See Appendix C for details on the setup and conduct of transition day fire.)

5-16. NBC FIRING

Since NBC plays an important part in our preparation for war on the modern battlefield, it is important that each soldier is prepared to accomplish the mission even if the area is contaminated and he must wear protective gear. (See Appendix C for details on setup and conduct of fire.)

5-17. NIGHT FIRE EXERCISE

The night fire exercise gives the soldiers the practical application of engaging targets using the AN/TVS-5 at night or during limited visibility. (See Appendix C for details on setup and conduct of fire.)

Section IV. ADVANCED GUNNERY

After firing 10-meter, day, NBC, and night, gunners need practice in applying what they have learned. They also need experience in engaging targets that depict realistic enemy formations. Advanced gunner exercises provide this experience in mounted, mounted NBC, and predetermined firing exercises.

5-18. OBJECTIVES

The objectives of the advanced gunnery phase are to prepare the gunners for combat. During this phase, training should be directed toward –

- Tracking and leading.
- Mounted target engagement techniques.
- Mounted NBC engagements.
- Predetermined fire techniques.

5-19. TRACKING AND LEADING EXERCISES

The gunner normally completes instruction in firing at stationary targets before he receives instruction in firing at moving targets. The technique of engaging a moving target differs from that of engaging a stationary target. The gun must be aimed ahead of the moving target a sufficient distance to cause the bullet and target to arrive at the aiming point at the same time. The distance is measured in target lengths. One target length as seen by the gunner is one lead. Leads are measured from the center of mass. The lead depends upon range, speed, and direction of movement of the target. To hit the target, the gunner aims at a point ahead of the target equal to the estimated number of leads, maintains the lead by tracking the target (manipulating the gun at the same angular speed as that of the target), and then fires. Fire is adjusted by observation of strike/tracer (Figure 5-36).

Figure 5-36. Lead technique.

a. **Tracking.** Tracking consists of maintaining correct alignment of the sights (with or without a lead) on a moving target by moving the gun at the same angular speed as that of the target.

(1) The gunner is required to aim at a prescribed point (center base to take advantage of the beaten zone) on the target and maintain that aim during uniform movement of the target. As instruction progresses, speeds used should differ for successive runs of the target. The speeds the 10-meter targets should move to represent speeds at various ranges are shown in Table 5-2.

TARGETS SPEEDS IN MILES PER HOUR	TARGET SPEEDS IN INCHES PER SECONDS CORRESPONDING TO–		
	300 M	500 M	900 M
7 1/2	6	4	2
15	12	8	4
30	24	15	9

Table 5-2. 10-meter moving target.

(2) The target handler must have practice in moving the target silhouette across the background at the varying speeds.

b. **Leading.** Mathematical computation or use of voluminous lead tables to obtain exact leads on a moving target are impractical in combat. The simple lead table shown in Table 5-3 gives the amount of lead necessary to hit a target moving at right angles (90 degrees) to direction to hit at speeds and ranges indicated.

SPEED IN MILES PER HOUR	RANGE OF TARGET		
	300 M	500 M	900 M
15	1/2 target length	1 target length	2 target lengths

Table 5-3. Lead table.

(1) The gunner must correct the lead as conditions change. If the target speed is 7 1/2 miles per hour, the amount of lead is half that shown in the table; at 30 miles per hour, double that shown, The angle at which the target is moving also alters the amount of lead taken. If the angle between line of fire and line of travel of the target is less than 45 degrees, use half the lead shown in the table.

(2) For targets moving directly toward the gun, the point of aim is placed on the center or the lower edge of the target, when possible. For targets moving directly away from the gun, the point of aim is placed on the center or upper edge of the target. Too much lead is better than too

little because the target runs into the fire; also, the observation of strike is easier. Intelligent use of the lead table includes immediate application of fire with estimated lead followed by necessary corrections based upon observation of strike/tracer.

c. **Tracking and Leading.** Combined tracking and leading exercises at 10 meters are used to gain proficiency in tracking the target. The gunner is required to repeat the tracking exercise while using a designated lead to simulate firing when his sights are properly aligned. As a further exercise in tracking and leading, the gunner may be required to track and lead moving targets at greater ranges. A vehicle can be run at right angles to the line of aim at ranges between 500 and 1,000 yards, and at varying speeds, averaging 15 miles per hour (Figure 5-37).

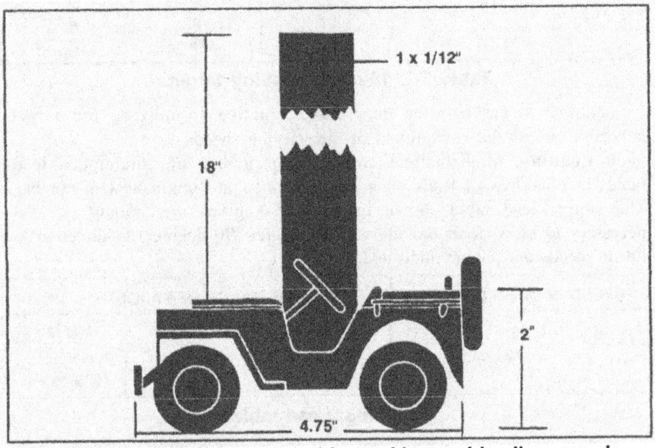

Figure 5-37. Aiming target used in tracking and leading exercise.

d. **Conducting the Lead Exercise (10-Meter).** The gunner is required to take a position at the gun, swing through the target's silhouette, and aim at a point ahead of the target equal to the prescribed lead from the center of mass. The gunner then directs the target handler to move the marking silhouette until the center of the target is at the point of aim. He repeats this procedure three times for each target lead announced. The target handler places his marking silhouette on the blank background, traces around it, and holds it in place for the gunner to aim, using the prescribed leads. Following the gunner's instructions, he moves the marking silhouette until the gunner commands HOLD. He then places a pencil dot at this point and returns the silhouette to the original position. This procedure is followed until the gunner has completed three tries for each target lead announced. The three pencil dots for each target should fit within a one-centimeter circle. The exercise should be conducted for varying left and right leads.

5-20. MOUNTED FIRING EXERCISE

The purpose of the mounted firing exercise is to teach the gunner techniques of firing the M2 MG from a mounted platform and to develop the gunner's ability to fire the M2 with it mounted on its primary carrier. (See Appendix C for details on the setup and conduct of fire.)

5-21. MOUNTED NBC FIRING EXERCISE

The probability of fighting mounted in a tactical environment that has been contaminated by NBC agents is very likely; therefore, gunners must be trained to engage targets while in a mounted NBC posture. (See Appendix C for details on the setup and conduct of fire.)

5-22. PREDETERMINED FIRING EXERCISE

The predetermined firing exercises are designed to instruct the gunners on preparing and using range cards during any visibility conditions. (See Appendix C for details on setup and conduct of fire.)

CHAPTER 6

COMBAT TECHNIQUES OF FIRE

Technique of fire is the method of delivering and controlling fire Each member of the machine gun crew must be trained in standard methods of applying fire, either as a crewmember or a gunner, and must perform his assigned task automatically and effectively. The simplest and most effective technique of delivering fire with the machine gun mounted on its ground or vehicular mount is to align the sights of the gun on the target and fire; this is called direct laying. At times, techniques of fire other than direct laying are more appropriate and effective. When delivering overhead fire or fire from position defilade, the gunner must use the appropriate technique.

Section I. FUNDAMENTALS

Before the machine gun can be employed to the best advantage using any firing technique, certain fundamentals must be understood and applied. These include:

- Characteristics of fire.
- Classes of fire.
- Fire control.
- Target types and methods of engagement.
- Overhead fire.
- Methods of laying the gun.
- Firing from position defilade.
- Final protective fires.

6-1. CHARACTERISTICS OF FIRE

The gunner's knowledge of his machine gun is not complete until he learns something of the action and effect of the projectiles when fired. This section discusses various characteristics of machine gun fire, including trajectory, cone of fire, and the beaten zone.

a. **Trajectory.** The trajectory is the curved path of the projectile in its flight from the muzzle of the weapon to its impact. The major factors that influence the trajectory are the velocity of the round, gravity, rotation of the round, and resistance of the air. The farther the round travels, the greater the curve of the trajectory. The highest point of the trajectory is called the maximum ordinate. This is a point approximately two-thirds of the distance from the gun to the target. The maximum ordinate increases as the range increases (Figure 6-1).

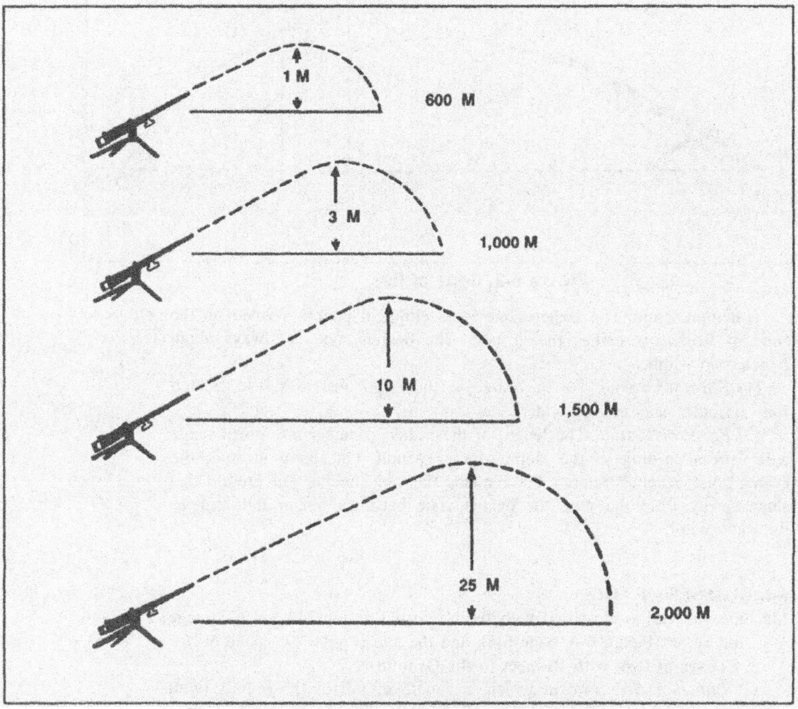

Figure 6-1. Maximum ordinates at key ranges.

b. **Cone of Fire.** When the weapon is fired automatically in bursts, all the rounds do not follow the same path. This is due to the vibrations of the gun and mount, variations in ammunition, and atmospheric conditions, which cause the rounds to follow a slightly different trajectory. This group of trajectories formed by a single burst is called the cone of fire (Figure 6-2).

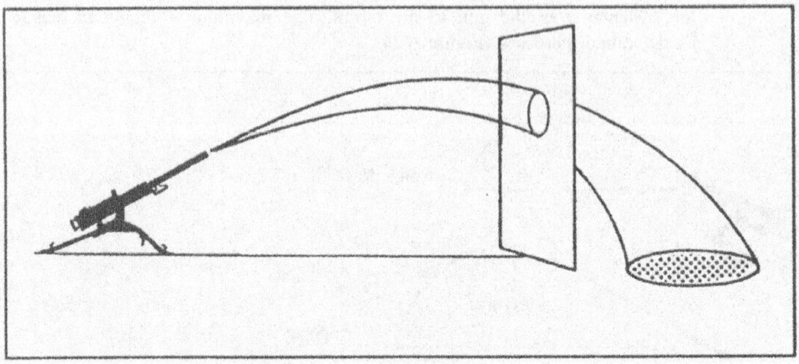

Figure 6-2. Cone of fire.

c. **Beaten Zone.** The beaten zone is an elliptical pattern formed by the cone of fire as it strikes the ground. The beaten zone is always about 2 meters in width.

(1) *Effect of range.* As the range to the target increases, the beaten zone becomes shorter and wider.

(2) *Effect of terrain.* The length of the beaten zone for any given range will vary according to the slope of the ground. On rising ground, the beaten zone becomes shorter but remains the same width. On ground that slopes away from the gun, the beaten zone becomes longer but remains the same width.

6-2. CLASSES OF FIRE

Machine gun fire is classified with respect to the ground (Figure 6-3, page 6-4), the target (Figure 6-4, page 6-5), and the gun (Figure 6-5, page 6-7).

a. **Classes of Fire with Respect to the Ground.**

(1) *Plunging fire.* Fire in which the angle of fall of the rounds (with reference to the slope of the ground) is such that the danger space is confined to the beaten zone, and the length of the beaten zone is materially shortened. Plunging fire is obtained when firing from high ground to low ground, when firing from low ground to high ground, and when firing at long ranges.

(2) *Grazing fire.* Grazing fire is fire in which the center of the cone of fire does not rise more than one meter above the ground. When firing over level or uniformly sloping terrain, the maximum extent of grazing fire obtainable is about 700 meters.

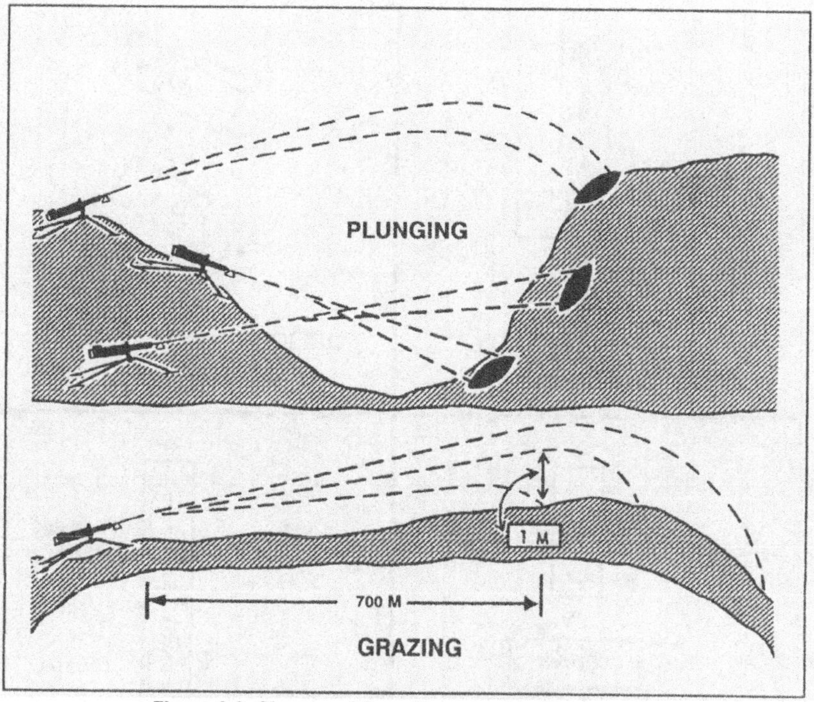

Figure 6-3. Classes of fire with respect to the ground.

b. **Classes of Fire with Respect to the Target.**

(1) *Frontal.* The long axis of the beaten zone is at a right angle to the long axis of the target.

(2) *Flanking.* Fire is delivered against the flank of a target.

(3) *Oblique.* The long axis of the beaten zone is at an angle (but not a right angle) to the long axis of the target.

(4) *Enfilade.* The long axis of the beaten zone coincides or nearly coincides with the long axis of the target. This class of fire is either frontal or flanking. It is the most desirable class of fire with respect to the target because it makes maximum use of the beaten zone.

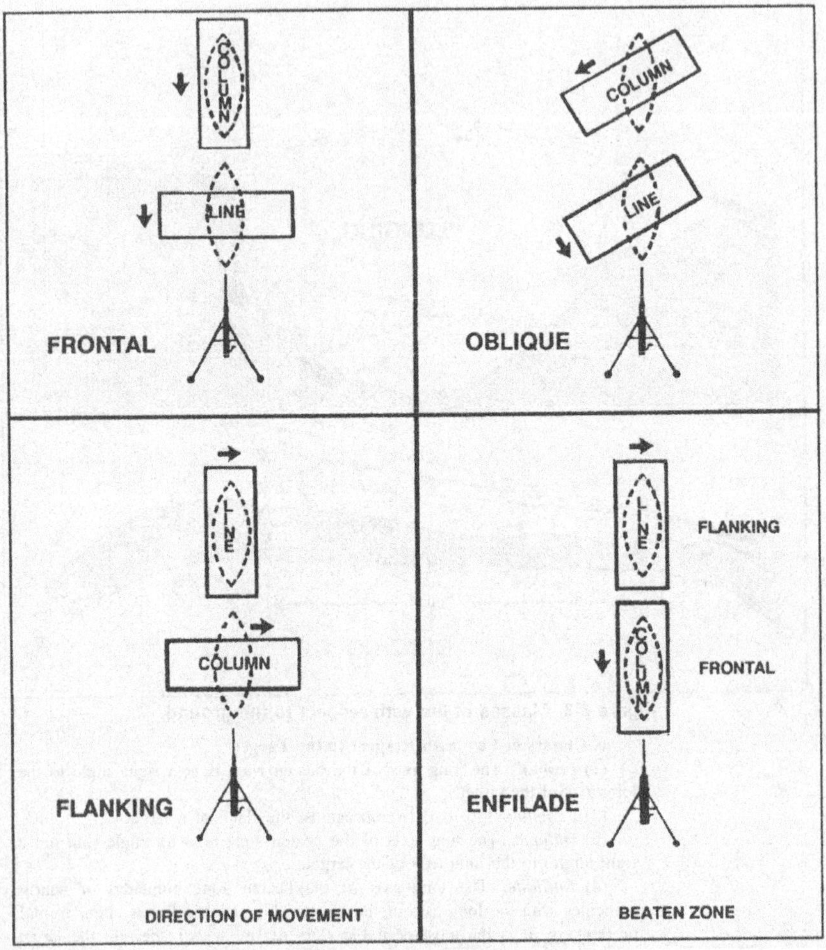

Figure 6-4. Classes of with respect to the target.

c. Classes of Fire with Respect to the Gun.

(1) *Fixed fire.* This is fire delivered on a point target with little or no manipulation needed. After the initial burst, the gunners will follow any change or movement of the target without command.

(2) *Traversing fire.* This is fire distributed against a wide target requiring successive changes in the direction of the gun. When engaging a wide target requiring traversing fire, the gunner should select successive aiming points throughout the target area. These aiming points should be close enough together to ensure adequate target coverage; however, they need not be so close as to be wasteful of ammunition by concentrating a heavy volume of fire in a small area. Two clicks on the traversing handwheel after each burst ensure coverage (2 clicks = 2 mils = constant width of beaten zone).

(3) *Searching fire.* This is fire delivered against a deep target or a target that has depth, requiring changes in elevation of the gun. The amount of elevation change depends upon the range and slope of the ground.

(4) *Traversing and searching fire.* This is fire delivered both in width and depth by changes in direction and elevation. It is employed against a target whose long axis is oblique to the direction of the fire.

(5) *Swinging traverse.* This is employed against targets that require major changes in direction but little or no change in elevation. Targets may be dense, wide, in close formations moving slowly toward or away from the gun, or vehicles or mounted troops moving across the front. The traversing slide lock lever is loosened enough to permit the gunner to swing the gun laterally.

(6) *Free gun.* This is when fire is delivered against moving targets that must be rapidly engaged with fast changes in both direction and elevation. Examples are aerial targets, vehicles, mounted troops, or infantry in relatively close formations moving rapidly toward or away from the gun position. To fire free gun on the M3 tripod mount, remove the T&E mechanism from the receiver and traversing bar and put it down. When firing swinging traverse and free gun, the weapon is normally fired at the cyclic rate of fire which is in excess of 450 rounds per minute. This consumes a lot of ammunition, and there is no beaten zone because each round seeks its own area of impact.

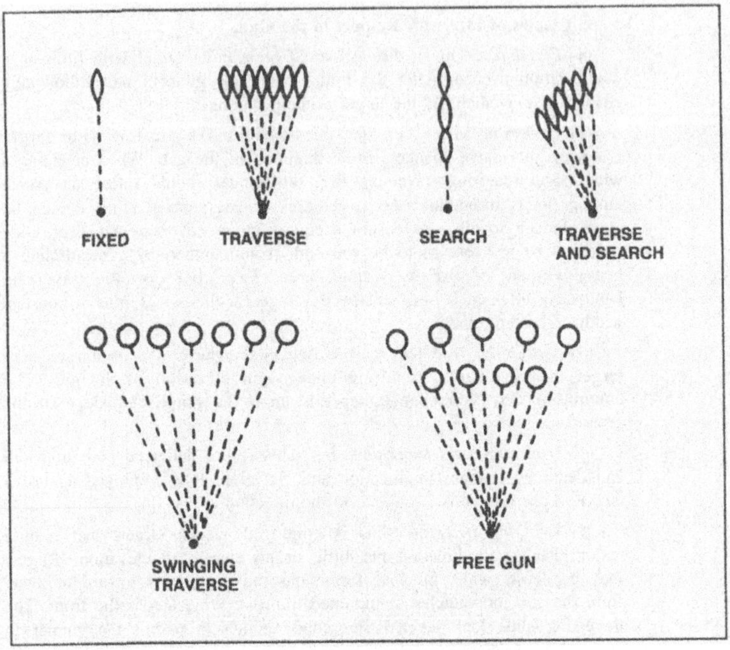

Figure 6-5. Classes of fire with respect to the gun.

Section II. FIRE CONTROL

Fire control of machine guns includes all operations connected with the preparation and actual application of effective fire on a target. It implies the-ability of the leader to-open fire at the instant he desires, to adjust the fire of the gun(s) on the target, to regulate the rate of fire, to shift from one target to another, and to cease firing. This ability to exercise proper fire control depends primarily on the discipline and the proper training of the crew. Failure to exercise fire control results in danger to friendly troops, loss of surprise, premature disclosure of positions, application of fire on unimportant targets, loss of time in adjusting fire, and waste of ammunition.

6-3. METHODS OF FIRE CONTROL

Of the several methods of controlling machine gun fire, the noise of battle will limit the use of some. Therefore, the leader must select the method or combination of methods that will best accomplish his purpose. The chain of fire control begins with the leader. He is responsible for both the technical and tactical employment of the gun and the training of the crew. He is responsible for passing on to the crewmembers all instructions and orders from his next higher leader regarding the situation and mission. He assigns sectors of fire and firing positions, designates targets to be engaged, adjusts fire, and ensures effective coverage of the targets.

a. **Oral.** Oral control is effective unless the leader is too far away from the gun crew(s), or the noise of battle makes it impossible for the gun crew(s) to hear him.

b. **Arm-and-Hand Signals.** This is an effective method when the gun crew(s) can see the leader. All crewmembers must understand the standard arm-and-hand signals used to control machine gun fire.

c. **Prearranged Signals.** These are either visual or sound signals such as pyrotechnics or blasts on a whistle. These signals should be included in the units' SOPs and must be clearly understood by all crewmembers.

d. **Personal Contact.** In many situations, the leader must move to individual crewmembers to issue orders. This method is used more than any other by small-unit leaders. The leader must use maximum cover and concealment to keep from disclosing the gun crew's position.

e. **Standing Operating Procedures.** SOPs cover actions the gun crews perform without command because they were developed and practiced during the training of the gun crews. Their application eliminates many commands and simplifies the leader's job of fire control.

6-4. TARGETS AND THEIR ENGAGEMENT

Targets presented to the machine gunners during combat will in most cases consist of enemy soldiers in various formations, which require distribution and concentration of fire. These targets have width and depth, and the application of machine gun fire is designed to completely cover the area in which the enemy is known or suspected to be. These targets may be easy to see or may be indistinct and difficult to locate.

a. When machine gun fire is under direct control of a leader, he designates the midpoint and flanks or ends of a target unless they are obvious to the gun crew(s). When a target other than a point target is engaged by two gunners, it is always divided. Each gunner applies his fire to that portion of the target corresponding to his position with relation to the other gun. Normally, each gunner engages one-half of the target;

however, gunners must be prepared to engage the entire target if necessary. Gunners continue to fire on the target until it is neutralized or until another signal is received from the leader.

b. The gunner's positions (including vehicular-mounted) should be numbered so each gunner will know which portion of a target he should engage. It should be emphasized that the positions are numbered – not the guns or gunners. To ensure that gunners react quickly and properly when they detect a target or when a target is designated by the leader, standard methods of applying fire to the various type targets are taught. These methods are the same for ground and vehicular-mounted guns. The following are the different types of targets and how they are engaged with the MG.

(1) Point targets are targets that require the use of a single aiming point. Enemy bunkers, weapon emplacements, vehicles, small groups of soldiers, and aerial targets such as helicopters or descending paratroopers are examples of point targets. A point target is engaged with fixed fire. If the target moves after the initial burst, the gun crew(s) keeps fire on the target by following its movement with the gun(s).

(2) Linear targets have sufficient width to require traversing fire and no more depth than can be effectively covered by the beaten zone. Linear targets are engaged with traversing fire.

(a) Two guns, normal division. The target is divided at the midpoint; the right gun engages the right half of the target, and the gun on the left engages the left half of the target. The point of initial lay and adjustment for both guns is at the midpoint of the target. After adjusting on the midpoint, the right gun traverses the right half of the target to include one aiming point beyond the last visible target flank and returns to the midpoint.

(b) Two guns, special division. If one portion of the target presents a greater threat than another, the target can be divided so fire is concentrated on that portion presenting the greatest threat. The special division of the target is accomplished by a subsequent fire command after firing begins. The gunners initially lay at the midpoint, regardless of the special division to be made, thus precluding confusion.

(c) One gun. A single gunner must engage the entire width of a linear target. The point of the initial lay and adjustment is on the midpoint, or that portion of the target presenting the greatest threat. The gunner traverses to either flank and then covers the remainder of the target (Figure 6-6, page 6-10).

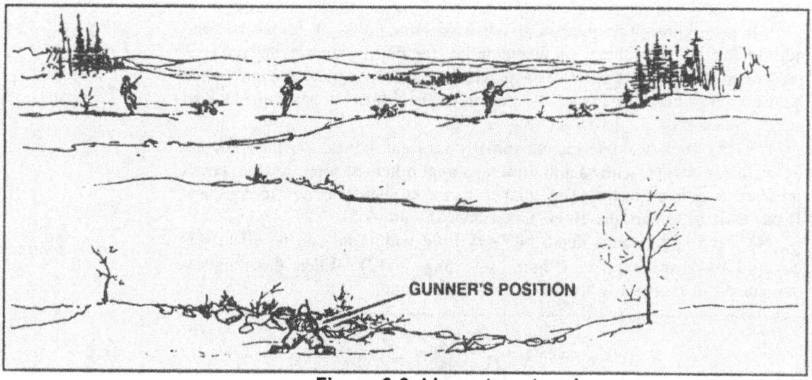

Figure 6-6. Linear target and one gun.

(3) Linear targets with depth are targets that have sufficient width to require traversing fire and depth which cannot be covered by the beaten zone. A combined change in direction and elevation (traversing and searching fire) is required to maintain effective fire on these targets (Figure 6-7). Linear targets with depth are engaged with traversing and searching fire. When range is announced, the range to the midpoint is given.

Figure 6-7. Linear target with depth.

(a) Two guns. The method of division, the point of initial lay and adjustment, and the extent of manipulation for both guns are the same as prescribed for linear targets. The gunners, however, apply enough search between each burst to ensure the center of the beaten zone is maintained at the center base of the target (Figure 6-8).

(b) One gun. A single gunner initially lays and adjusts on the midpoint of a linear target with depth unless some other portion of the target presents a greater threat. The gunner traverses and searches to the near flank, then he covers the entire target area (Figure 6-8).

(4) Deep targets have depth but very little width and can be effectively covered by searching fire (Figure 6-9, page 6-12). When the range is announced, it is given to the midpoint of the target.

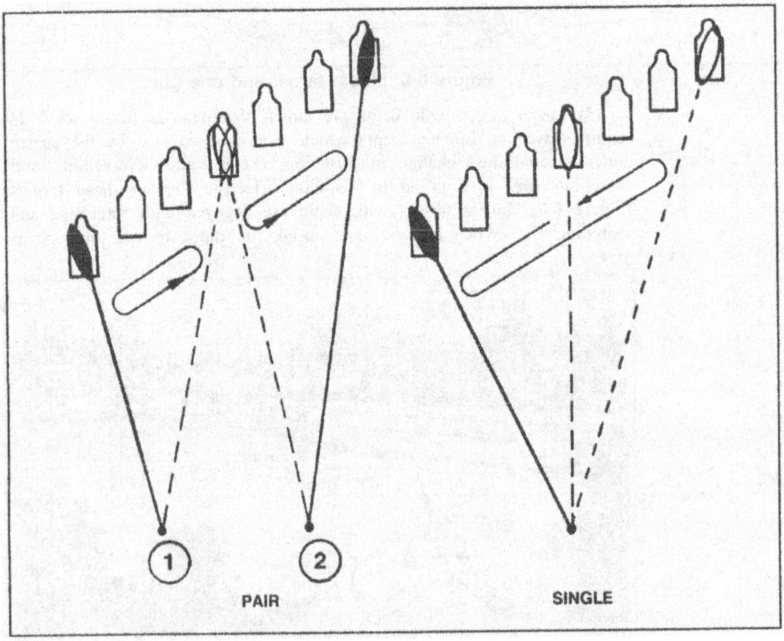

Figure 6-8. Engagement of linear targets with depth.

Figure 6-9. Deep target.

(a) Two guns. The point of initial lay of both guns is on the midpoint, which is also the point of division. Since enfilade fire is delivered, it is not necessary to adjust on the midpoint of the target because the long axis of the beaten zone will compensate for missing the midpoint. However, should the gunner's beaten zone be out of the lateral confines of the target, it will be necessary to adjust fires into the target area. After the initial bursts, the right gun searches to the near end of the target, and the left gun searches to the far end of the target. Both gunners then reverse their direction of search and return to the midpoint (Figure 6-10).

(b) One gun. A single gunner initially lays and fires at the midpoint of a deep target, unless another portion of the target presents a greater threat. The gunner immediately searches to the near end, then covers the entire target (Figure 6-10).

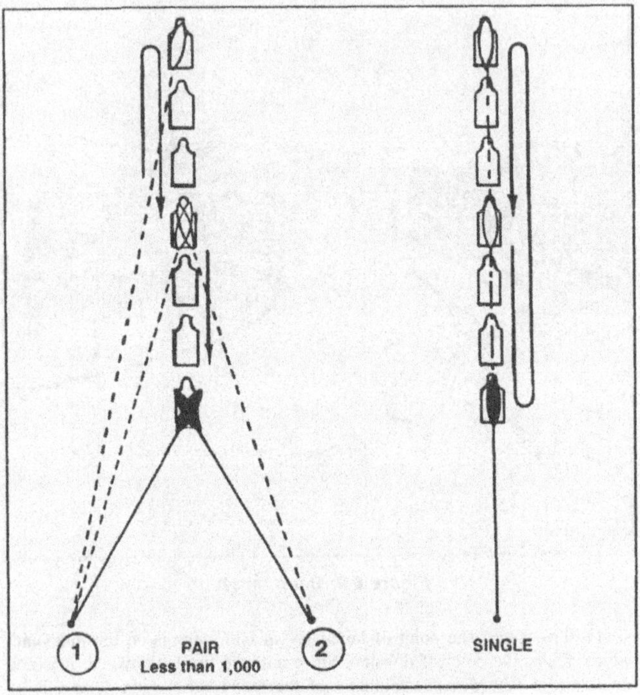

Figure 6-10. Engagement of deep targets.

(5) Area targets as discussed in this manual have considerable width and depth, and they require extensive traversing and searching fires. This type target exists when the enemy is known to be in a certain area, but his exact location is not known. A hilltop is a typical area target. The leader designates an area target by indicating to the gun crew(s) the width and depth of the target.

(a) Two guns. The target is divided at the center of mass; the right gun fires on the right half and the left gun fires on the left half. The point of initial lay and adjustment for both guns is on the center of mass. After adjusting on the center of mass, fire is distributed by determining the size of the beaten zones and applying direction and elevation changes that cause the most effective coverage of the target area. Both guns traverse and search their respective halves to the flanks, then return to the midpoint (Figure 6-11).

(b) One gun. A single gunner engages an area target by laying and adjusting on the center of mass, traversing and searching to either flank, then reversing the direction, traversing and searching to the other flank (Figure 6-11).

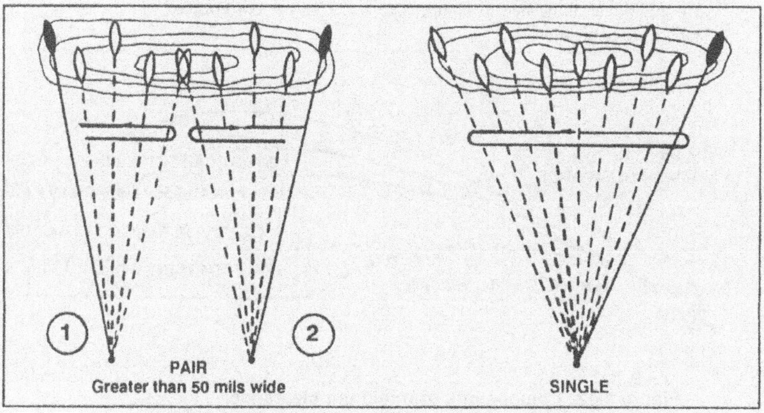

Figure 6-11. Engagement of area targets (objective).

NOTE: After the target is engaged in whatever formation it is in, the configuration of that target will change. The gunner must be trained to compensate for this change and still place effective fire on the target.

6-5. OVERHEAD FIRE

Overhead fire is fire delivered over the heads of friendly troops. A machine gun on a tripod is capable of delivering this type of fire because of the small and uniform dispersion of the cone of fire. In the attack, the use of overhead fire permits the machine gun to support the advance of rifle units. Sectors of fire allow the trainers to plan safe training while still incorporating the combat realities of overhead fires.

a. **Minimum Clearance.** The center of the cone of fire must clear the heads of the friendly troops by a prescribed distance (Figure 6-12). This distance, known as minimum clearance, is found by adding together the following elements:

- The height of a standing man, taken as 1.8 meters.
- Half the vertical dimension of the 100-percent cone of fire at the range to the troops.
- A margin of safety equal to the vertical distance which extends a 5-mil angle at the gun or 3 meters, whichever is greater.
- An additional allowance to compensate for a 15-percent error in range determination.

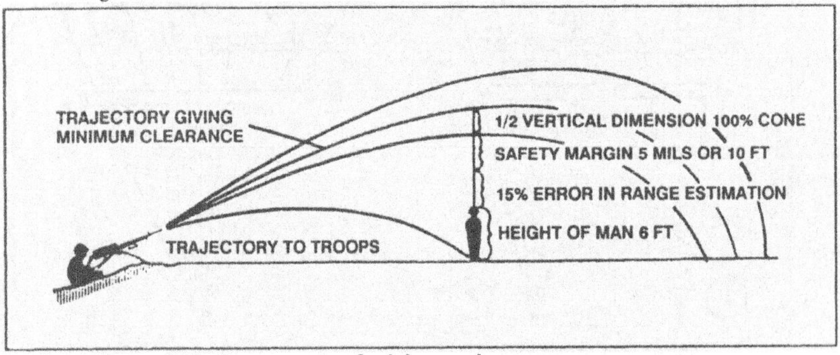

Figure 6-12. Components of minimum clearance.

b. **Safety Angles.** To obtain this minimum clearance, the gun is elevated so that the center of the cone of fire is raised from the feet-of the friendly troops to maintain clearance above their head. The amount of this elevation change is known as the safety angle. When the gun is fired from the tripod with the required safety angle, the center of impact determines the shortest range at which fire can be delivered over the heads of friendly troops. The range from the gun to the point of strike is called the corresponding range. When the ground is level or uniformly sloping between the gun and the target, the corresponding range for the safety angle used is obtained by converting the angle of elevation expressed in mils into range.

c. **Conditions.** Overhead fire is used only when the following conditions have been met:

(1) The safety limit has been determined and has been identified on the ground.

(2) The gun mount is firmly seated.

(3) Friendly troops have been notified, if at all possible, that fire is to be delivered over them.

(4) The rate of fire does not exceed 40 rounds per minute.

(5) The gun barrel is not badly worn. This condition is indicated by excessive muzzle blast.

d. **Uneven Terrain.** Level or uniformly sloping ground is seldom found in the field. This limits the use of firing tables and corresponding ranges in determining the limit of troop safety. In lieu of firing tables, a rule of thumb has been devised to give the gunner a simple method of checking for troop safety.

(1) The gunner's rule can be applied when the friendly troops are at least 350 meters in front of the gun position, and the range to the target is 850 meters or less (Figure 6-13).

(a) Lay the gun on the target with the correct sight setting to hit the target.

(b) Without disturbing the lay of the gun, set the rear sight at a range of 1,600 meters.

(c) Look through the sights and notice where the new line of aim strikes the ground. This is the limit of troop safety. When the feet of the friendly troops reach this point, fire must be lifted or shifted.

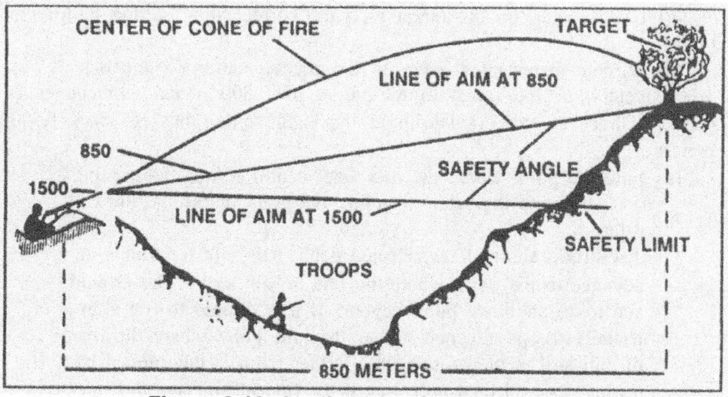

Figure 6-13. Application of gunner's rule.

(2) When the range to the target is greater than 850 meters, overhead fire should be delivered only in an emergency and then only out to a range in which either the tracers or the strike of the bullets can be seen by the gunner. In this situation the leader's rule applies (Figure 6-14).

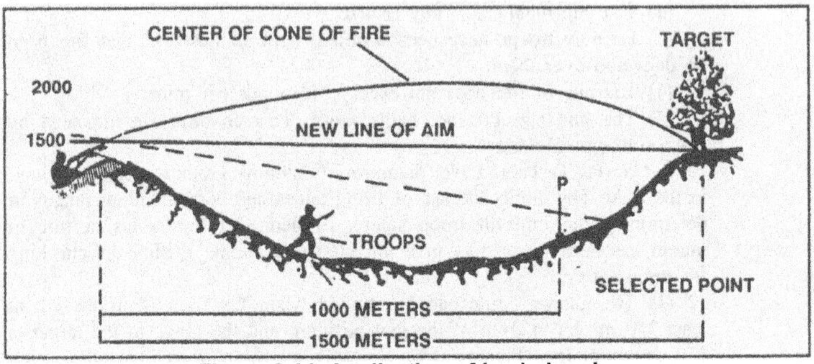

Figure 6-14. Application of leader's rule.

The platoon or section leader uses the leader's rule only when the target is greater than 850 meters. The rule is as follows:

(a) Select a point on the ground where it is believed friendly troops can advance with safety.

(b) Determine the range to this point by the most accurate means available.

(c) Lay the gun on the target with the correct sight setting to hit the target.

(d) Without disturbing the lay of the gun, set the rear sight to 1,600 meters, or the range to the target plus 500 meters, whichever is greater. Under no conditions should the sight setting be less than 1,500 meters.

(e) Note the point where the new line of aim strikes the ground.

- If it strikes at the selected point, that point marks the limit of safety.
- If it strikes short of the selected point, it is safe for troops to advance to the point where the line of aim strikes the ground and to an unknown point beyond. If it is desired to fire after friendly troops advance farther than the point where the line of aim strikes the ground, this farther point is determined by testing new selected points until the line of aim and the selected point coincide.

- If it clears the selected point, it is safe for the troops to advance to the selected point and to an unknown point beyond. If it is desired to have troops advance beyond the selected point, this farther point must be determined by testing new selected points until the line of aim and the selected point coincide. This point marks the line of safety.

e. **Precautions.** The following safety precautions *must* be observed in delivering overhead fire.

(1) Firmly emplace the tripod mount.

(2) Use depression stops to prevent the muzzle of the gun from accidentally being lowered below the safety limit.

(3) Do not deliver overhead fire through trees.

(4) Inform commanders of friendly troops when fire is to be delivered over their heads.

(5) Ensure that all members of the gun crew(s) are aware of the safety limit.

(6) Do not deliver overhead fire if the range from the gun to the target is less than 350 meters or more than 850 meters.

(7) Do not use a barrel that has excessive muzzle blast or is otherwise determined to be badly worn.

(8) Do not lay machine guns so their fire will cross at any point over the heads of friendly troops.

6-6. DEFILADE POSITIONS

To achieve maximum effectiveness, the machine gun must be employed using the technique of direct lay; however, at times it may be desirable to employ guns from defilade positions.

a. **Full Defilade.** A machine gun is in defilade when the gun and its crew are hidden from enemy ground observation by a land mass such as the crest of a hill. The position may be on the reverse side of the crest or the forward slope of the next higher ground (Figure 6-15). The gun must fire up and over the hill. Fire must be observed and adjusted by a crewmember who can observe the target from a position on a flank or to the rear of the gun (on higher ground). A defilade position allows little opportunity to engage new targets.

b. **Partial Defilade.** A machine gun is in partial defilade when a mask (usually the crest of a hill) provides the gun and gunner with some protection from enemy direct fire, but the gunner is able to engage the target using direct laying techniques. The gun is far enough up the slope so that the gunner can see the target through the sights but the lower portion of his body and lower portion of the gun are protected by the mask. Partial defilade positions are desirable when a fire mission cannot be accomplished from a defilade position.

c. **Advantages.** The gun and crew have cover and concealment from direct fire weapons. The crew has some freedom of movement in the vicinity of the gun position, and control and supply are facilitated. The characteristic smoke and flash of the gun are partially concealed from observation.

d. **Disadvantages.** Rapidly moving ground targets are not easily engaged because adjustment of fire must be made through an observer. Targets close to the mask usually cannot be engaged, and it is difficult to secure grazing fire for a final protective line.e.

e. **Position Selection.** The fire unit leader selects the location of the gun position. To select a position in partial defilade, he moves up the reverse side of the slope until he has the target in view above the mask when sighting at the height of the gunner's eye. To select a position in maximum defilade, he estimates the lowest point below the mask at which the gun can still engage the target without danger of hitting the mask.

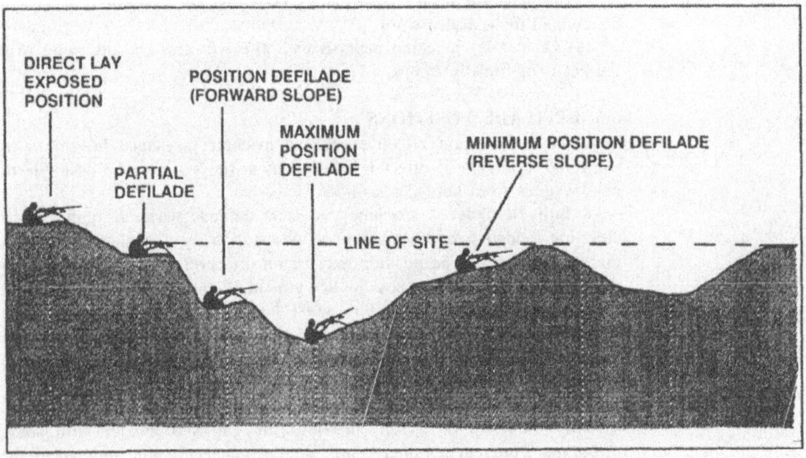

Figure 6-15. Minimum and maximum position defilade, partial defilade, and direct lay areas.

6-7. METHODS OF LAYING THE GUN FOR DEFILADE FIRING

The essential elements in engagement of a target from defilade position are direction, elevation, mask clearance, and adjustment of fire.

a. **Direction.** An observer places himself on the gun-target line in a position from which he can see the gun and the target. He aligns the gun approximately by having the gunner shift the mount. The gunner then loosens the traversing slide lock lever and, as directed by the observer, moves the gun right or left until it is aligned on the target; he then clamps it in that position. A prominent landmark, visible to the gunner through his sights, is selected as an aiming point. An aiming point on the gun-target line and at an equal or greater range than the target is desirable. However, an aiming point on the mask may be used. If the aiming point is on the gun-target line, the gun is laid on the aiming point and is thereby aligned for direction. If the aiming point is not on the gun-target line, the deflection is measured by binoculars or compass. This measured deflection is laid off with the gun.

b. **Elevation.** An aiming point visible from the gun position is selected (preferably a point at a greater range and at a higher elevation than the target) and the range to the target is determined. The leader, using binoculars, measures the vertical angle in roils from the aiming point to the base of the target. He then lays the gun on the aiming point with the sight set to hit the target. He directs the gunner to manipulate the gun through the number of mils measured. For example, in Figure 6-16, the range to the target is 1,300 meters. The angle read with the binoculars from the aiming point down to the base of the target is 12 mils. The sight is set at 1,300 meters, the gun laid on the aiming point, and the muzzle depressed 12 mils. If the aiming point is off the gun-target line, deflection in mils may be taken with the rear sight windage screw knob if it is not over 5 mils; otherwise, the deflection must be taken up on the traversing handwheel.

c. **Mask Clearance.** After the gun has been laid, determine if the entire cone of fire will clear the mask.

(1) *Visual method.* When the range to the mask is not more than 450 meters, mask clearance exists when the axis of the bore is elevated 7 mils or more above the gun-mask line. Mask clearance can be checked after the gun has been laid on the target by depressing the muzzle of the gun 2 mils and sighting along the bottom of the receiver and the barrel support. If this line of sight clears the mask, the clearance exists. Elevate 2 mils before firing.

(2) *Firing tables method.* Determine the range to the mask and obtain the corresponding angle of elevation for mask clearance from the firing

tables. The range corresponding to the angle of elevation is set on the gun sight. If the line of aim through the sight clears the mask, the clearance exists.

d. **Adjustment of Fire.** Under field conditions, even the most practical methods of laying the gun on the target quickly do not always result in the initial burst being on the target. For this reason, adjustment of fire on the target is essential. Creeping fire should be avoided. (See paragraph 6-10 for details on adjustment of fire.)

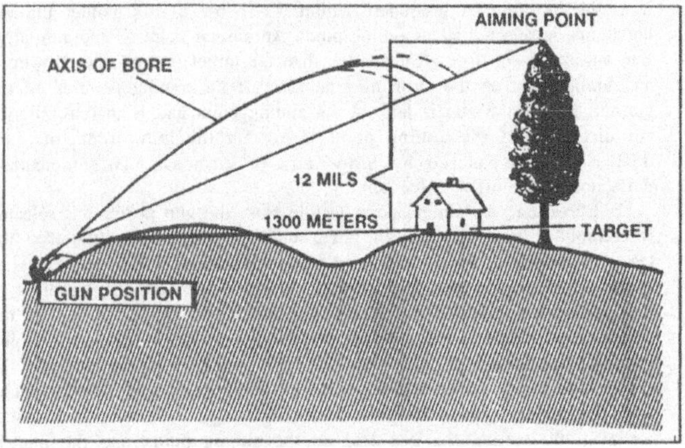

Figure 6-16. Aiming point method.

6-8. FINAL PROTECTIVE FIRES

These are types of fire that are placed on a predetermined line along which grazing fire is placed to stop an enemy assault. This fire is fixed in direction and elevation; however, a few roils of search are employed during firing to compensate for irregularities in the terrain. FPLs are always laid in using the extreme left or right of the tripod, causing the T&E to move to the extreme left or right on the traversing bar. The FPFs can be delivered in any visibility conditions. When terrain permits, final protective lines are assigned to machine guns along the forward line of troops as a part of the FPFs of the defending unit. The signal used to call for FPFs is normally prescribed in the company operation order. The authority to call for these fires may be delegated to the platoon leader of a forward rifle platoon. Final protective fires are ceased on order.

a. **Signals.** Arm-and-hand signals, voice commands, or pyrotechnic devices may be used in calling for these fires.

b. **Rates of Fire.** When firing FPFs, the rapid rate of fire is used unless it is obvious that a different rate is necessary to accomplish the mission. When engaging other preselected target areas, the rapid rate of fire is used until commanded to cease fire.

6-9. APPLICATION OF FIRE

To be effective, machine gun fire must be distributed over the entire target area. Improper distribution of fire results in gaps which allow the enemy to escape or use weapons against friendly positions without effective opposition.

a. The method of applying fire to a target is generally the same for either a single gun or a pair of guns. Direct laying is pointing the gun for direction and elevation so that the sights are aligned directly on the target. Fire is delivered in width, depth, or in a combination of the two. To distribute fire properly, the gunners must know where to aim, how to adjust their fire, and the direction to manipulate the gun. The gunner must aim, fire, and adjust on a certain point of the target. Binoculars may be used by the leader to facilitate fire adjustment.

b. The gunner ensures throughout his firing that the center of the beaten zone is maintained at the center base of the target for maximum effect from each burst of fire. When this is done, projectiles in the upper half of the cone of fire will pass through the target if it has height, and the projectiles in the lower half of the beaten zone may ricochet into the target (Figure 6-17).

c. The gunner must move his beaten zone in a certain direction over the target. The direction depends on the type of target and whether the target is engaged with a pair of guns or a single gun. When engaging targets other than point targets with a pair of guns, the targets are divided so that fire is evenly distributed throughout the target area. Fire delivered on point targets or a specific area of other target configurations is called concentrated fire.

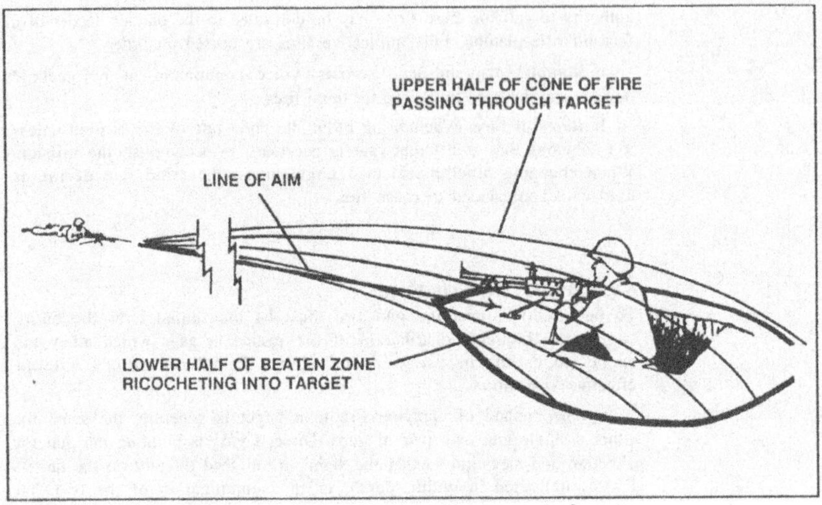

Figure 6-17. Line of aim and placement of center of beaten zone on target.

6-10. FIRE ADJUSTMENT

Machine gun fire is adjusted by observing the strike of the rounds, observing the flight of tracers, frequently re-laying the gun, or by a combination of these. Adjustment by observation of fire is the most important element of fire control if it is bold, aggressive, rapid, and continuous throughout the action.

a. The gunner is trained to observe and adjust his gun's fire without command. He is trained to anticipate the action of the enemy after the initial burst, and is prepared to shift his fire to cover any change in formation or movement of his target. If the gunner fails to accomplish this, the fire unit leader must promptly correct him by announcing or signaling subsequent fire commands. This responsibility to adjust fire continues through the chain of commands.

b. When subsequent fire commands are given, the gunner makes the required corrections and continues to engage the target without any further command to fire. If the gun is fired on the tripod mount, subsequent commands are given to make changes in direction, elevation, and the rate of fire. These changes are given orally as SHIFT RIGHT, SHIFT LEFT, ADD, or DROP. (For arm-and-hand signals see FM 21-60.) When making these announced changes,mils may be used to indicate the amount of desired shift; for example, SHIFT RIGHT 5 or SHIFT LEFT 7. When making changes in elevation, roils are not used, as it is normally difficult to determine just how high or low the center of the beaten zone is striking the ground in relation to the target.

(1) *Observation.* When firing on the 10-meter range, the strike of the bullets is visible on the target. When firing at greater distances, the strike of the bullets on the ground may cause dust to rise, which is visible to the gunner; however, during wet weather the strike cannot always be seen. In this event, the tracers will allow the gunner or crew leader to note the strike of the burst in relation to the target.

(2) *Adjustment.* Using the mil relation, one click of the traversing handwheel or elevating handwheel moves the strike of the bullet 1/2 inch on the target at a range of 10 meters.

(a) When firing on the 10-meter range, adjust by moving the shot group a required number of centimeters vertically or horizontally until the center of the group is on the aiming paster. Should the gunner's initial burst strike the target 2 centimeters to the left and 3 centimeters below the aiming paster, he adjusts his fire by traversing right 4 clicks and elevating 6 clicks before firing again.

(b) When firing on field targets, adjust by moving the burst into the target. One click (roil) on the traversing handwheel will move the strike 1/2 meter at 500 meters or 1 meter at 1,000 meters; however, the distance 1 click (roil) in the elevating handwheel will move the strike depends on the range to the target and the slope of the ground. The gunner determines the number of roils necessary to move the center of the strike into the target, and he manipulates the gun the required number of roils. This does not require the use of sights. For example, should the gunner fire on a target at 500 meters and observe the strike 10 meters to the right of the target and short about 50 meters, he would traverse the gun to the left 20 clicks (roils) and add one or more clicks (roils), depending on the slope of the ground.

(c) The gunner may use the adjusted aiming point method to adjust the fire. In this method the gunner must use his sights. He selects an aiming point that will place the next burst on the target. For example, should the gunner fire on a target at 500 meters and estimate that the strike is 20 meters short and 10 meters to the right of the target, he would rapidly select an aiming point approximately 20 meters beyond the target and 10 meters to the left of the target, lay on that aiming point, and fire.

6-11. ANTIAIRCRAFT GUNNERY

The MG can provide units with a self-defense capability against hostile low-flying, low-performance aircraft. These guns are employed in the air defense role as part of the unit's local defense. The MGs are not components of an integrated and coordinated air defense system. Unless otherwise directed, hostile aircraft within range of the gun (about 800 meters maximum effective range) should be engaged. The decision will be made by the commander. Typical targets are surveillance, reconnaissance, and liaison aircraft; troop carriers; helicopters; and drones.

a. **Engagement and Employment. The** mission is to impose maximum attrition upon the attacking enemy, such as low-flying, low-performance aircraft. Employment of MGs used for air defense is guided by the following defense design factors:

- Defense design should produce an equally balanced defense that is effective in all directions, unless a forced route of approach exists.
- Machine guns should be sited so that the maximum number of targets can be engaged, continuous fire can be delivered, and the most likely routes of approach are covered.
- Machine guns used to defend march columns should be interspersed in the convoy, with emphasis on the lead and rear elements (Figure 6-18).

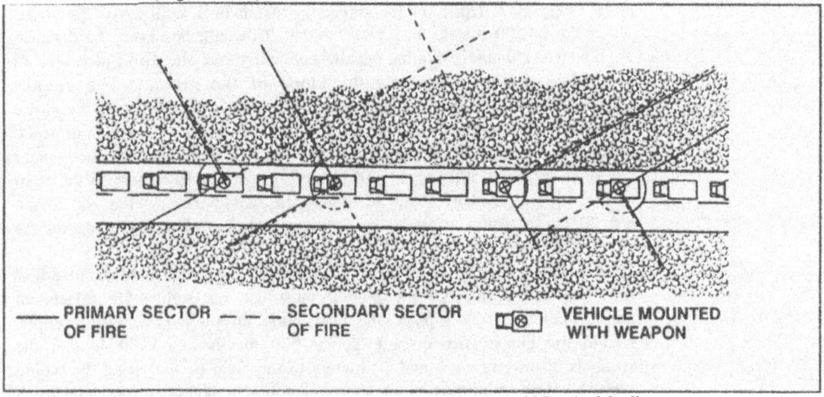

Figure 6-18. March column with four MGs (added).

b. **Target Selection and Engagement Control. These** actions depend upon visual means. The sites selected for the guns must provide maximum observation and unobstructed sectors of fire. Units furnished MGs in sufficient numbers should site them within mutual support distances of 90 to 360 meters. Each gun is assigned a primary and secondary sector of fire. Weapon crews maintain constant vigilance in their primary sectors of fire, regardless of the sector in which the guns are actually engaged.

Section III. LIMITED VISIBILITY CONDITIONS

The machine gun is provided with a stable tripod mount, M3, and a traversing and elevating mechanism. By manipulating the T&E mechanism, gun crews can record target data during good visibility and engage the same targets in poor visibility. This section provides guidance on machine gun firing techniques and terms used during limited visibility, which includes darkness, smoke, fog, rain, or snow.

6-12. DIFFICULTIES

Crewmembers encounter difficulties while defending during limited visibility, which preclude the use of many of the daylight techniques of engaging targets.

a. During limited visibility, the machine gunner's sector of responsibility cannot be observed in depth; therefore, targets are difficult or impossible to detect.

b. Visibility may be so limited that the leader cannot control the fires of his guns by selecting and directing fire on targets as he would during good visibility. Oral commands are not dependable, arm-and-hand signals may not be seen, and personal contact with the gunner is difficult.

c. At night, machine gunners have a tendency to fire indiscriminately at noises and suspected enemy locations.

To overcome these difficulties, special techniques must be developed for engaging targets and delivering preplanned fires by the use of range cards. (See Appendix E.)

6-13. TERMINOLOGY

The following terms must be familiar to MG crews for them to complete their missions in poor visibility.

a. **Sector of Fire.** An area (to be covered by fire) assigned to an individual or unit. Machine guns are normally assigned two sectors of fire, a primary and a secondary sector.

b. **Final Protective Line.** A predetermined line along which grazing fire is placed to stop an enemy assault. The FPL is fixed as to direction and elevation; however, a few roils of search are employed during firing to compensate for irregularities in the terrain. The FPL can be delivered regardless of visibility conditions. The FPL is always the inner limit of the primary sector, which is assigned close to the forward line of troops area. When terrain permits, FPLs are assigned to machine guns along the **FLOT** as a part of the final protective fires of the defending unit.

c. **Principal Direction of Fire.** A PDF is a priority direction of fire that marks a specific area assigned to a weapon. This area may extend from the gun position to the maximum effective range of the weapon and therefore is not fixed for elevation. Visible targets appearing in the PDF take priority over targets that may appear elsewhere in the sector. A PDF may be assigned to cover an area that provides good fields of fire, is a likely avenue of foot approach, or mutually supports an adjacent unit.

d. **Sector of Graze.** A wedge-shaped area formed by assigned sector limits that afford grazing fire (one meter high, maximum) from the muzzle of the weapon to the first major break in the terrain. The sector of graze is fired using swinging traverse in the primary sector of fire. It can be fired in the secondary sector in conjunction with field expedients by freeing the T&E mechanism and using the mount as a pivot. A sector of graze can be delivered regardless of the condition of visibility.

e. **Area of Graze.** This is an area, other than the sector of graze, within a sector of fire that is covered by grazing fire. Grazing fire need not be continuous from the muzzle of the weapon to the area over which grazing fire is desired.

6-14. TARGET ENGAGEMENT

A gunner's ability to detect and identify targets during limited visibility will vary, depending upon the amount of natural and artificial light and the types and numbers of sensors used. All tracer ammunition allows a gunner to more effectively engage visible targets during limited visibility; it should be used when possible. Gunners must be trained to fire low initially and adjust up when engaging targets during limited visibility. This helps them overcome the tendency to fire high during these conditions. The types of point targets machine gunners will be concerned with during limited visibility, particularly at night, are enemy automatic weapons and assaulting enemy personnel.

a. Point targets such as automatic weapons may be identified during limited visibility by their muzzle flashes. To effectively engage these targets, fire should be delivered in a heavy volume and adjusted by observing the tracer stream.

b. During the final stage of an enemy assault, machine guns normally fire at personnel on an FPL; they may be assigned a PDF. Both are considered as final protective fires and should be planned for and coordinated as such. If individual enemy soldiers are observed in the proximity of the gun position, they must be neutralized by someone other than the machine gunner (by the other crewmembers or by security forces of the supported unit). The FPFs are fired according to the order or SOP, and the machine gunner is not allowed to stop firing them except in accordance with those orders or SOP.

6-15. FIRE CONTROL

During limited visibility, the leader cannot direct the fires of his guns as effectively as with good visibility. Consequently, initiative is required of the gunners. When targets within their sectors become visible to gunners, they must engage such targets without command and continue to fire until the targets have been neutralized. Gun crews engage targets only when they can identify them, unless ordered to do otherwise. For example, if one gunner detects a target and engages it, the other gunners will observe the area in which fire is being placed. They will add their fire only if they can identify the target or are ordered to place fire in the area.

6-16. PREPLANNED FIRES

In addition to engaging appropriate visible targets, the gunner must be able to deliver preplanned fires during limited visibility. These fires are used to cover target areas of tactical significance (such as routes, avenues of approach, anticipated enemy supporting weapons positions, and probable enemy assault positions) and to establish sectors of graze and final protective lines. For maximum effect in all preplanned target areas, grazing fire should be obtained when possible.

a. **Obtaining Maximum Extent of Grazing Fire Over Level or Uniformly Sloping Terrain.** The machine gunner sets the rear sights at 700 meters; selects a point on the ground, which he determines to be at a range of about 700 meters; and lays, fires, and adjusts on this point. If the gunner cannot obtain 700 meters of grazing fire because of a major break in the ground at a range of less than 700 meters, he places the range to the break on his sight and lays, fires, and adjusts at that point.

b. **Determining the Extent of Grazing Fire on the Final Protective Line.** The extent of grazing fire on the FPL is determined using the techniques described above. Any intermediate breaks in the terrain along this line that cannot be covered by grazing fire from a gun firing along the line is considered dead space.

c. **Determining the Extent of Grazing Fire in the Sector of Graze.** The ranges to the extent of grazing fire in a sector of graze are determined by observing the terrain and by observing the tracer stream from behind or from a flank of the gun position. Normally, the extent of grazing fire within this area will be much less than on an FPL and will form an irregular pattern.

d. **Determining the Amount of Grazing Fire in an Area of Graze.** The same procedures used in paragraph 6-8a are used in determining the extent of grazing fire in an area of graze. The ranges to areas of grazing fire are determined by observing the flight of tracer ammunition from behind or from the flank of the gun position. The gunner determines the lateral extent of areas of graze by selecting and engaging successive aiming points in the area believed to afford grazing fire, using the same range setting as when determining the range to the extent of grazing fire.

6-17. NBC CONSIDERATIONS

During this phase of training, the gunner is introduced to firing the machine gun while in MOPP, keeping in mind that engagement of some targets in MOPP is a qualification requirement. Firing weapons is only part of the overall NBC training. Soldiers must first be familiar with the NBC equipment, its use, and proper wear before they progress to learning the techniques of MOPP firing. Although there is no different technique required to fire the MG, there are certain fundamentals that may be slightly impaired.

a. **Immediate Action.** Under normal conditions, a gunner should be able to clear a stoppage in two to four seconds; however, under full MOPP, this may take a few seconds longer. Dry-fire practice under these conditions is necessary to reduce time and streamline actions. When practicing with the hood/mask and gloves, care must be taken not to snag or damage the gloves or dislodge the hood/mask during movement. Trainers should apply immediate action to a variety of stoppages during dry fire until the gunners are able to instinctively do it without compromising their NBC environment.

b. **Target Detection.** Techniques and principles of target detection and target acquisition still remain valid during NBC conditions, but considerations must be made for limiting factors imposed by MOPP equipment. For example, vision is limited to what can be seen through the mask's lens/faceplate. Peripheral vision is severely restricted. The lens/faceplate may be scratched or partly fogged, thus further restricting vision. Gunners requiring corrective lenses must be issued insert lenses before training. Scanning movements may be restricted by the hood/mask. Any of these factors could adversely affect the gunner's ability to quickly and accurately detect targets. Extra skill practice should be conducted.

c. **Efficient Performance.** The trainer must keep in mind that although movements are slowed, tasks take longer, and function checks, loading, unloading, and cleaning are affected by MOPP, it is a must that the gunner avoid damaging MOPP gear and risk possible exposure to lethal agents. Because of the great difference between no MOPP and MOPP4, gunners must be trained in all aspects of operation and maintenance of the weapon while practicing at the highest MOPP level. Only through repeated training and practice can the soldier be expected to perform all tasks efficiently.

CHAPTER 7

TRAIN THE TRAINER PROGRAM

The information in this chapter tells how to train the trainer, how to advise the trainer, and how to assist the trainer in preparing the crew for the three phases of MG marksmanship.

7-1. CONCEPT

The Train The Trainer Program must be planned properly and conclude with a certification program. In the planning phase, the instructor teaches the trainer to be flexible and thorough. He must plan alternate exercises in case weather or other constraints prohibit the originally scheduled training. Training must not follow rigid timetables that inhibit training and learning; instead, schedules should be established that provide sufficient time to correct mistakes and ensure learning. The following are factors that the trainers are taught to consider:

a. **Support.** Training requires support. The trainers must ensure that the training is conducted within the resource levels and that the training received justifies the material used. Public address systems should be used if the group is larger than a platoon-size element.

b. **Time.** Ample time must be allocated for each phase.

c. **Participants.** Trainers must consider whether or not the groups or individuals to be trained are capable of benefiting from the phases selected. (See paragraph 7-2a.)

d. **Safety.** Safety is the most important factor and therefore will be the main consideration. Trainers as well as the personnel to be trained will ensure all safety precautions are met. Anyone observing any unsafe act will immediately call CEASE FIRE or HALT to any training. A good way to emphasize safety is to give the students a test on the procedures. An example of a written examination on safety is shown in Figure 7-1, page 7-2.

SAFETY REQUIREMENT EXAM		
INSTRUCTIONS: Mark each answer as being true or false.		
T	F	
	√	l. Firing can be done without a safety officer.
	√	2. Firing can be done without correct headspace.
	√	3. Firing can be done without correct timing.
√		4. No gun should be loaded until the command to do so.
√		5. Each gun is cleared and checked after it is fired.
√		6.No one is allowed in front of the gun for any reason unless authorized by the OIC/NCOIC.
	√	7. Guns can be fired outside safety areas.
	√	8. Guns can be pointed at other personnel while clearing them.
√		9. Firing stops immediately upon the command CEASE FIRE.
√		10. Personnel will stay in designated safety areas until the command is given to leave by OIC/NCOIC.
Answer key: l. F; 2. F; 3. F; 4. T; 5. T; 6. T; 7. F; 8. F; 9. T; 10. T.		

Figure 7-1. Safety examination.

7-2. TRAINER CERTIFICATION PROGRAM

The Trainer Certification Program is designed to build pride, confidence, and overall working/teaching knowledge of the functions, employment, and overall training of the MG. The trainer must know how to do all the tasks he is going to be teaching; the certification program determines his capability to do this. The program is conducted in three phases.

a. **Phase I: Basic Fundamentals of Marksmanship.** In this phase the instructor must ensure the trainers being trained perform dry fire techniques; prepare firing positions, fighting positions, and range cards; and manipulate the T&E mechanism. At the end of this phase, they will be given a hands-on examination to test their basic knowledge. This test may be developed by the OIC/NCOIC of the MG committee.

b. **Phase II: Basic Marksmanship.** In this phase, the trainer moves one step closer to becoming a skilled gunner. He must learn the correct procedures to zero the weapon; he must master the 10-meter target paster; and once this is accomplished, he must be able to engage targets on a transition range. He must show his mastery in this phase by engaging targets at different ranges. Upon completion, the trainer is ready to move to the advanced stage of marksmanship.

c. **Phase III. Advanced Gunnery.** This is the final phase that the trainer has to achieve to standard.

(1) First, he must understand the procedures using the techniques of advanced gunnery.

(2) He must then learn tracking, which consists of maintaining correct alignment of the sights on a moving target by moving the gun at the same angular speed as that of the target.

(3) He will learn fire distribution next, which means he must be able to distribute fire over the entire target area.

(4) The last stage the trainer must learn is the correct assault fire techniques. During this stage, the gunner will normally fire on an FPL after the enemy has assaulted. He may also be given a PDF to fire at. Another assault method the trainer learns is to fire the weapon while it is mounted on a moving M113.

d. **Recertification Program.** All MG instructors are required to be recertified on a semi-annual basis. This will consist of the trainer being required to present a selected period of instruction within the current POI. The selection will be made by the CDR/OIC/NCOIC.

NOTE: Documentation of the results of the Trainer Recertification Program and the requirements must be maintained on file.

7-3. RESPONSIBILITIES AND DUTIES OF THE TRAINER

The trainer must be present during the planning and during any instruction given. The success of the preparation and instruction of all training depends upon the thoroughness with which the trainer performs the following duties:

- He must assist the gunner in targeting the MG.
- He should require the gunner to inspect his equipment and MG.
- He should explain and require the gunner to explain the exercise that he is about to perform.
- He should ensure the gunner's sight picture is accurate.
- He must observe the gunner's position, grip, and manipulation during any firing.
- He must show the gunner how to adjust his fire by observation.
- He must point out errors and explain their effect on the exercise.

NOTE: Safety is a must during the entire certification program. The OICs/NCOICs must ensure that no unsafe acts are tolerated.

* APPENDIX A

SAFETY

Safety precautions must be observed during all marksmanship training. This appendix recommends safety precautions for the ranges described in this manual. Range safety requirements will vary with the different courses of fire.

A-1. TRAINING RANGES

The following safety precautions will be observed when using any training range.

a. A red flag will be displayed at the entrance or in some other prominent location on the range during firing.

b. Firing limits will be marked with red and white striped poles visible to all firers.

c. Firing lanes must be clearly marked for easy identification.

d. All communication equipment, such as microphones, PA systems, loudspeakers, and radios, must be in good working condition.

e. When not in use, all MGs will be kept in a prescribed area with proper safeguards.

f. No smoking will be permitted near ammunition, explosives, or flammables. Firefighting equipment must be installed in ammunition shed or area.

g. Hearing protection devices will be worn by all personnel during firing.

h. Obstructions will never be placed in the muzzles of guns about to be fired.

i. Always assume that guns are loaded until they have been thoroughly examined and found to contain no ammunition.

A-2. RANGE PROCEDURES

Safety requirements must be followed by all personnel on the range before, during, and after firing. It is the responsibility of the safety officer, OIC, and safety NCO to ensure all procedures are met. They are as follows:

a. **Before Firing**.

(1) All prescribed roadblocks and barriers will be closed and necessary guards posted.

(2) All guns will be checked to ensure that they are clear of ammunition and obstructions, and that the covers are up to show they are cleared.

(3) All individuals will be briefed on the firing limits of the range and firing lanes.

(4) Range clearance will be obtained from the installation range-control office.

(5) The downrange area will be checked before firing to ensure that all personnel and equipment are clear of the area.

(6) The NCOIC or range safety officer places a complete first aid kit on the range.

(7) The NCOIC or range safety officer places medical personnel on the range or nearby to enable quick contact.

* (8) Due to the potential injury hazards associated with improper headspace and timing, either the NCOIC or RSO will assume personal responsibility for ensuring that all weapons are set for proper headspace and timing. He will do this before emplacing weapons on the firing line. This applies to all manually fired weapons. It does not apply to mechanically fired weapons.

*** WARNING**

Before placing any manually fired weapon on the firing line, personally check each one for proper headspace and timing.

(9) Soldiers handle guns only on the command of the tower operator or the OIC.

(10) Soldiers draw ammunition only on command of the OIC. When using two or more lots of ammunition, the OIC keeps the lots separate. He identifies ammunition by lot numbers in case of accident or malfunction. Soldiers should retain original packaging for possible repacking of ammunition.

(11) Protect all ammunition from the rays of the sun. Keep oil away from ammunition.

(12) Avoid moving forward of the firing line except with permission of the tower operator, range safety officer, or OIC.

b. **During Firing**.

(1) Should anyone note an unsafe condition, that person will immediately call CEASE FIRE. Firing resumes only when the OIC so directs.

(2) All personnel know the danger of moving forward of the firing line to score their targets. Before the firing line is cleared, and before anyone can go forward, the OIC or range safety officer clears all machine guns.

(3) In the event of a runaway gun, the firer should keep the barrel pointed downrange. He should leave the cover latched.

c. **During Darkness**.

(1) The NCOIC or range safety officer checks the downrange area before firing. He ensures that the area is clear of personnel and equipment. To make sure, he asks three times over a public address system, "Is anyone downrange?" pausing long enough each time to permit a response.

(2) A blinking red light must be used in addition to the red flag. It should be displayed at the entrance to the range or at some other prominent location.

(3) Two red lights will be mounted on the striped poles marking the limits of fire. They must be visible to all firers.

(4) No one will move from his position until told to do so by the OIC.

d. **After Firing**.

(1) Safety personnel will inspect all guns to ensure that they are clear. All guns on the firing line will be rodded whether they were fired or not. A check will be conducted to determine if any brass, links, or live ammunition is in possession of the troops.

(2) When guns have been cleared, they will be kept in a prescribed area with the bolt lock to the rear, and the feed cover raised.

A-3. WEAPONS MAINTENANCE

Before disassembly can be conducted, the gun must be cleared as discussed in Chapter 1.

a. Make sure the bolt is forward before removing the backplate assembly.

b. Never attempt to cock the gun while the backplate is off and the driving spring assembly is in place. If the backplate is off and the driving spring assembly is compressed, the retaining pin on the driving spring rod can slip from its seat in the sideplate and could cause serious injury to anyone behind the gun.

c. When lifting the receiver, barrel, or the tripod, hold it with a firm grip to avoid dropping it and possibly injuring someone.

APPENDIX B

TRAINING AIDS AND DEVICES

The use of devices in marksmanship training programs are important factors because they may allow the gunner to get an idea of what actual combat is like. The training devices designed for the MG are the M19 blank firing attachment, the multiple integrated liner engagement system, the sighting bar, and the M3 recoil amplifier barrel used with plastic ammunition.

B-1. BLANK FIRING ATTACHMENT

The M19 BFA was developed to permit the MG to fire the M1A1 blank cartridge in the automatic fire mode. The BFA is an easy to install, reliable device that allows the MG to be used more realistically during an FTX. The M19 BFA weighs 15.5 pounds and can be installed using either a crescent wrench or the blank/live round discriminator (a component of the BFA that has been designed to serve as a wrench). The design of the M19 will not allow a live round to be loaded while the blank/live round discriminator cover is in position. Normal headspace and timing must be made when firing blank rounds using the M19 (Figure B-1).

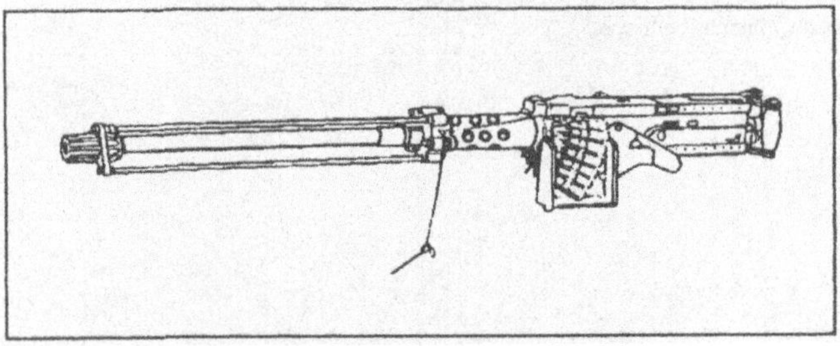

Figure B-1. M19 BFA mounted on MG.

WARNING

The flash of the M1A1 round fired with the M19 BFA at night is extremely bright. It could cause temporary night blindness, and night vision devices may be temporarily disabled.

B-2. MULTIPLE INTEGRATED LASER ENGAGEMENT SYSTEM

The MILES transmitter has been developed for the MG as well as for many other weapons. It consists of a lightweight laser transmitter that sends coded hit/kill messages whenever blank rounds are fired. The MILES transmitter is normally used as part of an M113 APC MILES kit; however, it can also be used on truck- or ground-mounted machine guns with modifications of the detector straps and the combat vehicle kill indicator (CVKI) light (Figure B-2).

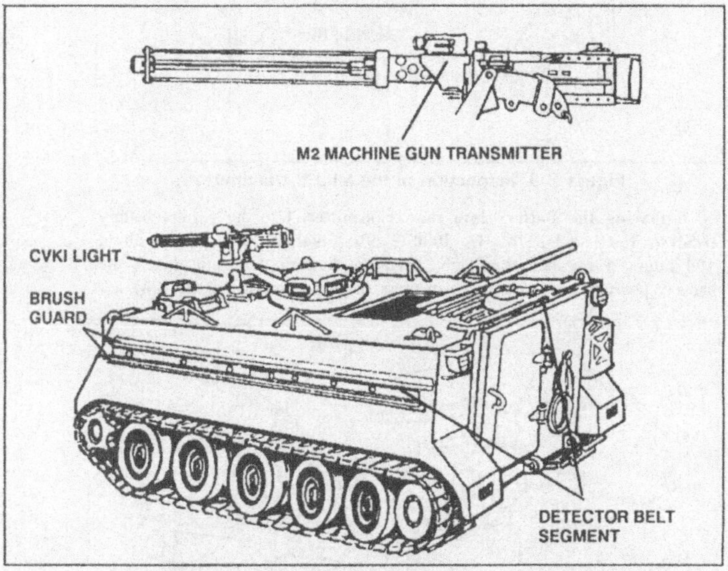

Figure B-2. Components of the M113 APC MILES kit.

a. Inspecting and Servicing the MILES Transmitter. Remove any dirt or oil from the lens using lens paper or a soft, dry cloth. Take care not to scratch the lens. Inspect the foam microphone cover; if it is wet or caked with dirt or blank firing residue, clean or replace it. Inspect the transmitter for indications of damage that would prevent normal operations (Figure B-3). Clean all surfaces. Ensure that the orange key is present. Use DA Form 2404 to report any damage to the transmitter.

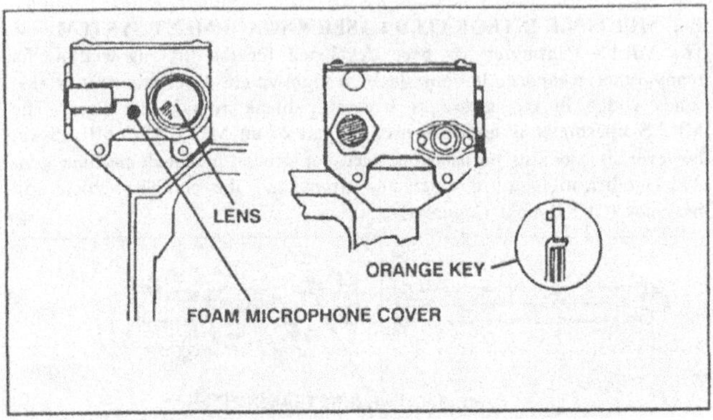

Figure B-3. Inspection of the MILES transmitter.

b. **Placing the Battery into the Transmitter.** Use the correct battery (NSN 6135-01-063-1978, BA 3090/U, 9V, alkaline); it will last about 100 hours. Flip open the latch; open the battery door and insert the battery; press the door closed; then press the latch closed (Figure B-4).

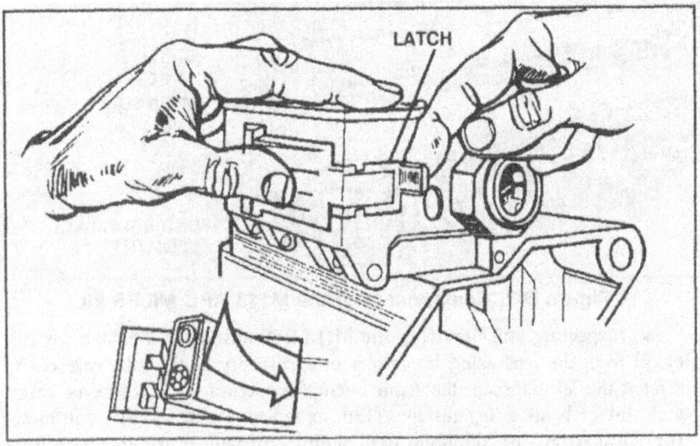

Figure B-4. Battery insertion.

c. **Attaching the Transmitter to the Machine Gun.** Attach the machine gun to its mount. Attach the M19 BFA to the machine gun. Unscrew the knob from the side of the transmitter bracket; swing the bottom plate down; and place the transmitter on the barrel support, flush against the receiver. The lip on the back of the mounting bracket must sit behind the barrel cooling bracket. Swing the bottom plate back up against the transmitter mounting bracket and tighten the knob securely with your hand (Figure B-5).

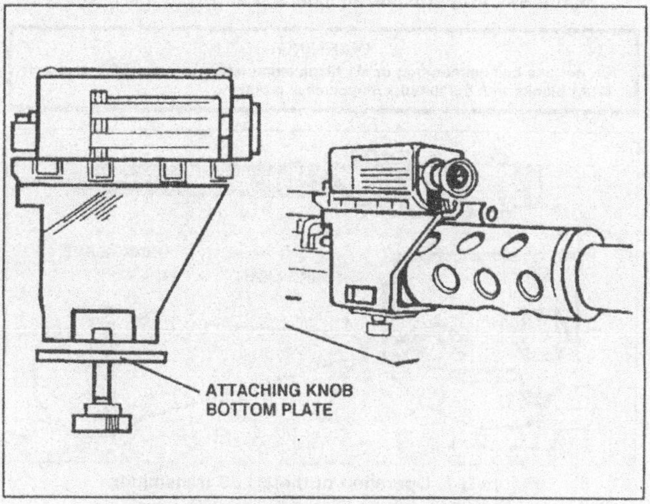

Figure B-5. MILES transmitter attachment.

d. **Operating the Transmitter.** Ensure that the M19 BFA is correctly mounted. Turn the orange key to WEAPON ON. Load the machine gun with M1A1 blank ammunition only (Figure B-6). Fire normally. The sound of the blanks firing will trigger the transmitter. The transmitter will work only as long as the supply of blank ammunition lasts. Once a day, use a clean cloth to remove blank fire residue from the transmitter lens. To test the transmitter, fire a shot burst and watch the firing lamp (Figure B-7). If no light, replace the battery and test again. If still no light, replace the transmitter. Use DA Form 2404 to report the information.

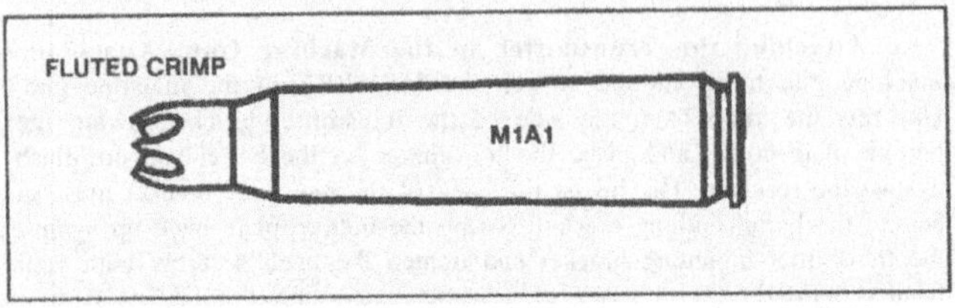

Figure B-6. Proper blank ammunition to use with M19 BFA.

WARNING

Do not use ball ammunition or M1 blank ammunition with the BFA. Use only M1A1 blanks with the fluted, crimped case mouth.

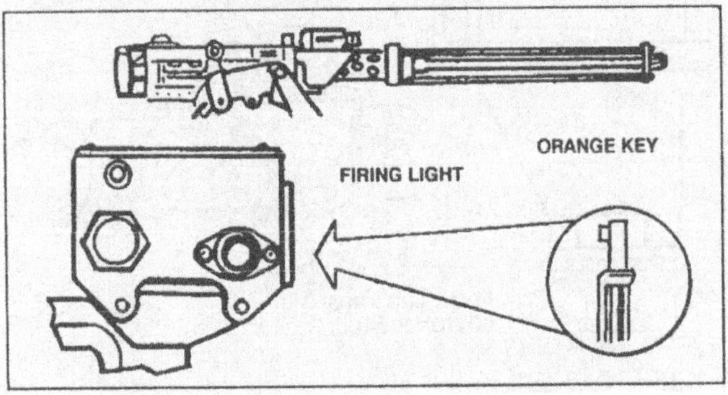

Figure B-7. Operation of the MILES transmitter.

e. **Aligning the Transmitter to the Machine Gun.** The transmitter must be aligned in the dry fire mode, using the dry fire trigger cable. To align the transmitter, you need a soldier wearing a MILES helmet and torso harness. The gun should already be mounted on a tripod and have the T&E attached.

(1) Position the soldier 50 to 75 meters away from the machine gun. Have him place a green controller key in his MILES receptacle and turn the key to the CONTROLLER position.

(2) Connect the trigger cable assembly to the receptacle on the rear of the MILES transmitter. Use the controller key to reset the system.

(3) Insert an orange weapon key into the transmitter receptacle and turn to WEAPON ON.

(4) Fire at the soldier by pressing the trigger cable button. Traverse the gun left and right, up and down, until the soldier's buzzer signals a "near miss." Hold the gun in that position.

(5) While he is being fired on, have the soldier move to his right until his buzzer stops. Have him mark that location.

(6) Have the soldier move to his left until the buzzer stops again. Have him mark that location.

(7) Have the soldier estimate the center between the two marks, and stand there.

(8) Cease fire. Without moving the barrel of the machine gun, adjust the sights until the soldier is seen through the rear sight aperture.

(9) Begin firing again and move the barrel up and down, noticing where the buzzer stops. Estimate the center of this up and down area and adjust the sight elevation to that point.

(10) Remove the trigger cable assembly and screw the protective cover onto the transmitter receptacle.

f. **Resetting the Transmitter After a Kill.** Remove the orange key from the transmitter. Use it to silence the "kill" indicator and the CVKI light. The transmitter will not fire with the key removed. When the controller resets the "kill" indicator, turn the transmitter to WEAPON ON and reinsert the orange key. Continue the mission.

g. **Using Precautions.** The following safety considerations apply to the M1A1 blank cartridge:

(1) Never fire blanks directly at personnel within 20 meters of the gun.

(2) Never fire blanks without wearing hearing protection.

(3) Never fire blanks when the temperature is below 0 degrees or above 120 degrees Fahrenheit.

(4) Never stand directly to the side of the weapon muzzle when blanks are fired.

(6) Never fire blanks in bursts in excess of 20 rounds.

(7) Always allow a cooling period between bursts to avoid overheating the gun.

B-3. SIGHTING BAR

The sighting bar is a locally fabricated device used for practicing the sighting and aiming exercise (Figure B-8). Paragraph 5-9 explains its use. This exercise requires the soldiers to look through the peephole and center on the front sight blade. The gunner adjusts the peephole until he is able to align the front sight blade in the center of it. The instructor will then review and critique the gunner on sight alignment. The device is also equipped with a scaled target for practicing sight picture. Once sight alignment is mastered, the gunner will be required to show a correct sight picture. This is done by adjusting the scaled target until the front sight plate is bottom center of it. Again the instructor will review and critique the gunner. This exercise will continue until the gunner has mastered the technique of sighting and aiming; this requires a lot of practice.

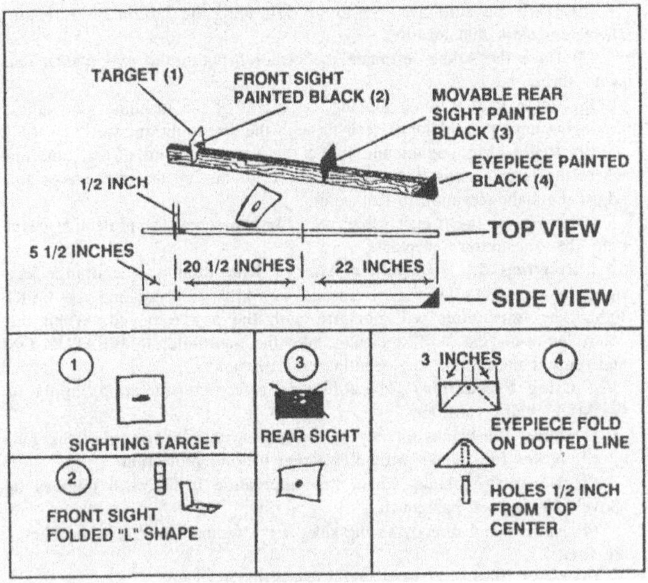

Figure B-8. Sighting bar.

B-4. SHORT RANGE TRAINING AMMUNITION

To augment the use of the caliber .50 ammunition, a new generation of SRTA (the M858 ball and the M860 tracer) has been developed.

a. The caliber .50 ball and tracer cartridges have configurations and dimensions similar to those of the corresponding service rounds; however, the base of the cartridge, which encloses the primer, is made of aluminum. The rest of the case is-made of a blue-colored plastic material. The tip of the tracer projectile is colored red. The complete round weighs about 1.05 ounces and is 5.19 inches long, which is slightly shorter than the service round. The cartridges are linked with M9-type metallic links for use with the MG. Muzzle velocity of the projectile is about 4,000 feet per second. The light weight of the projectile, about 50 grains, and its low sectional density cause its velocity to drop rapidly.

b. The SRTA allows training in small local training areas without fixed training facilities, in MOUT facilities, and in combat training theaters. With its 150-meter usable range, the SRTA can be used on grouping, zeroing, and 10-meter scaled silhouette firing.

WARNING

The SRTA projectile causes considerable damage out to 150 meters and is considered dangerous out to its maximum range of 700 meters.

B-5. M3 RECOIL AMPLIFIER BARREL ASSEMBLY

An M3 RABA (Figure B-9) is used with the machine gun for firing the training ammunition. It is assembled to the gun in the same manner as the standard barrel. The RABA provides the means to boost the power that is required to recoil the barrel and effectively cycle the weapon when relatively underpowered M858 SRTA or M860 SRTA-T are fired in the machine gun.

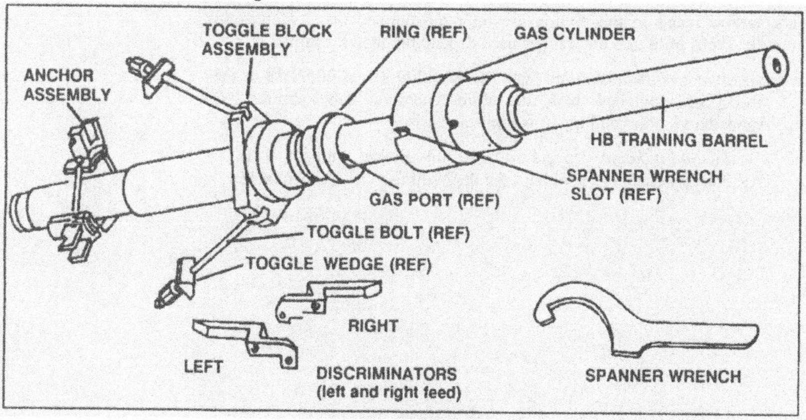

Figure B-9. M3 RABA.

APPENDIX C

RANGES AND TRAINING EXERCISES

This appendix provides guidance for US Army machine gun exercises. It is designed to be a working guide for trainers and commanders by explaining range layouts, conduct of firing exercises, and qualification standards. The two basic firing ranges will be the multipurpose MG transition range and the multipurpose range complex.

* Section I
MULTIPURPOSE MACHINE GUN TRANSITION RANGE

All machine gun firing can be conducted on the multipurpose MG transition range. It allows for complete firing of the 10-meter course as well as transition, day, night, and NBC firing. The firing area has 10 lanes; its layout is shown in Figure C-1. Details on the setup and target configuration are described in FM 25-7. The minimum personnel required to operate the range are as follows: OIC, NCOIC, safety officer, ammunition NCO, tower operator, lane NCOs, primary instructor, and concurrent training instructors. Local policy may require more personnel.

* The targets listed in this section are only recommendations. Units may alter the types of targets and ranges used depending on availability.

- *For night qualification*, units with the AN/PEQ-2A, ANPAS 13 or the AN/TVS-5 upgraded with the third-generation tube should use Appendix G tables and qualification standards.
- *For day qualification*, leaders will use this section until the STRAC (AR 350-38) changes, at which time they will use only Appendix G.

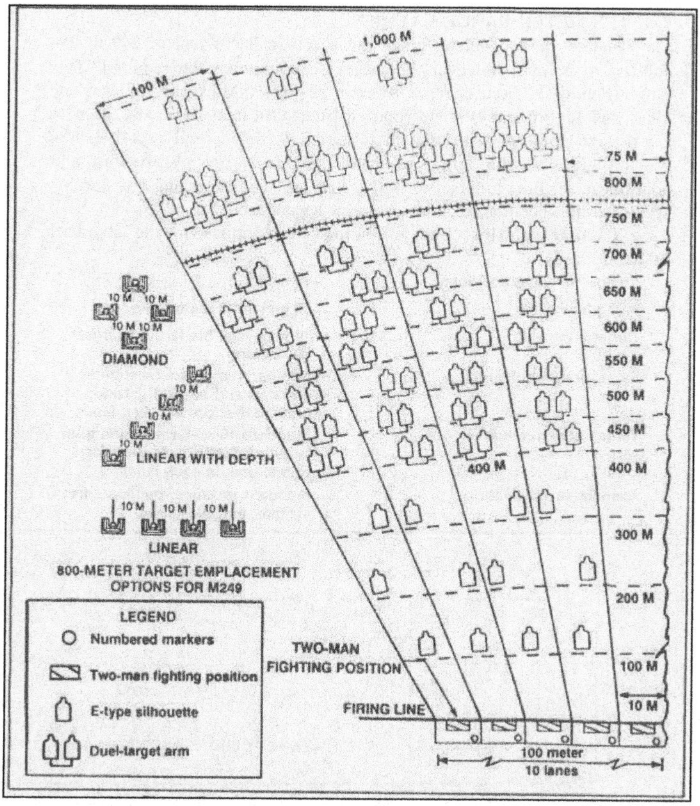

Figure C-1. Multipurpose machine gun transition range layout.

C-1. TEN-METER RANGE LAYOUT

The standard, basic 10-meter range can accommodate a unit of 200 to 250 soldiers at a time; however, concurrent training may be required. This range (Figure C-2) can be used to zero the M249, M60, and M2 machine guns, and to fire the 10-meter portion of qualification. It is also used to familiarize soldiers with the characteristics, noise, and recoil of the weapon. This is the range used to practice target observation and adjustment of fire, to practice MG traversing and searching, to develop speed during operation, and to obtain an accurate burst.

a. **Characteristics.** The following data applies to the standard 10-meter range.

Number of firing positions	20 lanes.
Firing lane width	60 meters (3 meters per lane).
Target area width	60 meters at the farthest targets (10 meters).
Firing point configuration	Numbered markers on slightly elevated and sodded ground; brass deflectors between lanes.
Target configuration	Standard 10-meter machine gun target stretched over a wooden frame, one on each lane.
Associated facilities	Standard facilities, public address system, and bleachers.

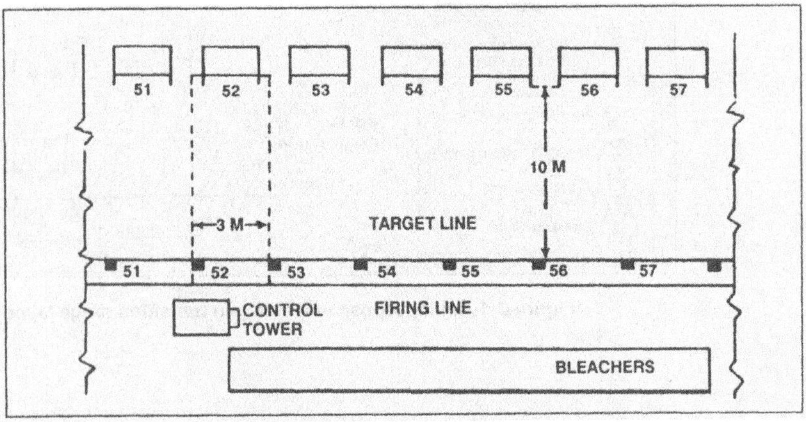

Figure C-2. Ten-meter range layout.

b. **Setup of Range.** The 10-meter range should meet the following requirements:

(1) The firing line should be long enough to emplace 20 MGs, allowing 3 meters between positions. For control, each position should be numbered.

(2) The target line should be 10 meters in front of the firing line. The MG marksmanship target is a paper target pasted onto target cloth that is stretched over a wooden frame. One target is set up for each position and is numbered to correspond with the numbered position.

(3) The instruction site for this range is the bleachers, which should be to the rear of the firing line.

(4) The control tower is located to the immediate rear and center of the firing line.

c. **Personnel Required.** Each range must be staffed with the following personnel:

• One officer in charge.
Ž One safety officer.
Ž One principal instructor.
• One assistant instructor for every 10 students.
• One ammunition NCO.
• One tower operator.
• Medical personnel.

d. **Equipment Required.** Although more equipment may be required by local range regulations, safety regulations, or unit SOPs, the minimum equipment required to operate the range is as follows:

• One public address system.
• One MG for each firing lane and one extra gun for every five lanes as a backup.
• One cleaning rod per assistant instructor.
Ž One scorecard per soldier.
Ž One asbestos glove for every two weapons.
Ž One M3 tripod for each lane.
• One caliber .50 pintle for each lane.
Ž One caliber .50 T&E mechanism for each lane.
• Medical evacuation capability.
• Communication equipment (wire or radio as required to operate the range).

* C-2. TEN-METER FIRING EXERCISES

The purpose of 10-meter firing is to develop skills in the delivery of initial bursts on target. When conducted properly, it will train gunners in the basic skills of zeroing, controlling the burst, traversing, and traversing and searching techniques. This paragraph discusses firing the 10-meter exercise on the multipurpose transition range; however, it is conducted the same on a basic 10-meter range except that the groups will be broken down to fit the firing lanes. The course is designed to fire the tripod-mounted M2. The basic MG target is used for all 10-meter firing and it allows four gunners to use it for practice or two gunners for practice and qualification. Those units wanting to fire qualification right after practice should use section A for practice and B for qualification for the first gunner, and sections C and D for the second gunner. Ten-meter firing will be conducted IAW Firing Table I (Figure C-3).

TASK	CONDITIONS TARGET/SITUATION	AMMO	STANDARD_____
1. Zero the M2 machine gun at 10 meters.	Instruct the gunner to engage scoring paster 1 in his assigned section with 6 single rounds in two 3-round shot groups and then paster 2 with the remaining rounds.	12 single rounds of 12.7-mm	Gunner must impact 4 of 6 rounds of the 3-round shot groups within paster 2 of his assigned section._____
2. Engage a 10-meter target using controlled-burst techniques.	Instruct the gunner to engage paster 3 in his assigned section using controlled bursts, then engage paster 4 in the same manner. Gunner will use 5- to 7-round bursts for this engagement.	two 7-round belts of 12.7-mm	Gunner must impact one controlled burst on pasters 3 and 4 in his assigned section. (No score required.)_____
3. Engage scoring pasters 5 through 6 using the traverse and search technique of fire.	Instruct the gunner to engage pasters 5 through 6 using traverse and search technique of fire. Gunner will use 5- to 7-round bursts for this engagement.	35-round belt of 12.7-mm	Gunner must impact one round on each scoring paster of 5 through 6 in his assigned section. (No score required.)_____
4. Engage scoring pasters 7 through 8 using the traverse and search technique of fire.	Instruct the gunner to engage pasters 7 through 8 using traverse and search technique of fire. Gunner will use 5- to 7-round bursts for this engagement.	56-round belt of 12.7-mm	Gunner must impact one round on each scoring paster of 7 through 8 in his assigned section _____. (No score required.)_____
5. Engage linear target at 10 meters using traverse fire technique.	Instruct the gunner to engage pasters 1 through 4 in his assigned section using traverse fire. Gunner will use 5- to 7-round bursts for this engagement.	28-round belt of 12.7-mm	Gunner must impact at least four rounds on each scoring paster 1 through 4 in his assigned section _____.
6. Engage scoring pasters 7 through 8 using the traverse and search technique of fire.	Instruct the gunner to engage pasters 7 through 8 using traverse and search technique of fire. Gunner will use 5- to 7-round bursts for this engagement.	56-round belt of 12.7-mm	Gunner must impact at least four rounds on each scoring paster 7 through 8 in his assigned section _____.
7. Engage scoring pasters 5 through 6 using the traverse and search technique of fire.	Instruct the gunner to engage pasters 5 through 6 using traverse and search technique of fire. Gunner will use 5- to 7-round bursts for this engagement.	35-round belt of 12.7-mm	Gunner must impact at least four rounds on each scoring paster 5 through 6 in his assigned section _____.

*** Figure C-3. Ten-meter firing table (Firing Table I).**

a. **Objective**. In order for the M2 gunner to become proficient with the machine gun, the following objectives must be considered:

- Obtain an accurate initial burst.
- Master the technique of controlled burst firing.
- Traverse and search the machine gun effectively.
- Observe and adjust fire.
- Operate with speed.

b. **Organization**. The unit is organized into groups of 10. Each group is assigned a firing order number. One order becomes gunners and the other assistant gunners. The gunners and assistant gunners are assigned lanes and required to set up their guns and perform pre-fire checks.

(1) Ten-meter practice fire (Tasks 1 through 4).

(a) The gunner is required to set his rear sight at 500 yards. On command, the first order zeros using scoring pasters 1 and 2 of their assigned section.

(b) At the completion of zeroing, the gunners are required to engage the next two aiming pasters (scoring pasters 3 and 4 of the same section), using controlled bursts, and then area targets 5 through 6 and 7 through 8, using traverse and search techniques. Time is allowed between bursts to permit gunners to observe and adjust their fire.

(c) After the first order has fired, the second order fires the course in the same manner using the next assigned section. After both orders have fired, the guns are cleared.

(2) Ten-meter qualification (Tasks 5 through 7).

(a) The gunner is then required to set his rear sight at 550 yards. On command, the first order fires at scoring pasters 1 through 4 of their assigned section.

(b) At the completion of their linear engagement, commands are given to require the gunners to engage area targets 5 through 6 and then 7 through 8, using traverse and search techniques. Time is allowed between bursts to permit gunners to observe and adjust their fire.

(c) After the first order has fired qualification, the second order fires the course in the same manner using the next assigned section. After both orders have fired, the guns are cleared.

(2) *Ten-meter qualification (Tasks 5 through 7).*

(a) The gunner is then required to set his rear sight at 550 yards. On command, the first order fires at scoring pasters 1 through 4 of their assigned section.

(b) At the completion of their linear engagement, commands are given to require the gunners to engage area targets 5 through 6 and then 7 through 8, using traverse and search techniques. Time is allowed between bursts to permit gunners to observe and adjust their fire.

(c) After the first order has fired qualification, the second order fires the course in the same manner using the next assigned section. After both orders have fired, the guns are cleared.

c. **Target Analysis.** The exercises use the basic machine gun target (FSN 6920-078-5128) (Figure C-4). The following explanation of the target, including the size of the aiming pasters and scoring spaces, will aid in zeroing the M2 and will facilitate control during firing exercises.

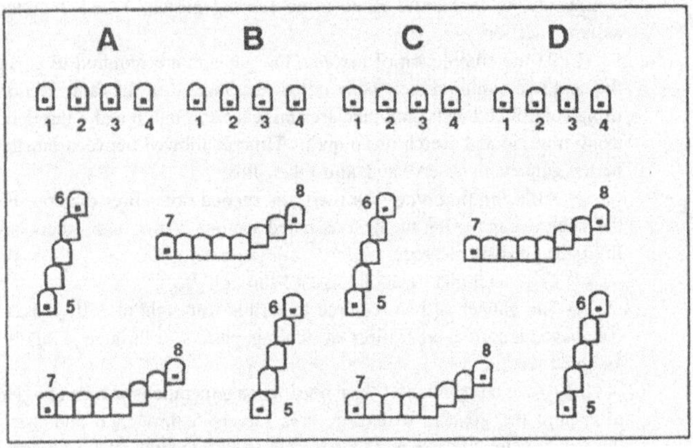

Figure C-4. Basic machine gun target.

(1) The target consists of four sections lettered A, B, C, and D. Each section has four point targets numbered 1, 2, 3, and 4; and two sets of area targets numbered 5 through 6 and 7 through 8. Each space is 4 centimeters wide and 5 centimeters high. The black aiming pasters within some numbered scoring spaces are 1-centimeter squares. Firing at targets 1 through 4 allows the gunner to use the traverse-fire technique

targets 5 through 6 and 7 through 8 give them practice in traverse and search techniques.

(a) Point targets. Point targets on the basic machine gun target are considered to be pasters 1 through 4 of section A, B, C, and D. Firing at point targets exposes the gunner to zeroing techniques and controlled-burst fire techniques. Targets 1 through 4 will also be used as a linear target for qualification.

(b) Area targets. Area targets on the basic machine gun target are considered to be pasters 5 through 6 and 7 through 8 of section A, B, C, and D. Target group 5 through 6 exposes the gunner to T&E manipulation when using the tripod mount and body position changes to engage targets in depth (elbow position changes). This causes him to use a series of aiming points to disburse fire across the target when using the tripod. Target group 7 through 8 exposes the gunner to position changes to engage linear targets with depth. It causes him to control the burst length from the weapon, use a series of aiming points, and disburse fire across the target.

(2) Targets are analyzed and scored to determine the gunner's proficiency and to see if more training is needed in any of the fundamentals of M2 gunnery. During firing with a properly zeroed weapon, a target is best analyzed by considering the common errors of M2 gunnery (Figure C-5).

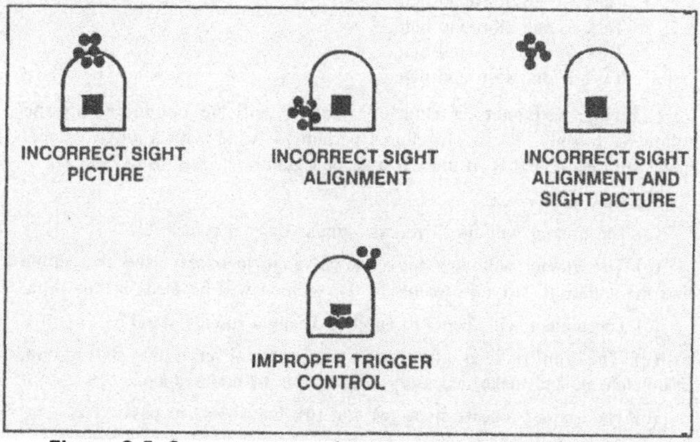

Figure C-5. Common errors found on the basic MG target.

NOTE: Large shot groups are usually caused by incorrect position and grip; small shot groups outside of the scoring space are usually caused by incorrect sight alignment, sight picture, or zero.

d. **The 10-Meter Zero.** Ten-meter zeroing is accomplished by adjusting the rear sight until the strike of the projectile coincides with the point of aim at a given range. On the 10-meter range, the elevation scale must be set at 500 yards so that it can impact on the desired aiming point. Before zeroing the weapon, the gunner must first center the rear sight for windage by aligning the index with the center of the windage scale.

e. **Controlled Burst Firing.** This firing exercise reinforces the dry-fire experience and allows the gunner practice in firing of controlled bursts and in providing the most accurate and tightest shot groups obtainable with the M2.

f. **Ammunition Issue.** The ammunition for the firing of this exercise will be broken down at the ammunition point and issued to each assistant gunner as he is assigned a firing point. The total ammunition for this exercise is 236 rounds (117 for practice and 119 for qualification). Issue will consist of the following rounds or belts:

- Task 1, 12 single rounds
- Ž Task 2, two 7-round belts.
- Task 3, one 35-round belt.
- Task 4, one 56-round belt.
- Task 5, one 28-round belt.
- Task 6, one 56-round belt.
- Task 7, one 35-round belt.

g. **Firing Sequence.** Firing of Table I will be conducted in the following manner. These procedures pertain to firing with a single gunner using sections A and B; if there is a second gunner, he will use C and D.

(1) *Task 1, Zero.*

(a) The gunner will fire 3 rounds, single shot, at paster A1.

(b) The gunner will then move down range to observe the shot group and triangulate it. No adjustments to the weapon will be made at this time.

(c) The gunner will then fire another 3 single rounds at A1.

(d) The gunner then goes down range to observe the shot group, triangulate it, and make necessary adjustments to his weapon.

(e) The gunner repeats steps (c) and (d), but shoots at paster A2.

NOTE: If the gunner is able to zero his weapon using 9 rounds, have him use the remaining 3 to confirm his zero. If the gunner is unable to zero within 12 rounds, remove him from the firing line and give him remedial training.

(2) *Task 2, Controlled Burst Firing.*

(a) The gunner will fire the first burst of 5 to 7 rounds at paster 3.

(b) The gunner then goes down range to observe, marks the projectile holes, and analyzes his burst.

(c) The gunner repeats steps (a) and (b), but fires at paster 4 of same section.

NOTE: The primary purpose of this task is to get the gunner to fire 5- to 7-round bursts. Impact on the target is desired, but not necessary to complete this task. The gunner will fire a 5- to 7-round burst at each paster.

(3) *Task* 3, *Traverse and Search Fire.*

(a) The gunner will engage pasters 5 through 6, section A, firing a 5- to 7-round burst for each paster, using the traverse and search technique.

(b) The gunner will then move down range to observe and analyze his targets.

NOTE: These tasks are fired from the tripod firing position. If the gunner shows difficulty in manipulating the weapon, remove him from the firing line and give him remedial training.

(4) *Task 4, Traverse and Search Fire.*

(a) The gunner will engage pasters 7 through 8, section A, firing a 5- to 7-round burst at each paster, using the traverse and search technique.

(b) The gunner will then move down range to observe and analyze his targets.

(5) *Task 5, Traverse Fire.*

(a) The gunner will engage pasters 1 through 4, section B, firing a 5- to 7-round burst at each paster, using the traverse technique.

(b) The gunner will then move down range to observe and analyze his targets.

(6) *Task 6, Traverse and Search Fire.*

(a) The gunner will engage pasters 7 through 8, section B, firing a 5- to 7-round burst at each paster, using the traverse and search technique.

(b) The gunner will then move down range to observe and analyze his targets.

(7) *Task 7, Traverse and Search Fire.*

(a) The gunner will engage pasters 5 through 6, section B, firing a 5- to 7-round burst at each paster, using the traverse and search technique.

(b) The gunner will then move down range to observe and analyze his targets.

DANGER

Before any gunner goes down range, the lane NCO must ensure that the weapon has been properly cleared. Anyone observing any unsafe acts should immediately call a cease fire and notify range personnel.

h. **Scoring Procedures.** When scoring the 10-meter target, all scoring spaces are scored (1 through 4, 5 through 6, and 7 through 8). One point is given for each round impacting within each space. Rounds touching the boundary of a scoring space are considered hits, but they can be counted in only one scoring space. When firing at 1 through 4 with 28 rounds, the maximum score is 28 points. In group 5 through 6, five scoring spaces are engaged with 35 rounds with a possible score of 35 points. When firing groups 7 through 8, eight scoring spaces are engaged with 56 rounds with a possible score of 56 points.

(1) The total possible score of both groups is 119 points. A minimum of 84 points is required to pass the 10-meter course of fire.

(2) Soldiers failing to achieve minimum standards must be retrained and retested in a dry-fire mode until proficiency is demonstrated. They may not move on to transition firing. Soldiers should then re-fire the 10-meter portion with close supervision and coaching to ensure that the fundamentals are applied properly during live fire.

C-3. TRANSITION FIRING EXERCISES

Transition firing of the M2 machine gun will teach the gunner some techniques of fire that he may encounter in combat situations. The gunner will field zero his weapon and also incorporate the techniques of fire during limited visibility and NBC environments. The gunner needs to be aware that during certain situations his capabilities will be degraded. He is expected to compensate for these situations by applying all available techniques. Within this training, the gunner will be required to apply all the fundamentals of gunnery learned in preparatory gunnery training and 10-meter firing. Instructors should encourage gunners to perform immediate action if a stoppage occurs during fire. This procedure may be modified if local policies require the gunners to notify the range personnel first.

a. **Objectives.** The objectives of transition fire are —

- To engage targets at long ranges with the tripod-mounted machine gun.
- To understand the characteristics of fire.

Ž To field zero the M2 machine gun.

Ž To use range estimation to determine the distance to targets.

Ž To apply the method of adjusted point of aim.

b. **Organization.** Organization of a unit for transition firing is conducted the same as for 10-meter firing. Field zeroing is the first firing task of transition day-fire phase. The transition day-fire portion of the course must be fired to become a qualified M2 gunner.

c. **Ammunition Issue.** The ammunition for the firing of this exercise will be broken down at the ammunition point and issued to each assistant gunner as he is assigned a firing point. The total ammunition for this exercise is 182 rounds, and issue will consist of two belts; one 28-round belt and one 154-round belt.

d. **Firing Sequence.** Firing of Table II (Figure C-6) will be conducted in the following manner. These procedures pertain to firing with a single gunner using a tripod-mounted M2. If there is a second gunner, he will use the same procedures.

(1) *Task 1, Field Zero.*

(a) The gunner must first center the rear sight in the same manner as sight setting for 10-meter firing. He selects the appropriate range mark with his elevation knob.

(b) The gunner will load one 28-round belt of ammunition and fire a burst of 5 to 7 rounds at the 550-meter, double E-type silhouette.

(c) If the impact of the rounds (beaten zone) is over the target, the gunner has probably fired with an incorrect sight picture (too high on the target). He needs to relay the gun on the original point of aim and then he relays back on the target and fires again. If the gunner again fails to zero with a proper sight picture, he should have the unit armorer inspect the weapon before continuing the zeroing procedure.

(d) The gunner will observe the beaten zone and make adjustments to the sights so the round will impact on the target.

(e) After adjustments are made, the gunner will then relay on the center base of the target and fire another 5- to 7-round burst.

(f) The gunner will repeat steps c, d, and e with the remaining rounds.

NOTE: If the gunner is unable to zero within 28 rounds, he is removed from the firing line and given remedial training.

(2) *Task 2 through Task 8, Single and Multiple Engagements.*

(a) The gunner will load one 154-round belt of ammunition and fire bursts of 5 to 7 rounds at double E-type silhouettes at ranges of 800, 400, 700, and 1,000 meters at vehicular targets. He will also engage multiple double E-type silhouettes targets at 400 to 700; 550 and 800; and 400, 550, and 1,000 meters respectively.

(b) The gunner will observe the beaten zone and make adjustments as necessary to hit the targets.

(c) After firing is completed, weapons will be cleared and the gunner critiqued. If there is a second gunner to fire, the entire procedure will be repeated.

NOTE: After both gunners have fired for practice, they will then be assigned new lanes and issued more ammunition for record firing. If possible, the gunner should keep the same weapon he used for practice fire. The only change in the above procedures will be in step (c); here, the gunner will be scored along with the critique. If the gunner is unable to meet standard, he is removed from the firing line, given remedial training, and re-fired.

DANGER

Before any gunner changes lanes, the lane NCO must ensure that the weapon has been properly cleared. Anyone observing any unsafe acts should immediately call a cease fire and notify range personnel.

TASK	CONDITIONS TARGET/SITUATION	AMMO	STANDARD
1. Zero the M2 machine gun.	Gunner is in stationary tripod firing position and engages a 550-meter, double E-type silhouette. Gunner will use 5-to 7-round bursts for this engagement.	28-round belt of 12.7-mm.	Gunner must impact one burst on the 550-meter target.
2. Engage a single, double E-type silhouette.	Gunner is in stationary tripod firing position and engages an 800-meter, double E-type silhouette. Gunner will use 5-to 7-round bursts for this engagement.	14 rounds of 12.7-mm.	Gunner must impact one burst on the 800-meter target within 20 seconds.

Figure C-6. Practice/Qualification Table II.

TASK	CONDITIONS TARGET/SITUATION	AMMO	STANDARD
3. Engage a single. double E-type silhouette.	Gunner is in stationary tripod firing position and engages a 400-meter, double E-type silhouette. Gunner will use 5-to 7-round bursts for this engagement.	14 rounds of 12.7-mm	Gunner must impact one burst on the 400-meter target within 20 seconds.
4. Engage a single. double E-type silhouette.	Gunner is in stationary tripod firing position and engages a 700-meter, double E-type silhouette. Gunner will use 5-to 7-round bursts for this engagement.	14 rounds of 12.7-mm	Gunner must impact one burst on the 700-meter target within 25 seconds.
5. Engage a single. double E-type silhouette.	Gunner is in stationary tripod firing position and engages a 1,000-meter, vehicle type silhouette. Gunner will use 5-to 7-round bursts for this engagement.	14 rounds of 12.7-mm	Gunner must impact one burst on the 1,000-meter target within 25 seconds.
6. Engage multiple double E-type silhouettes.	Gunner is in stationary tripod firing position and engages 400-and 700-meter, double E-type silhouettes. Gunner will use 5-to 7-round bursts for this engagement.	28 rounds of 12.7-mm.	Gunner must impact one burst on each target within 35 seconds.
7. Engage multiple double E-type silhouettes.	Gunner is in stationary tripod firing position and engages 550-and 800-meter, double E-type silhouettes. Gunner will use 5-to 7-round bursts for this engagement.	28 rounds of 12.7-mm.	Gunner must impact one burst on each target within 35 seconds.
8. Engage multiple double E-type silhouettes.	Gunner is in stationary tripod firing position and engages 400-, 550-and 1,000 meter, double E-type and vehicle silhouettes. Gunner will use 5-to 7-round bursts for this engagement.	42 rounds of 12.7-mm.	Gunner must impact one burst on each target within 45 seconds.

Figure C-6. Practice/Qualification Table II, continued.

e. **Scoring Procedures.** One point is given for each target hit with an additional 2-point bonus if the target is hit on the first burst. The total possible points for day fire is 33. A minimum 7 points (7 out of 11 exposures) is required to meet standard on the day-fire course.

C-4. NBC FIRING

Since NBC plays an important part in our preparation for war on the modern battlefield, it is important that each soldier is prepared to accomplish the mission even if the area is contaminated and he must wear protective gear.

a. **Objectives.** The objectives of this training are:

- To teach the gunners to take appropriate action when notified of a contaminated area.
- To identify targets while in a contaminated area.
- To engage targets while in appropriate NBC gear.

b. **Organization.** The unit is organized the same as in paragraph C-2b, except both the gunner and assistant gunner must wear protective masks. Protective clothing is optional.

c. **Ammunition Issue.** The ammunition for the firing of this exercise will be broken down at the ammunition point and issued to each assistant gunner as he is assigned a firing point. The total ammunition for this exercise is 182 rounds, and issue will consist of two belts; one 28-round belt and one 154-round belt.

d. **Firing Sequence.** Firing of Table III (Figure C-7, pages C-16 and C-17) will be conducted in the following manner. These procedures pertain to firing with a single gunner using a tripod-mounted M2. If there is a second gunner, he will use the same procedures.

(1) *Task 1, Field Zero.*

(a) The gunner must first center the rear sight in the same manner as setting the sights for 10-meter firing. He selects the appropriate range mark with his elevation knob.

(b) The gunner will load one 28-round belt of ammunition and fire a burst of 5 to 7 rounds at the 550-meter, double E-type silhouette.

(c) If the impact of the rounds (beaten zone) is over the target, the gunner has probably fired with an incorrect sight picture (too high on the target). He should re-fire concentrating on the proper point of aim and sight picture. If the impact is still off, have the unit armorer correct the calibration and the gunner repeat the zeroing procedure.

(d) The gunner will observe the beaten zone and make adjustments to the sights so the round will impact on the target.

(e) After adjustments are made, the gunner will then relay on the center base of the target and fire another 5- to 7-round burst at the same target.

(f) The gunner will repeat steps c, d, and e with the remaining rounds.

NOTE: If the gunner is unable to zero within 28 rounds, he is removed from the firing line and given remedial training.

(2) *Task 2 through Task 8, Single and Multiple Engagements.*

(a) The gunner will load one 154-round belt of ammunition and fire bursts of 5 to 7 rounds at double E-type silhouettes at ranges of 800,400,

700, and a 1,000-meter vehicular target. He will also engage multiple, double E-type silhouettes targets at 400 and 700; 550 and 800; and 400, 500, and 1,000 meters, respectively.

(b) The gunner will observe the beaten zone and make adjustments as necessary to hit the targets.

(c) After firing is completed, weapons will be cleared and the gunner critiqued. If there is a second gunner to fire, the entire procedure will be repeated.

NOTE: After both gunners have fired for practice, they will then be assigned new lanes and issued more ammunition for record firing. If possible, the gunner should keep the same weapon he used for practice fire. The only change in the above procedures will be in the last step; here, the gunner will be scored along with the critique. If the gunner is unable to meet standard, he is removed from the firing line, given remedial training, and re-fired.

DANGER

Before any gunner changes lanes, the lane NCO must ensure that the weapon has been properly cleared. Anyone observing any unsafe acts should immediately call a cease fire and notify range personnel.

e. **Scoring Procedures.** One point is given for each target hit with an additional 2-point bonus if the target is hit on the first burst. The total possible points for day fire is 33. A minimum of 7 points (7 out of 11 exposures) is required to meet standard on the NBC fire course.

TASK	CONDITIONS TARGET/SITUATION	AMMO	STANDARD
1. Zero the M2 machine gun.	Gunner is in stationary tripod firing position and engages a 550-meter, double E-type silhouette. Gunner will use 5-to 7-round bursts for this engagement.	28-round belt of 12.7-mm.	Gunner must impact one burst on the scaled, 550-meter target.
2. Engage a single, double E-type silhouette at 800 meters while wearing a protective mask.	Gunner is in stationary tripod firing position and engages a 800-meter, double E-type silhouette. Gunner will use 5-to 7-round bursts for this engagement.	14 rounds of 12.7-mm.	Gunner must impact one burst on the 800-meter target within 20 seconds.

Figure C-7. NBC Qualification Table III.

TASK	CONDITIONS TARGET/SITUATION	AMMO	STANDARD
3. Engage a single, double E-type silhouette at 400 meters while wearing a protective mask.	Gunner is in stationary tripod firing position and engages a 400-meter, double E-type silhouette. Gunner will use 5-to 7-round bursts for this engagement.	14 rounds of 12.7-mm.	Gunner must impact one burst on the 400-meter target within 20 seconds.
4. Engage a single, double E-type silhouette at 700 meters while wearing a protective mask.	Gunner is in stationary tripod firing position and engages a 700-meter, double E-type silhouette. Gunner will use 5-to 7-round bursts for this engagement.	14 rounds of 12.7-mm.	Gunner must impact one burst on the 700-meter target within 20 seconds.
5. Engage a single, double E-type silhouette at 1,000 meters while wearing a protective mask.	Gunner is in stationary tripod firing position and engages a 1,000-meter, vehicle type silhouette. Gunner will use 5-to 7-round bursts for this engagement.	14 rounds of 12.7-mm.	Gunner must impact one burst on the 1,000 meter target within 25 seconds.
6. Engage multiple, double E-type silhouettes at 400 and 700 meters while wearing a protective mask.	Gunner is in stationary tripod firing position and engages 400-and 700-meter, double E-type silhouettes. Gunner will use 5-to 7-round bursts for this engagement.	28 rounds of 12.7-mm.	Gunner must impact one burst on each target within 35 seconds.
7. Engage multiple, double E-type silhouettes at 550 and 800 meters while wearing the protective mask.	Gunner is in stationary tripod firing position and engages 550-and 800-meter, double E-type silhouettes. Gunner will use 5- to 7-round bursts for this engagement.	28 rounds of 12.7-mm.	Gunner must impact one burst on each target within 35 seconds.
8. Engage multiple, double E-type silhouettes at 400, 500, and 1,000 meters while wearing the protective mask.	Gunner is in stationary tripod firing position and engages scaled 400-,500-, and 1,000-meter, double E-type and vehicle type silhouettes. Gunner will use 5-to 7-round bursts for this engagement.	42 rounds of 12.7-mm.	Gunner must impact one burst on each target within 45 seconds.

Figure C-7. NBC Qualification Table III, continued.

C-5. NIGHT-FIRE EXERCISE

The night-fire exercise gives the soldiers the practical application of engaging targets using the AN/TVS-5 at night or during limited visibility.

a. **Objectives.** The objectives of this training are:

- To teach the gunners to zero the AN/TVS-5 to the M2.
- To engage targets at different ranges using the AN/TVS-5.
- To detect targets using the AN/TVS-5.

b. **Organization.** Organization of a unit for night firing is conducted the same as day fire. AN/TVS-5 zeroing is the first firing task of the night-fire phase. The night-fire portion of the course is also a requirement to become a qualified M2 gunner.

c. **Ammunition Issue.** The ammunition for the firing of this exercise will be broken down at the ammunition point and issued to each assistant gunner as he is assigned a firing point. The total ammunition for this exercise is 180 rounds, and issue will consist of the following rounds or belts:

- Task 1, 12 single rounds and two 7-round belts
- Task 2 through Task 8, one 154-round belt.

d. **Firing Sequence.** Firing of Table IV (Figure C-8) will be conducted in the following manner; these procedures pertain to a single gunner using a tripod-mounted M2 with an AN/TVS-5 If there is a second gunner, he will use the same procedures.

(1) *Task 1, Zero the AN/TVS-5.*

(a) The gunner must first mount the AN/TVS-5 to the weapon and place it into operation. Once the device is mounted, the gunner will fire a 7-round burst to seat the device, then fire another 7-round burst and tighten the device to make sure it is settled.

(b) The gunner must then center the reticle pattern in the field of view of the device. He then places the reticle aiming point on the 50-meter zero target aim point and fires three single rounds.

NOTE: Each click of the azimuth or elevation adjustment actuator moves the strike of the round 1/2 inch at 50 meters. One click of adjustment moves the reticle one square of the target at 50 meters.

(c) If the impact of the round (beaten zone) is over the target, the gunner has probably fired with an incorrect sight picture (too high on the target). He should re-fire concentrating on the proper point of aim and AN/TVS-5 sight picture. If the gunner again fails to zero with a proper device sight picture, he should inspect the sight calibration of the device.

(d) The gunner will observe the beaten zone and make adjustments to the sights so the strike of the round impacts on the target.

(e) After adjustments are made, the gunner will then fire another round at the same target until zero is obtained.

NOTE: If the gunner is able to zero his weapon using 12 rounds, he is then ready to continue the course. If not, he is removed from the firing line and given remedial training.

(2) *Task 2 through Task 8, Single and Multiple Engagements.*

(a) The gunner will load one 154-round belt of ammunition and fire bursts of 5 to 7 rounds at double E-type silhouettes at ranges of 800, 400, 700, and a 1,000-meter vehicular target. He will also engage multiple, double E-type silhouettes targets at 400 to 700; 550 and 800; and 400, 500, and 1,000 meters, respectively.

(b) The gunner will observe the beaten zone and make adjustments as necessary to hit the targets.

(c) After firing is completed, weapons will be cleared and the gunner critiqued. If there is a second gunner to fire, the entire procedure will be repeated.

NOTE: After both gunners have fired for practice, they will then be assigned new lanes and issued more ammunition for record firing. If possible, the gunner should keep the same weapon he used for practice fire. The only change in the above procedures will be in the last step; here, the gunner will be scored along with the critique. If the gunner is unable to meet standard, he is removed from the firing line, given remedial training, and re-fired at the commander's discretion.

DANGER

Before any gunner changes lanes, the lane NCO must ensure that the weapon has been properly cleared. Anyone observing any unsafe acts should immediately call a cease fire and notify range personnel.

e. **Scoring Procedures.** One point is given for each target hit with an additional 2-point bonus if the target is hit on the first burst. The total possible points for night fire is 33. A minimum 7 points (7 out of 11 exposures) is required to meet standard on the night-fire course.

NOTE: Firing Table IV is set up for gunners to engage targets out to 1,000 meters under moonlight. However, if visibility is limited by other conditions, then the commander may use his discretion to alter the ranges for better visibility.

TASK	CONDITIONS TARGET/SITUATION	AMMO	STANDARD
1. Zero the AN/TVS-5 to the M2 machine gun.	Gunner is in stationary tripod firing position at 50 meters and engages an M16A1/A2 zero target. Gunner will use single shots for this engagement. Two bursts will be used to spot the device.	12 single 12.7-mm rounds and two 7-round belts.	Gunner must impact 4 of 6 rounds of two 3-round shot groups within a 4-cm circle IAW FM 23-28.
2. Engage a single, double E-type silhouette at 800 meters while using the AN/TVS-5.	Gunner is in stationary tripod firing position and engages an 800-meter, double E-type silhouette. Gunner will use 5-to 7-round bursts for this engagement.	14 rounds of 12.7-mm.	Gunner must impact one burst on the 800-meter target within 20 seconds.
3. Engage a single, double E-type silhouette at 400 meters while using the AN/TVS-5.	Gunner is in stationary tripod firing position and engages a 400-meter, double E-type silhouette. Gunner will use 5-to 7-round bursts for this engagement.	14 rounds of 12.7-mm.	Gunner must impact one burst on the 400-meter target within 20 seconds.
4. Engage a single, double E-type silhouette at 700 meters while using the AN/TVS-5.	Gunner is in stationary tripod firing position and engages a 700-meter, double E-type silhouette. Gunner will use 5-to 7-round bursts for this engagement.	14 rounds of 12.7-mm.	Gunner must impact one burst on the 700-meter target within 25 seconds.
5. Engage a single, double E-type silhouette at 1,000 meters while using the AN/TVS-5.	Gunner is in stationary tripod firing position and engages a 1,000-meter, double E-type silhouette. Gunner will use 5- to 7-round bursts for this engagement.	14 rounds of 12.7-mm.	Gunner must impact one burst on the 1,000-meter target within 25 seconds.
6. Engage multiple, double E-type silhouettes at 400 and 700 meters while using the AN/TVS-5.	Gunner is in stationary tripod firing position and engages 400-and 700-meter, double E-type silhouettes. Gunner will use 5- to 7-round bursts for this engagement.	28 rounds of 12.7-mm.	Gunner must impact one burst on each target within 35 seconds.
7. Engage multiple, double E-type silhouettes at 550 and 800 meters while using the AN/TVS-5.	Gunner is in stationary tripod firing position and engages 550- and 800-meter, double E-type silhouettes. Gunner will use 5- to 7-round bursts for this engagement.	28 rounds of 12.7-mm.	Gunner must impact one burst on each target within 35 seconds.
8. Engage multiple, double E-type silhouettes at 400, 550, and1,000 meters while using the AN/TVS-5.	Gunner is in stationary tripod firing position and engages 400-, 550-, and 1,000-meter, double E-type silhouettes. Gunner will use 5- to 7-round bursts for this engagement.	42 rounds of 12.7-mm.	Gunner must impact one burst on each target within 45 seconds.

Figure C-8. Night Fire Qualification Table IV.

* C-6. QUALIFICATION STANDARDS

To qualify on the M2, the gunner must achieve the minimum standards on the firing tables.

a. **Firing Table I**. On this table, a gunner can earn one point for each round that hits in each space. He can earn up to 119 points on this table, but must earn at least 12 points (12 out of 17 pasters).

b. **Firing Tables II, III, and IV**. On each of these tables, the gunner can earn one point for each target hit plus a two-point bonus if he hits each target with the first burst. He can score up to 33 points on each of these tables, but to qualify, he must score at least 23 points on each.

c. **All Tables**. The gunner firing tables I through IV can earn up to 218 points. However, he must earn at least 12 points on Table I plus at least 23 points on each of the other three firing tables. Thus, he needs at least 12 + (3 x 23) = 12 + 69 = 81 points to qualify:

FIRING TABLE	POINTS NEEDED TO QUALIFY
Firing Table I	12
Firing Table II	23
Firing Table III	23
Firing Table IV	+ 23
	81 TOTAL POINTS NEEDED TO QUALIFY ON M2

d. **Ratings**. The grader rates each gunner as follows:

MAXIMUM	218
EXPERT	196 TO 217
FIRST CLASS	174 TO 195
SECOND CLASS	153 TO 173
UNQUALIFIED	152 OR LESS

e. **Scorecard**. This change prescribes a new edition of DA Form 7007-R. This blank, reproducible form may be copied from the back of this manual onto 8 1/2 by 11-inch paper. It is also available on the Army Electronic Library (AEL) CD-ROM (EM0001) and at the USAPA website located at (http://www.usapa.army.mil).

MACHINE GUN SCORECARD FOR M2

For use of this form, see FM 23-65, Change 1, or the back of this form. The proponent agency is TRADOC.

DATA REQUIRED BY THE PRIVACY ACT OF 1974

AUTHORITY: 10 USC 3012(g)/Executive order 9397.

PRINCIPAL PURPOSE: To aid individual training on targets at various ranges.

ROUTINE USE: To evaluate individual proficiency. SSN is used for positive identification purposes only.

DISCLOSURE: Voluntary. However, mess rating and scoring require some tracking method. Therefore, only those soldiers who provide their SSNs can receive scores or qualify on weapons or equipment.

1a. LAST NAME	1b. FIRST NAME	1c. MI	2. DATE *(YYYYMMDD)*
LEMAY	CURTIS	X	20021211

3. SSN	4. UNIT	5. LANE
222-22-2222	2/29TH INF	4

	6a. FIRING TABLE I			7a. FIRING TABLE II					8a. FIRING TABLE III					9a. FIRING TABLE IV				
TSK	RANGE (M)	HIT	PTS	RANGE (M)	TIME	HIT	PTS	BON	RANGE (M)	TIME	HIT	PTS	BON	RANGE (M)	TIME	HIT	PTS	BON
1	10	NA	NA	550	None	NA	NA	NA	550	None	NA	NA	NA	550	None	NA	NA	NA
2	10	NA	NA	800	20 Sec	X	1	2	800	20 Sec	X	1	2	800	20 Sec	X	1	2
3	10	NA	NA	400	20 Sec	X	1	2	400	20 Sec	X	1	0	400	20 Sec	X	1	2
4	10	NA	NA	700	25 Sec	X	1	2	700	25 Sec	X	1	2	700	25 Sec	X	1	2
5	10	X	20	1,000	25 Sec	X	1	2	1,000	25 Sec	X	1	2	1,000	25 Sec	X	1	2
6	10	X	30	400	35 Sec	X	1	2	400	35 Sec	X	1	2	400	35 Sec	X	1	2
				700		X	1	2	700		X	1	2	700		X	1	2
7	10	X	45	550	35 Sec	X	1	2	560	35 Sec	X	1	0	550	35 Sec	X	1	2
				800		X	1	2	800		X	1	2	800		X	1	2
8	NA	NA	NA	400	45 Sec	X	1	2	400	45 Sec	X	1	2	400	45 Sec	X	1	2
				550		X	1	0	550		X	1	2	550		X	1	2
				1,000		X	1	2	1,000		X	1	2	1,000		X	1	2

6b. SUBTOTAL FIRING TABLE I	7b. SUBTOTAL FIRING TABLE II	8b. SUBTOTAL FIRING TABLE III	9b. SUBTOTAL FIRING TABLE IV
95	31	29	33

10. TOTAL SCORE	11. RATING
188	GUNNER FIRST CLASS

12a. OIC's PRINTED OR TYPED NAME	13a. GRADER'S PRINTED OR TYPED NAME
SSG DWIGHT X. EISENHOWER	SSG EDWIN X. ROMMEL

12b. OIC'S SIGNATURE	13b. GRADER'S SIGNATURE
Dwight X. Eisenhower	Edwin X. Rommel

DA FORM 7007-R, JAN 2002 — DA FORM 7007-R, JUN 1991, IS OBSOLETE. — USAPA V1.00

Figure C-9. Example completed DA Form 7007-R (front).

The following procedure will be used to fill out the scorecard.

1. NAME: Enter last name , first name, and middle initial.

2. SSN: Enter gunner's social security number.

3. UNIT: Enter unit designation.

4. DATE: Enter date of firing.

5. LANE: Enter the lane number for the firing point of your gunner.

6. HIT: When firing Table I, enter the number of rounds impacting within the scoring space in task numbers 5, 6, and 7. For Tables II, III, and IV, enter an X when target is hit.

7. POINTS: Enter 1 point for each round impacting within the scoring space in task numbers 5, 6, and 7 in Table I. Gunners will receive 1 point for each target hit in Tables II, III, and IV.

8. BONUS: Entry is used only for Tables II, III, and IV. Two bonus points are awarded when gunner hits the target on initial burst of each target exposure.

9. TOTAL SCORE: Enter the total of combined hits, points, and bonus scores.

NOTE: It is possible that a gunner may score the required points outlined for qualification. However the gunner must also score 70 percent on each firing table before he is considered a qualified gunner.

MAXIMUM	218
EXPERT	196-218
GUNNER FIRST CLASS	174-195
GUNNER SECOND CLASS	153-173
UNQUALIFIED	152 and below

Back, DA Form 7007-R, Jun 91

Figure C-10. Example DA Form 7007-R, back.

Section II. MULTIPURPOSE RANGE COMPLEX

The multipurpose range complex is designed to conduct armor and mobilized machine gun firing. Detailed setup and target configuration is described in FM 25-7; its layout is shown in Figure C-11. This section also discusses the mounted firing exercise, requirements for crews, rating procedures, and standards.

C-7. DESIGN SPECIFICS

The tower and administrative facilities must be positioned so they do not impede tactical maneuver onto or off of the range.

a. Hardened emplacements are for aerial gunnery exercises in addition to tank and BFV gunnery.

b. System calibration targets must be provided at ranges of 950, 1,200, and 1,500 meters, as shown in Figure C-11.

c. Gunnery tasks that require the use of dud-producing ammunition cannot be fired on the range proper. Provisions for these tasks must be made in impact areas adjacent to the range.

d. Double target mechanisms are recommended for use in the first six target groupings for BFV training, or for positioning throughout the range as needed to support local training requirements.

e. The administrative area is not shown in exact location or scale.

f. Additional defilade positions maybe required for BFV gunnery.

g. The addition of battle positions and positioning of personnel targets closer to the baseline may be necessary to permit BFV dismounted infantry training.

Carrier 2 will upload and wait at the ammunition point. When the course is clear, carrier 2 will move to the initial fire point followed by carrier 4 and proceed as stated. Carrier 3 should now be at the ammunition point for uploading.

(1) *Task 1, Conduct Prefire Inspection.*

(a) The crew will move forward to the initial firing position. Once the carrier is in position, the crew will perform a prefire inspection by setting the headspace and timing, and load a 7-round belt of ammunition.

(b) The gunner will then fire the 7-round belt at a 450-meter target to ensure the machine gun is operational.

NOTE: If the machine gun fails to fire or fires sluggishly, recheck the headspace and timing and attempt to fire again. If the situation continues, clear the weapon and have the carrier move off the firing course and report to the armorer to have the problem corrected.

(c) Upon completion of Task 1, the gunner will clear the weapon and await further instructions.

NOTE: To accomplish Task 2 and Task 3, the gunner will load one 28-round belt of ammunition.

(2) *Task 2, Engage Stationary Target From a Mounted Stationary Firing Position.*

(a) The gunner will engage a 450-meter stationary personnel target from the initial firing location.

(b) The gunner will use a 5- to 7-round burst for this engagement.

(c) The gunner will observe the beaten zone and make adjustments as necessary to hit the target.

(d) Upon completion of Task 2, the gunner will await further instructions.

(3) *Task 3, Engage Stationary Target From a Stationary Firing Position.*

(a) From the initial firing location, the gunner will engage an 850-meter stationary vehicle target.

(b) The gunner will use a 5- to 7-round burst for this engagement.

(c) The gunner will observe the beaten zone and make adjustments as necessary to hit the target.

(d) Upon completion of Task 3, the gunner will await further instructions.

NOTE: On command from the grader, the crew will move the carrier to the next predetermined firing location, load one 28-round belt, and conduct Task 4.

(4) *Task 4, Engage Moving and Stationary Targets From a Mounted Stationary Firing Position.*

(a) From the second predetermined firing location, the gunner will engage one moving vehicle target at a range of 800 meters and one stationary vehicle target at a range of 1,000 meters.

(b) The gunner will use a 5-to 7-round burst for this engagement.

(c) The gunner will observe the beaten zone and make adjustments as necessary to hit the targets.

(d) Upon completion of firing Task 4, the gunner will await further instructions.

NOTE: When conducting Task 5, the gunner will load one 28-round belt and engage targets from a moving carrier along a predetermined route.

(5) *Task 5, Engage Stationary Targets While Firing From a Moving Carrier.*

(a) The carrier will be moving along the predetermined firing route, at which time the gunner will engage stationary vehicle targets at ranges of 300 and 500 meters.

(b) The gunner will use a 5- to 7-round burst for this engagement.

(c) The gunner will observe the beaten zone and make adjustments as necessary to hit the targets.

(d) Upon completion of Task 5, the gunner will await further instructions.

NOTE: On command from the grader, the crew will move the carrier to the third predetermined firing location, load one 28-round belt, and conduct Task 6.

(6) *Task 6, Engage a Stationary and a Moving Target From a Mounted Stationary Firing Position.*

(a) From the third predetermined firing location, the gunner will engage one 600-meter personnel target and one 800-meter moving vehicle target from a stationary firing position.

(b) The gunner will use a 5- to 7-round burst for this engagement.

(c) The gunner will observe the beaten zone and make adjustments as necessary to hit the targets.

(d) Upon completion of Task 6, the gunner will await further instructions.

NOTE: From the same firing location as Task 6, the crew will be placed in a chemical environment and be required to go to MOPP4.

(7) *Task 7, React in Chemical Environment.*

(a) Upon command, the crew will be required to react to a chemical agent.

(b) Upon completion of Task 7, the gunner will await further instructions.

NOTE: To accomplish Task 8 through Task 10, the gunner will load one 84-round belt of ammunition.

(8) *Task 8, Engage Stationary Target From a Mounted Stationary Firing Position While in MOPP4.*

(a) From the same firing location as Task 6, the gunner will engage a stationary target at a range of 500 meters from a mounted stationary firing position while in MOPP4.

(b) The gunner will use a 5- to 7-round burst for this engagement.

(c) The gunner will observe the beaten zone and make adjustments as necessary to hit the target.

(d) Upon completion of Task 8, the gunner will await further instructions.

(9) *Task 9, Engage Moving and Stationary Targets From a Mounted Stationary Firing Position While in MOPP4.*

(a) From the same firing location, the gunner will engage one 800-meter moving vehicle target and personnel targets at 1,000 meters.

(b) The gunner will use a 5- to 7-round burst for this engagement.

(c) The gunner will observe the beaten zone and make adjustments as necessary to hit the targets.

(d) Upon completion of Task 9, the gunner will await further instructions.

(10) *Task 10, Engage Multiple Stationary Targets in a Mounted Stationary Firing Position While in MOPP4.*

(a) From the same firing location, the gunner will engage one 300-meter personnel target and two stationary vehicle targets at ranges of 500 and 700 meters.

(b) The gunner will use a 5- to 7-round burst for this engagement.

(c) The gunner will observe the beaten zone and make necessary adjustments to hit the target.

(d) Upon completion of firing Task 10, the gunner will clear his machine gun, have it inspected by the safety officer, move from the firing line, and turn in any excess ammunition. Then the crew will be critiqued.

TASK	CONDITIONS TARGET/SITUATION	AMMO	STANDARD
1. Conduct prefire inspection.	At a stationary position given a 450-meter stationary target, headspace and timing gauge, unit's organic vehicle, and M2 machine gun.	7-round belt of 12.7-mm.	Conduct prefire inspection and fire 7 rounds to confirm operation of machine gun.
2. Engage stationary target from a stationary firing position.	Gunner is in a stationary mounted firing position and engages a 450-meter stationary personnel target. Gunner must use 5- to 7-round bursts for this engagement.	14 rounds of 12.7-mm.	Gunner must impact on target.
3. Engage stationary target from a stationary firing position.	Gunner is in a stationary mounted firing position and engages a 850-meter stationary vehicle. Gunner must use 5- to 7-round bursts for this engagement.	14 rounds of 12.7-mm.	Gunner must impact on target.
4. Engage moving and stationary targets from another stationary firing position.	Gunner is in a stationary mounted firing position and engages a moving vehicle at 800 meters and stationary vehicle at 1,000 meters. Gunner must use 5- to 7-round bursts for this engagement.	28-round belt of 12.7-mm.	Gunner must impact on each target.
5.Engage stationary targets while firing from a moving vehicle.	While the gunner's vehicle is moving, the gunner engages stationary vehicles at 300 and 500 meters. gunner must use 5- to 7-round bursts for this engagement.	28-round belt of 12.7-mm.	Gunner must impact on each target.
6. Engages a stationary and moving target from a stationary firing position.	Gunner is in a stationary firing position and engages a 600-meter personnel and 800-meter moving vehicle targets. gunner will use 5- to 7-round bursts for this engagement.	28-round belt of 12.7-mm.	Gunner must impact on each target.
7. React to chemical agent.	While in a stationary position, given protective mask and protective clothing, crew must react to chemical agent.	N/A	In accordance with FM 3-3.
8. Engage stationary target from a stationary firing position while in an upgraded NBC posture.	Gunner is in a stationary firing position and engages a 500-meter vehicle target. gunner will use 5- to 7-round bursts for this engagement.	14 rounds of 12.7-mm.	Gunner must impact on target.

Figure C-12. Mounted Firing Exercise Table VI.

TASK	CONDITIONS TARGET/SITUATION	AMMO	STANDARD
9. Engage moving and stationary targets while in a stationary firing position in an upgraded NBC posture.	Gunner will engage 800-meter moving vehicle target and personnel targets at 1,000 meters. Gunner will use 5- to 7-round bursts for this engagement.	28 rounds of 12.7-mm.	Gunner must impact on targets.
10. Engage multiple stationary targets while in a stationary firing position in an upgraded NBC posture.	Gunner will engage 300-meter personnel targets, 500-and 700-meter vehicle targets. Gunner will use 5- to 7-round bursts for this engagement.	42 rounds of 12.7-mm.	Gunner must impact on each target.

Figure C-12. Mounted Firing Exercise Table VI, continued.

e. **Scoring Procedures.** The mounted firing exercise is graded on a Go/No Go basis. Commanders may increase the difficulty of any or all tasks to align the exercise with the unit's mission.

C-9. PREDETERMINED FIRING EXERCISE

The predetermined firing exercise is for the gun crew that has demonstrated proficiency during the basic phase of gunnery. Emphasis will be on developing range cards and confirming range-card data during day and night firing.

a. **Objectives.** The objective of this training is to reinforce what was developed in the fundamental gunnery phases. It is designed to increase the effectiveness of the M2 MG crew by building their confidence to quickly and accurately deliver a large volume of fire on a prescribed target.

b. **Organization.** The unit is assembled in the bleachers, given instructions, and briefed on training that will be conducted while they are on the range. After briefing, they will be organized into gun crews and moved to firing lanes. Lanes will be used IAW local range policies.

c. **Ammunition.** This exercise requires 168 rounds of 12.7-mm linked ammunition. The gunner is allotted two bursts per target during the day phase and one burst per target during the night phase. Each gunner will be issued two belts of ammunition; one 112-round belt for the day phase and one 56-round belt for the night phase.

d **Firing Sequence.** The sequence of firing will be conducted IAW Firing Table VII (Figure C-13, page 33). The suggested sequence of firing is as follows:

(1) *Day phase.*

(a) Task 1, Prepare a Range Card. Once the gunner is assigned his firing point, he must prepare a range card for that position. Range cards must be prepared IAW paragraph E-2.

(b) Task 2, Obtain Direction and Elevation Readings for the Final Protective Line. During this task, the gunner is allotted 28 rounds to obtain and record the direction and elevation reading of his FPL.

(c) Task 3, Obtain Direction and Elevation Readings for Point Targets. The gunner will be required to obtain and record direction and elevation readings for point targets located at 400, 600, 800, and 1,000 meters. He is allotted 56 rounds for this task.

(d) Task 4, Obtain Direction and Elevation Readings for Linear Targets. The gunner will be required to obtain and record direction and elevation readings for linear targets located at 600 and 800 meters. He is allotted 28 rounds for this task.

(2) *Night phase.*

NOTE: Each crew is scored as a group in the night phase. (The day phase is not scored.) The crew applies the data obtained during the day and engages their targets. Each crew receives 10 points for each target engagement. A minimum of 40 out of a possible 70 is required. This exercise is not a requirement for qualification; however, commanders can use this training to test their gun crews' proficiency.

(a) Task 5, Engage Point Targets Using Range-Card Data. The gunner must engage point targets located at 400, 600, 800, and 1,000 meters using range-card data. He is allotted 28 rounds for this task. The grader will announce the sequence of engagements.

(b) Task 6, Engage Area Targets Using Range-Card Data. The gunner must engage linear targets located at 600 and 800 meters using range-card data. He is allotted 14 rounds for this task. The grader will announce the sequence of engagements.

(c) Task 7, Fire Final Protective Line. The gunner will fire his FPL to obtain grazing fire. He is allotted 14 rounds for this task.

e. **Conduct of Firing.** The gunner, assistant, and leader will prepare a range card for that position. Once complete, each crew will be given 168 rounds as prescribed in a predetermined firing table.

(1) Each gunner initially lays on his target by using the dry-fire technique. Each crew is issued 168 rounds of ammunition, and the gunner is ordered to load by the group NCO. When all the gunners of a group are ready, the NCO announces, "Up," to the officer in charge.

(2) When all groups are ready to fire, the OIC announces, "Give me an 'Up' when you have engaged all targets and have obtained the correct data to all targets. You are clear to fire."

(3) When the gunner has correctly engaged his target (FPL), the leader records the information from the T&E mechanism and traversing bar onto a range card. Then, the assistant gunner becomes the gunner, the gunner becomes the leader, and the leader becomes the assistant gunner.

(4) As each member of the crew becomes the gunner, he fires at a different preselected target (order number 2, a linear target; order number 3, a point target).

(5) After each crew member has been a gunner, assistant gunner, and leader and all data on the three targets have been obtained, the group NCO clears and checks the machine guns of his group and announces to the OIC, "Group cleared and checked."

TASK	CONDITIONS TARGET/SITUATION	AMMO	STANDARD
1. Prepare a range card for the M2 machine gun.	The gunner is given a range card, a selected firing position, M2 machine gun, M3 tripod, and silhouette targets located from 400 to 1,000 meters.	N/A	Gunner must prepare a range card IAW Appendix E, FM 23-65.
2. Determine and record direction and elevation for final protective line.	The gunner is allotted 12.7-mm linked ammo, and a sector of fire with limits.	28 rounds of 12.7-mm.	Gunner must record the direction and elevation for his range card.
3. Determine and record direction and elevation for point targets.	The gunner is allotted 12.7-mm linked ammo, four E-type silhouettes located at 400, 600, 800, and 1,000 meters.	56 rounds of 12.7-mm.	Gunner must impact one burst on each target.
4. Determine and record direction and elevation for area targets.	The gunner is allotted 12.7-mm linked ammo, two E-type silhouettes located at 600 and 800 meters.	28 rounds of 12.7-mm.	Gunner must impact one burst on each target.
5. Engage point targets using range card data.	The gunner is allotted 12.7-mm linked ammo, the same firing position, and his prepared range card.	28 rounds of 12.7-mm.	Gunner must impact one burst on each target.

Figure C-13. Predetermined Firing Table VII.

TASK	CONDITIONS TARGET/SITUATION	AMMO	STANDARD
6. Engage area targets using range card data.	The gunner is allotted 12.7-mm linked ammo, and his prepared range card.	14 rounds of 12.7-mm.	Gunner must impact one burst on each target.
7. Fire final protective line.	The gunner is allotted 12.7-mm linked ammo, and his prepared range card.	14 rounds of 12.7-mm.	Gunner must obtain grazing fire when firing his FPL.

Figure C-13. Predetermined Firing Table VII, continued.

APPENDIX D

FIGHTING POSITIONS

Fighting positions are stable platforms from which a gunner can engage the enemy. Fighting positions are dug as often as possible, but for the most part only when a unit is in the defense and has the time. This appendix discusses the construction of only two fighting positions: the one-, two-, or three-man, tripod-mounted position; and the M63 antiaircraft emplacement.

D-1. CONSTRUCTION CONSIDERATIONS

When constructing a fighting position, the leader and gunner should consider factors that affect the position, such as cover, concealment, fields of fire, size, and improvements.

a. The cover in a machine gun fighting position protects the gun crew from frontal small-arms fire and from fragments of high explosive shells impacting within 3 to 5 meters of the position. The frontal parapet must be at least one M16 rifle length thick and high enough to hide the helmets of the soldiers in the position. Overhead cover built out of sandbags stacked on logs will protect from shell fragments. The logs should be at least 6 inches thick, with at least two layers of sandbags over them (Figure D-1).

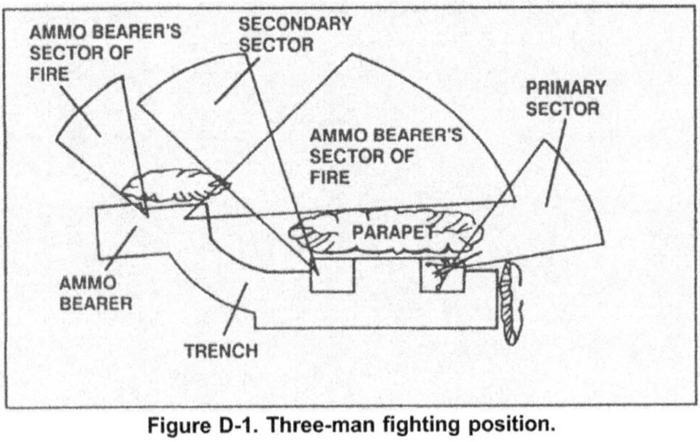

Figure D-1. Three-man fighting position.

b. Concealment hides the machine gun fighting position from direct observation by the enemy, who will make every attempt to locate and destroy heavy machine gun positions early in the battle. The position must be made to blend with its surroundings with both natural and man-made camouflage. Excess dirt from the initial digging should be moved away from the position, and the sod used to re-cover the overhead protection. From the enemy side, the position should not be visible at 35 meters or less (hand grenade range). It should never be obvious even at close range. Use sod or dampen the ground in front of the muzzle to reduce the dust cloud caused by firing. Consider what the position looks like from the air and camouflage it to blend with the surrounding terrain.

c. Selective clearing that does not destroy natural camouflage may be necessary to provide good observation and fields of fire in both primary and secondary sectors of fire.

d. The position should be armpit deep, be wide enough to allow two soldiers with load-bearing equipment to move freely, and have two distinct firing platforms.

e. Improvements to the position may include a grenade sump, a sloping floor with shallow trench for drainage, and a rear parapet for protection against shell fragments and small-arms fire from the rear (other friendly positions or supporting fire from armored personnel carriers).

D-2. CONSTRUCTION OF A TRIPOD-MOUNTED POSITION

After being assigned a sector of fire with an FPL or a PDF, the gun crew begins constructing the fighting position. The tripod is placed in position first and marked so that the weapon will be pointed in the general direction of the target area. A preliminary sketched range card is drawn to show the limits of the sector. The gun crew then outlines the shape of the platform and hole to include the area for the frontal cover in the ground (Figure D-2). The crew then starts digging out the platforms. When they get about 4 to 6 inches deep, the MG is put in place to cover the primary sector of fire until construction is complete.

a. When assigned an FPL, emplace the gun by locking the traversing slide to the extreme left or right of the traversing bar, depending on which side of the primary sector the FPL is on. Align the barrel on the FPL by shifting the tripod.

NOTE: No direction entry is needed in the data section of the range card for the FPL.

b. When assigned a PDF, emplace the gun by locking the traversing slide at the center of the traversing bar. Shift the tripod and gun until the barrel is aimed at the center of the sector. Check coverage of the sector limits by traversing the gun fully left and right.

Figure D-2. Planning the fighting position.

NOTE: In the data section of the range card, record the direction and elevation data of the PDF and the sector limits from the T&E mechanism.

c. The crew digging the hole uses the dug-up dirt to build up the cover – first for frontal cover and then for flank and rear cover. They dig the hole deep enough to protect the crew and still let the gunner shoot with comfort (usually about armpit deep) (Figure D-3). They fix the tripod legs in place by digging, sandbagging, or staking them down. This will ensure that the gun does not shift during firing, which would render the range card data useless.

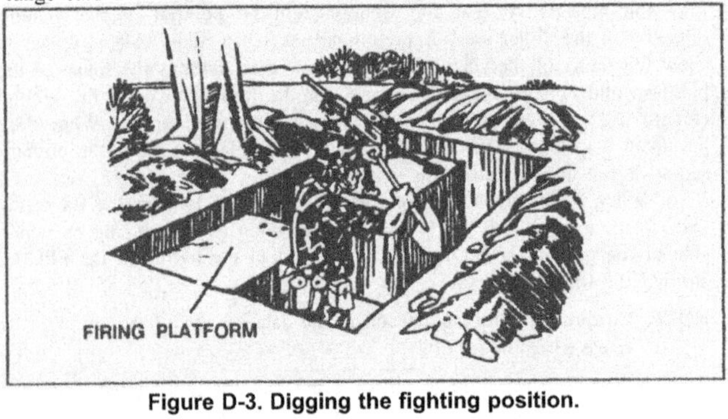

Figure D-3. Digging the fighting position.

(1) The crew digs three trench-shaped grenade sumps at various points where the crew can kick grenades into them (Figure D-4).

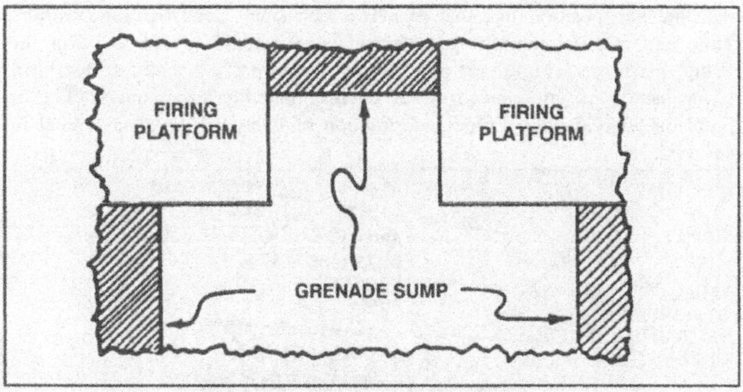

Figure D-4. Digging grenade sumps.

(2) When a position does not have a secondary sector of fire, the crew digs only half of the position (Figure D-5).

Figure D-5. Half of a position.

(3) When a position has both a primary and secondary sector, the crew prepares two firing platforms. The crew prepares overhead cover for a machine gun position like that of a two-man small arms fighting position. Time and material permitting, overhead cover should extend to cover the firing platforms (Figure D-6). Failure to properly construct overhead cover can result in reduced fields of fire, inability to mount NVDs, or problems in reloading. Proper construction of overhead cover is critical to survival.

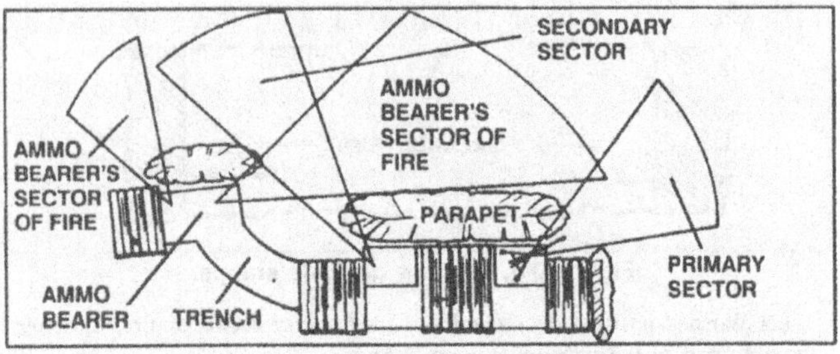

Figure D-6. Two firing platforms with overhead cover.

(4) When there is a three-man crew for the machine gun, the ammunition bearer digs a one-man fighting position to the flank of the gun position so that he can see and shoot to the oblique. This will allow him to cover the front of the machine gun's position (Figure D-7).

Figure D-7. Ammo bearer covering the front.

D-3. CONSTRUCTION OF AN ANTIAIRCRAFT EMPLACEMENT

Place the machine gun on the M63 mount alongside the designated location of the gun position. If needed, it can fire at any time during the construction process. Construct a circular position, with sufficient room to traverse the gun 360 degrees (normally, three and a half M16 lengths in diameter and one M16 length deep). Use the dirt dug out of the circular hole to construct a wall of sandbags around the position. Using sandbags, build a stable platform for the mount. Stake down the legs of the mount and place sandbags over them. This will prevent the mount from tipping backward when low-level targets are engaged. Include the M63-mounted machine gun's fire in the plan for defending the unit against ground attack (Figure D-8).

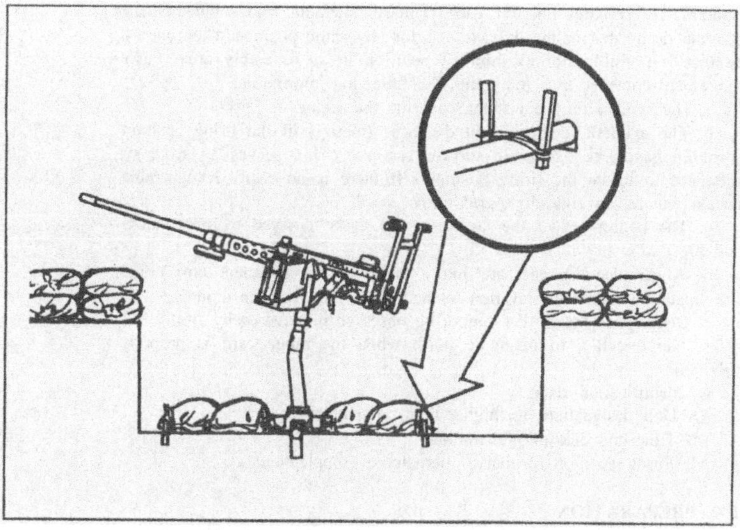

Figure D-8. Open pit-type antiaircraft position.

APPENDIX E

RANGE CARDS

A range card is a sketch or diagram of the terrain that a weapon is assigned to cover by fire. It shows possible target areas and terrain features plotted in relation to a firing position. The information on a range card is used for planning and controlling fire, for rapidly detecting and engaging targets, and for orienting replacement personnel or units. Therefore DA Form 5517-R, should be used to record the information.

E-1. RANGE CARD SECTIONS

Each machine gunner normally prepares range cards for his fighting position. He prepares one for each primary, alternate, and supplementary position designated in the defense and for any static position when enemy contact is possible; for example, a position in an assembly area. Each range card contains, as a minimum, the following information:

a. The symbol for the weapon covering the sector.

b. The azimuth (degrees) and distances (meters) of the firing position from an easily recognizable terrain feature. (This serves as an easy reference to locate the firing position.) If there is no easily recognizable terrain feature, an eight-digit grid maybe used.

c. The boundaries of the area assigned to be covered by observation and fire.

d. Areas where targets are likely to appear (engagement areas) and the range, azimuth, and elevation to them from the firing position.

e. Dead space (areas that cannot be observed or covered by fire).

f. The direction of magnetic north when the range card is properly oriented.

g. Identification data.

- Unit designation (no higher than company).
- Time and date of preparation.
- Firing position (primary, alternate, or supplementary).

E-2. PREPARATION

The following steps are taken to prepare a range card. (See Figure E-1, page E-3, for an example of a completed range card).

a. Draw the symbol for the MG in the lower center of the range card.

b. Show the sector of fire by drawing solid lines from the weapon symbol to the left and right limits. Sketch any easily recognizable terrain features that can be used to identify the sector. In the data section, indicate the magnetic azimuth and the range to the far limits of the sectors of fire.

NOTE: The left and right limits are labeled one and two, respectively, in the sector sketch and a circle drawn around each number. The left and right limits should be the first items drawn and labeled in the sector sketch and the data section of the range card. (The azimuth and range are not placed on the solid lines for left or right limit.)

c. Place target reference points at the location designated by the platoon leader or unit SOP, and at any other locations where a target is likely to appear. Number each TRP and likely target locations in the sketch section of the range card.

d. Draw a maximum engagement line across the sector of fire for the MG. These lines depict the maximum ranges at which a target can be effectively engaged.

e. Show dead space or areas where targets cannot be engaged with direct fire by drawing diagonal lines across the areas and writing the words DEAD SPACE.

f. Draw a magnetic north arrow on the range card to orient it with the terrain. Then add identification data – unit designation (no higher than company level), time and date of preparation, and type of position (primary, alternate, or supplementary).

g. Enter the information for the weapon reference point in the remarks block on the range card.

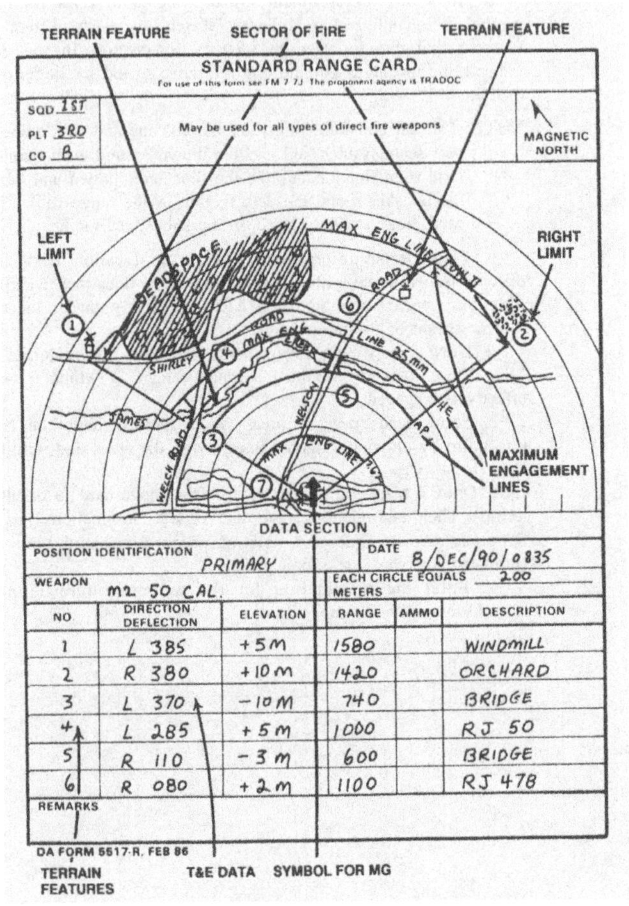

NO	DIRECTION DEFLECTION	ELEVATION	RANGE	AMMO	DESCRIPTION
1	L 385	+5m	1580		WINDMILL
2	R 380	+10m	1420		ORCHARD
3	L 370	-10m	740		BRIDGE
4	L 285	+5m	1000		RJ 50
5	R 110	-3m	600		BRIDGE
6	R 080	+2m	1100		RJ 478

Figure E-1. Range card.

* APPENDIX F

AERIAL DEFENSE

The two methods for defending against aerial attacks are passive air defense and active air defense. Passive measures are taken to avoid air detection; active measures combat air attacks.

F-1. PASSIVE AIR DEFENSE

Passive air defense measures are a first line of defense. They include troop or vehicle dispersion, concealment and camouflage, and observation and reporting. Attack avoidance is based on the reasoning that what can be seen from the air can be destroyed and what cannot be seen will probably not be attacked.

F-2. ACTIVE AIR DEFENSE

Active air defense comes into play when a unit has been detected by hostile aircraft or when ordered to interdict hostile aircraft. Volume of fire is the key to its effectiveness. Because the action is so fast, the response must be coordinated and tightly controlled, usually at platoon level. At company level, too many are involved, making response time too slow. At section and squad level, too few are involved to achieve the volume of fire needed for an effective defense.

NOTE: All aircraft engagement firing is done at the cyclic rate.

F-3. AERIAL ENGAGEMENT

In engaging aerial targets, there are two requirements that must be considered: lead and line. Superelevation (compensation made for the pull of gravity on the projectile) is another consideration for some weapons, but the caliber .50 MG projectile is basically flat out to 800 meters. Therefore the gunner does not have to worry about it. He can instead observe the tracers crossing the target course line and make the required adjustment from them.

a. **Lead.** Leading the aircraft is the compensation made for its speed when aiming the weapon so the projectile and the aircraft will arrive at the same spot at the same time. The required lead for a given aircraft is always calculated as a certain number of lengths of that aircraft. A high performance aircraft will need a greater lead factor than a slow one. The gunner should be taught to estimate leads by projecting images of the fuselage in the front of the aircraft.

(1) If the gunner is able to track an aircraft and lead the air craft with a constant number of leads, there will be two points along the target course line where he will attain hits. As long as the gunner understands this principle, he can bring effective fire on an aircraft.

* (2) There are many different types of hostile aircraft (FM 44-80). To simplify things for the gunner, all aircraft are classified into two types: high performance and low performance. Any aircraft that has an estimated speed of more than 150 knots is classified as high performance and anything slower is low performance.

b. **Line.** Line is the requirement that the projectile must intersect the target course line of the aircraft. If the projectiles do not pass through this target course line, the aircraft cannot be engaged.

F-4. FLY-THROUGH TECHNIQUE

As an aerial target moves along its target course line, the lead required to engage the target changes because the range and angle between the aircraft and the gun position are changing. The lead required increases from the initial sighting (A in Figure F-1) to midpoint (B in Figure F-1). The required lead is at its maximum at midpoint.

a. As the aircraft moves beyond the midpoint on the target course line, the lead needed begins to decrease. Therefore, by leading an aerial target by less than the maximum required lead, there will be two points along the target course line where the constant lead will be the correct lead and the aircraft will fly through the projectiles.

b. In instructing the fly-through method of engagement, do not teach the gunners to bring continuous fire from points A through E, teach them to bring fire at points A through B and again before point D through the second fly through. As the gunner becomes proficient in this method, he will learn to adjust his lead and bring longer and more accurate fire on the target.

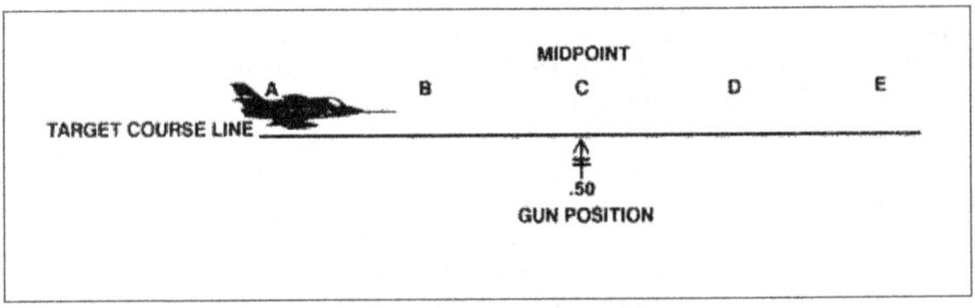

Figure F-1. Aerial target engagement.

* APPENDIX G

NIGHT OPTICS

Weapons-mounted night optics are less versatile than goggles with regard to situational awareness. However, they offer greater magnification and resolution at crew-served weapons ranges.

The graduated reticles on night optics offer the flexibility of engaging at various ranges. Laser pointers fix on one range, so the firer must "aim off" before engaging targets at a different range.

Night optics weigh more than goggles--they are awkward to move with and employ. Using night optics requires that the firer make specific adjustments to firing positions. Also, night optics might work poorly in some ambient light or thermal conditions.

Despite their drawbacks, properly boresighted night optics used in the right conditions offer units an extraordinary option: to engage targets beyond the range of opposing force weapons at night.

Section I. AN/PAS-13 (V3) HEAVY WEAPON THERMAL SIGHT

The AN/PAS-13 (V3) heavy weapon thermal sight (HWTS) (Figure G-1, page G-2) is a silent, lightweight, compact, durable, battery-powered, infrared imaging sensor that consumes little battery power. The self-contained infrared (IR) imaging sensor in this sight helps the firer acquire the target in low visibility conditions.

G-1. EQUIPMENT DESCRIPTION AND OPERATION

The HWTS works well at night and in the daytime. The telescope's IR sensor receives infrared light, converts it to digital data, processes it, and then displays it digitally as an infrared image for the user.

a. **Components**. The HWTS has two functional groups: the basic sensor and the telescope. See Figure G-1, page G-2, for equipment data.

(1) ***Basic Sensor***. The scanner reflects the IR light it receives from the telescope onto the detective (sensor) assembly. The assembly senses IR light, converts it into a video image, then conditions the video for display on the LED array. The LED array illuminates both the IR image and the reticle. This light image reflects off the scanner, which forms the actual image the firer sees in the eyepiece.

(2) ***Telescope***. The telescope receives IR light from an intended target and its surroundings, and then magnifies and projects this light onto the scanner on the basic sensor.

b. **Compatibility**. The heavy weapon thermal sight fits the weapons shown in Figure G-2:

- M2 .50-caliber machine gun.
- M24 sniper rifle.
- MK 19 grenade machine gun.
- Squad leader's M16.

c. **Operational Modes**. To place the HWTS in operational mode--

(1) If you have an AN/PAS-13 (V3), see TM 11-5855-302-12&P.

(2) If you have an AN/PAS-13 (OMNI), see TM 11-5855-312-10.

Weight:	4.5 pounds
Field of View (FOV):	
Narrow	3 degrees
Wide	9 degrees
Telescopic Magnification:	
Narrow FOV	10X
Wide FOV	3.3X
Dimensions:	
Length:	18 inches
Width:	6.25 inches
Height:	6.25 inches
Power Consumption:	
ON mode	12 watts
STANDBY mode	5.5 watts
Battery Life:	
71% STANDBY mode	10 hours
EMERGENCY mode	3.5 hours

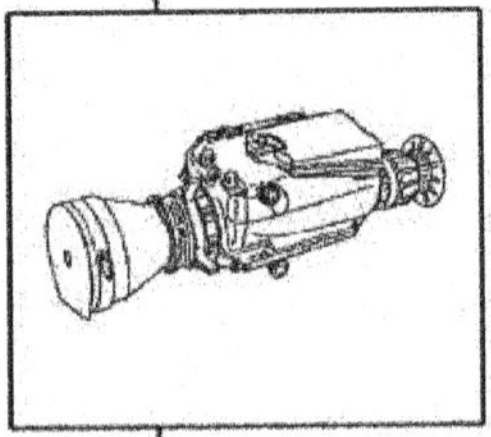

Figure G-1. Heavy weapon thermal sight.

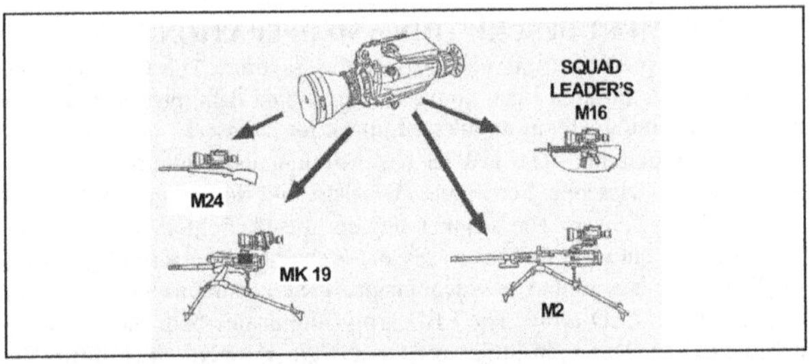

Figure G-2. Weapons that use the AN/PAS-13 (V3) HWTS.

G-2. FUNDAMENTALS OF MARKSMANSHIP

The fundamentals of marksmanship remain about the same for the AN/PAS-13 (V3) heavy weapon thermal sight (HWTS) as they do for the

AN/PVS-4. (TM 11-5855-302-12&P discusses HWTS reticle selection and point of aim.)

a. **Firing Position**.

(1) ***Sitting***. When using the tripod in the low or high position, sit directly behind the gun between the trail legs of the tripod. Extend your legs under the tripod, cross them, or brace your feet on the tripod. Regardless of which you choose, place your elbows on the inside of your thighs for support. Place your right eye in the HWTS eyepiece.

(2) ***Standing***. With the M2 mounted on a vehicle, stand with both hands on the control grips and your thumbs resting on the trigger. Keep your elbows against your body, your body forward, and your chest against your hands to brace the gun. Place your right eye in the HWTS eyepiece.

(3) ***Kneeling***. When using the M2 in a fighting or hasty tripod-mounted position, kneel and grasp the control grips with your thumbs on the trigger. Place your eye in the HWTS eyepiece.

(4) ***Prone***. Use the prone position when firing from a tripod that is sitting in a low position. Lie on the ground directly behind the gun and spread your legs a comfortable distance apart, with your toes pointed outward. Rest your left elbow on the ground and grasp the elevating handwheel of the T&E mechanism with your left hand. Grasp the right spade grip with your right hand, with the right thumb in position to press the trigger.

b. **Aiming**.

(1) Determine the range to the target. The aiming point of the M2 reticle depends on the range to the target. To determine range, you can use a range card, estimate range, use the reticle's aiming box, or use TRPs. The HWTS has two M2 reticles--a wide and a narrow field of view.

(2) Use the wide or narrow field of view to scan, to help determine range, and to engage targets.

(3) Use the indicators on the reticle to determine range, or use other methods of range determination.

(4) Set the mounting bracket range selector to NEAR for any target less than 1,400 meters and to FAR for any target beyond 1,400 meters.

(5) Using the T&E mechanism, sight the reticle aiming point on the target (bottom center).

(6) Take the slack out of the T&E by holding the M2 down and to one side. A loose T&E mechanism can move the HWTS reticle off the target's aiming point.

NOTE: Squad leader: observe the impact of the round to help the gunner adjust the sight to bring the strike of the round onto the target.

c. **Breath Control**. This fundamental of marksmanship does not change.

d. **Trigger Squeeze**. This fundamental of marksmanship does not change.

> **WARNING**
>
> ***Do not* press your eye hard against the HWTS eyepiece while engaging a target. The M2 kicks back slightly and could harm you. Secure the M2 as well as you can to increase accuracy and to help prevent injury.**

e. **Boresighting**. This paragraph provides guidance on boresighting. For zeroing procedures, see TM 11-5855-302-12&P and TM 11-5855-312-10. (This stays in the pocket of the sight's soft storage case.) Boresight as follows:

(1) Place the M2 in the ready-to-fire position 10 meters from the boresight target offset. To avoid a failed boresight zero, level the target and weapon before you adjust the laser.

NOTE: The HWTS will not focus in on a silhouette copied to plain paper. To create a boresight target offset, use a 25-meter M16 zero target and attach to it a copy of the boresight target offset.

(2) Mount the HWTS and boresight the weapon. Set the M2 mount to the NEAR position.

(3) Use the reticle select switch to change to the M2 reticle. The reticle type displays in the upper right corner on the raster.

(4) Make adjustments with the T&E until the bore light moves onto the bore light aiming point on the boresight target offset.

(5) Adjust the reticle with the reticle adjuster on the right side of the HWTS. Remember to hold the adjuster down for a count of five. Press and hold the adjuster up to move the reticle up, down to move the reticle down, forward to move the reticle left, and back to move the reticle right.

(6) Use the vertical and horizontal reticle adjuster to move the 1,000-meter aiming point of the reticle to the HWTS aiming point on the boresight target offset.

(7) Recheck the bore laser point of aim and make sure the reticle remains in the center mass of the HWTS aim point on the boresight target offset.

TRAINING STRATEGIES:	When using night optics, M2 gunners use night initial training strategies. Additional nonfiring exercises also apply.

f. **Qualifications**. Changes to M2 qualifications accommodate new night vision sights and laser pointers. Using the new tables reduces the time and ammunition required to train and qualify gun crews. Lack of ranges and other resources previously limited qualification options. The new tables increase the opportunities for qualification and the flexibility in the process of qualification. Simpler scoring and grading procedures allow gunners to maintain a high level of proficiency.

(1) With the availability of night vision sights and laser pointers, the current night standards are not challenging. Taking into consideration the "own the night" concept, gunners need to qualify at night to the same or a greater standard than we do during the day. Using these proposed qualification tables and training strategies, the gunner can do so. Each night practice and qualification table has three variations, to reflect the different range, weapon, and sight combinations. Each day table has a variation for each type target.

(2) Use the training strategies and preliminary marksmanship instruction previously described to prepare for day practice and qualification tables. Graders need to prepare, and to ensure ranges are prepared, in accordance to the latest published manuals on the subject. After the gunner trains, he should use Firing Table I then Firing Table II. If he fails to meet the standards in Firing Table I, he must repeat the training and fire the table again. Gunners *must* pass the practice table before moving on to the qualification table.

(3) The gunner must pass day qualification before proceeding to night qualification training.

g. **Nonfiring Exercise**. Use this exercise to train the gunner to use the T&E in order to sight the laser on target.

(1) Place the M2 in the ready-to-fire configuration, with the sight mounted on the M2, and with the M2 mounted on the M3 tripod.

(2) Have the gunner close his left eye and look into the sight with his right eye.

(3) Tell the gunner to focus on a target in either the narrow or wide field of view.

(4) Tell the gunner to close both eyes. After he closes his eyes, you, the trainer, must turn and change the setting on the sight.

(5) Tell the gunner to achieve the same sight picture he had before you reset it.

(6) Look through the sight to ensure the gunner has acquired a clear sight picture. If not, show him how to get one.

(7) Repeat the exercise until the gunner achieves proficiency with the sight.

h. **T&E Manipulation and Aiming Exercise**. Use this exercise to train the gunner to manipulate the T&E, acquire targets, and properly sight the target based on the target range. This exercise also trains the gunner to use the mounting bracket to adjust for range.

(1) Set up two to three targets that the gunner can acquire with the HWTS using the T&E.

(2) Regardless of the actual range to the target, tell the gunner to sight in on the first target and give him the range you want it to be.

(3) Ensure the gunner sets the mounting bracket to the correct setting and uses the T&E correctly.

(4) Time the gunner as he conducts this exercise, and create a competition between the gunners in training.

i. **Practice Qualification.** Use practice qualification, modeled after the day qualification scenario, to transition the gunner into engaging multiple timed targets at various ranges.

j. **Qualification.** Use qualification to assess the gunner's ability to engage targets on a timed scenario. If he does not meet the standard during the night practice table, train him again. He must meet the standard on the practice table before firing the qualification table. Conduct the nonfiring exercise as follows:

(1) Place the M2, with HWTS mounted, in the ready-to-fire configuration on the M3 tripod.

(2) Have the gunner close his left eye and look into the HWTS with his right eye.

(3) Tell the gunner to focus on a target in either the narrow or wide field of view.

(4) Tell the gunner to close both eyes. After he closes them, change the setting on the sight.

(5) Tell the gunner to open his eyes and to achieve the same sight picture he had before you reset it.

(6) Look through the sight to ensure the gunner has acquired a clear sight picture. If not, show him how to get one.

(7) Repeat the exercise until the gunner reaches proficiency with the sight.

Section II. AN/TVS-5

Gunners use the AN/TVS-5 (Figure G-3, page G-7) a portable, battery-operated electro-optical instrument, to observe and aim weapons fire at night. The AN/TVS-5 amplifies reflected light such as moonlight, starlight, and sky glow, making the scene clearly visible to the operator. The sight *does not* emit visible or infrared light (except from the eyepiece) that the enemy can detect. By using this device, the gunner can observe the area, and then detect and engage any suitable target. The quality of the zero determines the usefulness of the weapon. Accurate zeroing requires practice. Figure G-3 also provides equipment data for the AN/TVS-5.

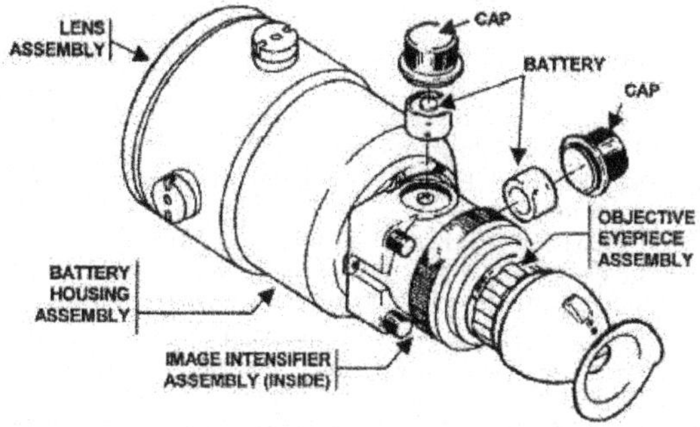

EQUIPMENT DATA

Magnification	5.6X
Field of view	9 degrees
Eyepiece focus	+3 to -6 diopters
Objective lens focus	25 meters to infinity
Range	1,000 to 1,200 meters for vehicle targets (under ideal conditions)
Reticle adjustment	+/-2.5 degrees (in ¼-mil increments)
Operating conditions	-65 F (-54C) to +125 F (+52C)
Battery type	BA-5567/U (2 each)
Voltage	3.0 volts DC
Shelf life	One year at 70 F (21C)
Operating life	About 50 hours at normal temperature

Figure G-3. AN/TVS-5 components and data.

CAUTION

Handle the AN/TVS-5, a precision electro-optical instrument, carefully at all times.

G-3. OPERATION

The AN/TVS-5 has the following controls and indicators:

a. Use the ON-OFF/TUBE BRIGHTNESS control to apply power to the sight and to control the brightness of the image-intensifier tube. This control also enables the ON-OFF RETICLE BRIGHTNESS control to function.

b. Use the ON-OFF RETICLE BRIGHTNESS to apply power to the reticle and to control the brightness of the reticle.

c. Use the OBJECTIVE FOCUS RING to adjust the range focus from 25 meters to infinity.

d. Use the DIOPTER FOCUS RING to focus the eyepiece.

e. Use the DIOPTER INDICATOR to learn the current direction of rotation of the DIOPTER FOCUS RING for plus and minus diopters.

f. Use the RETICLE ELEVATION ADJUSTMENT ACTUATOR to adjust the reticle up and down. Each click moves the strike of the round 1 inch at 100 meters.

g. Use the RETICLE AZIMUTH ADJUSTMENT ACTUATOR to adjust the reticle right or left. Each click moves the strike of the round 1 inch at 100 meters.

WARNING

- ***Do not*** **press the eye guard except with your eye area and then only to operate the sight. Used improperly, the sight emits illumination that the enemy can detect.**
- ***Do not*** **use the sight without the eye guard attached, or you may receive an injury when the weapon recoils.**
- **Use care when discarding the batteries. Their contents cause extreme irritation to the eyes and to oral and nasal passages. To prevent explosion, avoid burning the batteries.**
- ***Do not*** **short-circuit the batteries.**
- ***Do not*** **recharge the batteries.**
- **Remove batteries before storing the night vision sight.**
- **Always replace both mercury batteries at the same time.**

G-4. MOUNTING AND DISMOUNTING PROCEDURES

The AN/TVS-5 mounts and dismounts as follows:

a. **Mounting Procedures**. If the mounting bracket (Figure G-4) has not been installed already, install it now as described in TM 11-5855-214-10.

(1) Align the scribe line on the sight with the scribe line on the bracket.

(2) Place the sight in the groove of the bracket and tighten the lever screw clockwise.

(3) Secure the lever screw with lacing wire or tape to ensure the sight does not vibrate loose.

(4) Seat the device. After firing the initial burst, retighten the lever screw to ensure a secure mount for the sight. If unable to fire at this time, lightly shake the sight to ensure it is mounted correctly.

b. **Dismounting Procedures**.

(1) Remove the lacing wire or tape from the lever screw.

(2) Loosen the lever screw until the sight comes free, and then lift it off the bracket.

(3) Remove the batteries and place the sight in its carrying case.

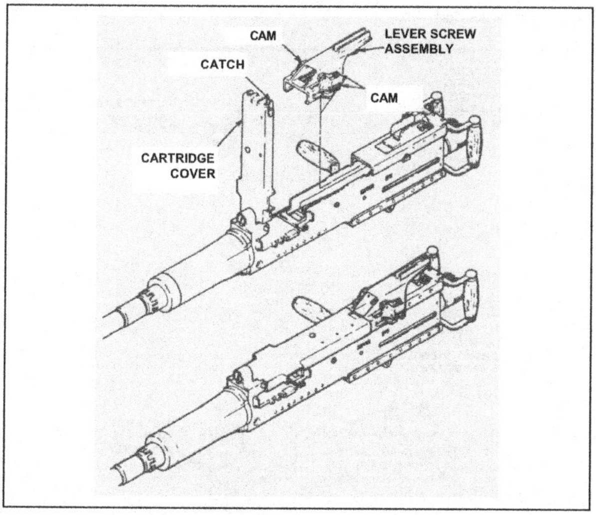

Figure G-4. Installation of M2 mounting bracket assembly.

G-5. FUNDAMENTALS OF MARKSMANSHIP

Except for aiming, the fundamentals of marksmanship for the AN/TVS-5 are the same as those for the AN/PAS-13 heavy weapon thermal sight.

a. Determine the range to the target (Figure G-5, page G-10). The range setting on the HWTS depends on the range to the target. To determine range to target, use any of several methods: range cards, range estimation techniques, the upper portion of the AN/TVS-5 reticle, or TRPs.

b. After determining range, use the aiming point designated for that range on the M2 reticle (Figure G-6, page G-12 shows the old reticle; Figure G-7, page G-13 shows the new reticle). If using the HWTS bracket and the M2 reticle, adjust the bracket to the nearest settings. Zero the sight on this setting and use the appropriate dot for the distance to the target as indicated in the reticle. For ranges beyond 1,200 meters, change the bracket to the far setting.

- Change "4" to 1,400 meters.
- Change "6" to 1,500 meters.
- Change "8" to 1,600 meters.
- Change "10" to 1,700 meters.
- Change "12" to 1,800 meters.

c. Bring the aiming point of the AN/TVS-5 onto the target using the T&E as explained in Section V.

d. Hold the weapon tight against the T&E, press the trigger, and adjust fire as needed.

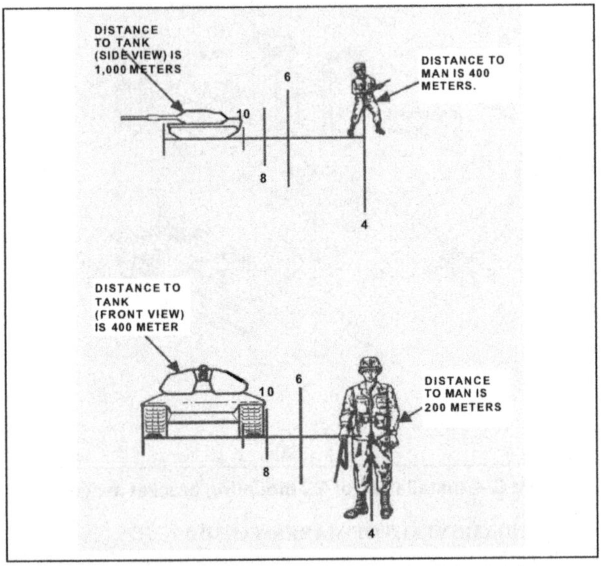

Figure G-5. Range estimation for M2 (old reticle).

G-6. BORESIGHTING PROCEDURES

Boresighting the AN/TVS-5 to the M2 requires the following steps:

a. Place the M2 in the ready-to-fire configuration, with the AN/TVS-5 mounted, 25 meters from the bore-light-offset zero targets. Ensure the weapon and target sit level before making any adjustments.

b. Adjust the T&E until the bore light moves onto the bore light aiming point on the boresight target offset.

c. Use the vertical and horizontal reticle adjusters to move the 1,000-meter aiming point on the reticle to the AN/TVS-5 aiming point.

d. Recheck the bore light aiming point and the AN/TVS-5 aiming points to ensure they remain center mass of the target.

G-7. AMMUNITION, NIGHT TRAINING STRATEGY

Table G-1 shows STRAC changes effective with this change publication.

EVENT	CURRENT STRAC BALL/MIX	REC'D CHGS	TOTAL BALL/MIX
10M ZERO	12 BALL	0	12 BALL
10M RECORD	119 BALL	0	119 BALL
TRANSITION ZERO, PRACTICE	50 MIX	ELIMINATE	0
TRANSITION RECORD	154 MIX	ELIMINATE	0
NIGHT ZERO, PRACTICE, RECORD	182 MIX	ELIMINATE	0
AG INSTR FIRE	54 MIX	ELIMINATE	0
TABLE I	0	140 MIX	140 MIX
TABLE II	0	140 MIX	140 MIX
TABLE III	0	140 MIX	140 MIX
TABLE IV	0	140 MIX	140 MIX
TOTAL	131 / 440 = 571	560 MIX	131/560 = 691

NOTE: The above STRAC supports one qualification. Allocations increase with the frequency of qualifications required. Check the firing unit's force activity designator (FAD).

Table G-1. Changes to STRAC effective with this publication.

Section III. TABLES AND QUALIFICATION, NIGHT COURSE OF FIRE

Improved firing tables and qualification methods allow the gunner to track his performance while he fires the tables. Requalifying gunners need not repeat the entire course of fire. This section standardizes the four tables to lead the gunner on a gradual path to qualification. The gunner must pass each firing table before he moves on to the next one. The tables in this section require 691 rounds to qualify a gunner--93 more than the old tables required, and 120 more than the STRAC previously authorized. The guidance in this appendix supersedes the STRAC. Though the new tables allow more rounds, they cut the time allowed to qualify. However, they recommend a band of ranges rather than a specific range for each target. This adds considerable flexibility. The ten-meter firing exercise is only a means of training--it is not a qualification table.

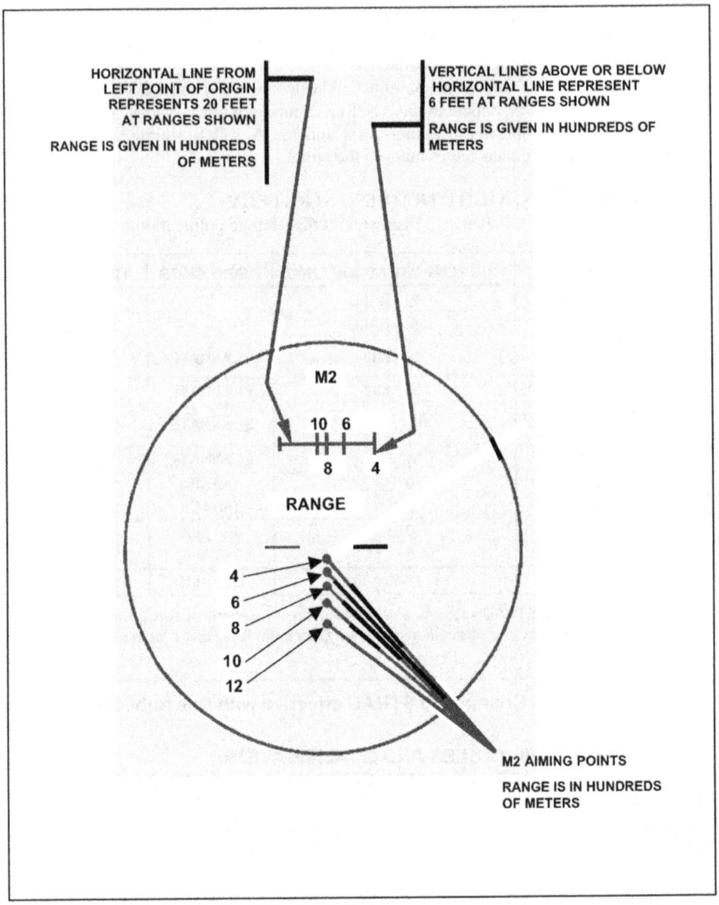

Figure G-6. Aiming points for the M2 (old reticle).

G-8. DIFFERENCES BETWEEN PRACTICE AND QUALIFICATION TABLES

Practice tables allow thirty extra seconds for each engagement. Also, infantry gun crews can practice in the daytime, but must qualify in both the daytime and at night.

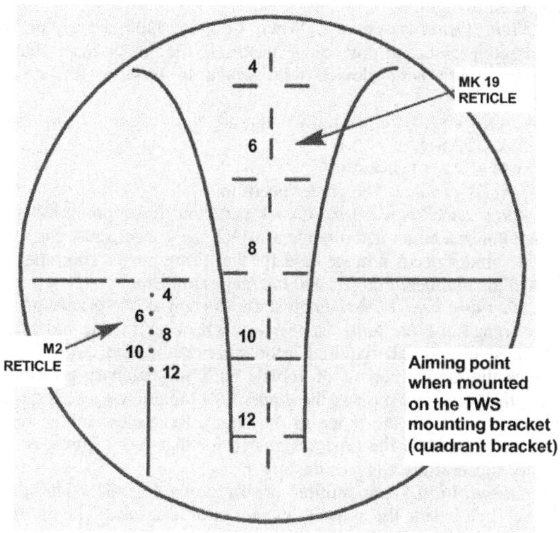

Figure G-7. Aiming points for the M2 (new reticle).

a. **Scoring**. For point target engagements (lightly armored vehicle targets, such as BRDMs and BTRs), give full credit (GO) when the gunner hits the target.

b. **Range Setup**. Set up targets and ranges as follows. Select target ranges IAW the scorecard for the applicable firing table. When choosing or placing targets, make sure no dead space falls within 100 meters of any of them. Any dead space near the target could keep the gunner from seeing the round impact, which he has to do in order to adjust fire:

(1) ***Hull Targets***. You need not modify hull targets on an impact range.

(2) ***Popup Silhouettes***. Set up a thermal source on popup silhouette targets so the gunner can acquire them with the HWTS. If the gunner will be using the AN/TVS-5, place a light source on each target.

- Targets between 400 and 600 meters--two chemical lights.
- Targets between 600 and 900 meters--three chemical lights.

c. **Grading**. Each firing point requires one grader.

(1) ***Grading Equipment***. While firing the day tables, the grader needs a set of binoculars. His other equipment needs vary according to the type of range used at night.

(a) *Night Vision Assistance*. When firing at hull targets, or when popup targets provide no downrange feedback, the grader must obtain an appropriate night vision device with which to observe the strike of the round.

- AN/PVS-14, 7B, with 3X magnifier.
- AN/TAS-4.
- AN/PAS-13 (heavy).

(b) *Target Scenario*. The grader needs this.

(c) *Range and Range Card*. At each particular firing point, the grader gives the gunner a target and a range at which the gunner must engage the target. The grader needs a range card for that firing range. The range card should include numbered targets and the ranges to them.

(2) ***Start and End Times***. Time starts as soon as the grader provides the target range and the target appears--which occur in this order, since graders must provide all required information before the target appears. Sometimes the targets consist of vehicle hulls in stationary positions. In this case, time starts as soon as the grader tells the gunner which target to engage and gives him the range to the target. Regardless of the type of target, time ends when the time designated for that task expires, or when the gunner successfully engages the target.

(3) ***Ammunition***. Ammunition breaks down by task. The gunner places each belt beside the assistant gunner in order of use. The number of rounds authorized for each task equals the number of rounds in each belt. For example, eight tasks require eight belts of ammunition, which should all sit within reach of the assistant gunner in the order the gunner plans to fire them.

(4) ***Fire Control***. Controlling M2 firing presents no problem when the range setup includes a firing lane for each firing point. However, this ideal situation seldom occurs with hull targets. In fact, some ranges must use the same target for each lane. This can cause confusion. When, inevitably, more than one gunner fires at the same target, no one can tell who hit it. The 400-meter target usually presents this problem. To prevent it, make sure that only one gunner at a time engages each target. If you want to let more than one gunner to fire at the same time, and then mix up the tasks so the gunners are firing at different targets. For example, have Point One fire at the 1,500-meter target while Point Two engages the 600-meter target, and so on. This requires careful coordination and communication between the graders.

d. **Firing Tables**. Figures G-8 and G-9 (page G-16), Figure G-10 (page G-17), and Figure G-11 (page G-18) show completed examples of the recommended day and night practice and qualification tables for the infantry M2 gunner and crew. Blank, reproducible forms are included at the back of the manual for users to copy onto 8 1/2 by 11-inch paper. They are also available on the Army Electronic Library (AEL) CD-ROM (EM0001) and at the USAPA website: http://www.usapa.army.mil.

(1) The first task allows evaluation of field zeroing. The gunner confirms the zero, even if he boresighted the weapon. If the gunner fails to zero within 14 rounds, graders remove him from the line and train him some more before letting him refire the table.

(2) All tasks have point targets. Graders should change one or two targets between the ranges of 600 and 900 meters to area targets, but should not change the round count or the time. They base their changes on range resources and the commander's guidance.

(3) Graders match the correct table to each target and weapon-sight configuration.

(4) Graders use range finders to predetermine the range to each target from each firing point.

(5) Graders set up M2s mounted or dismounted (in the tripod configuration), based on the range constraints and the commander's guidance.

(6) Leaders can have the unit fire the tasks in any order, and should do so to avoid redundancy in practice and qualification.

(7) For NBC qualification, the leader uses the practice tables.

e. **Ammunition Requirements for Night Training Strategy**. Table G-1 shows the ammunition requirements for a night training strategy.

G-9. SPECIAL MOUNTING PROCEDURES

This paragraph explains the special mounting and dismounting procedures required to mount a night vision device on the M2 heavy barrel .50 caliber machine gun.

a. **Mounting Procedure**. Before mounting the sight, the gunner ensures the bolt is forward and the rear sight is in the down position.

(1) Release and raise the top cover assembly to the upright position.

(2) Place the mounting bracket over the breach and slide it rearward until it stops.

(3) Push the three locking cams rearward until the bracket is secured (lock the side-locking cam first), and close the top cover assembly.

M2 CALIBER .50 HEAVY BARREL MACHINE GUN FIRING TABLE I
DAY PRACTICE SCORECARD

For use of this form, see FM 23-65, Change 1; the proponent agency is TRADOC.

DATA REQUIRED BY THE PRIVACY ACT OF 1974

AUTHORITY: 10 USC 3012(g)/Executive order 9397.
PRINCIPAL PURPOSE: To aid individual training on targets at various ranges.
ROUTINE USES: To evaluate individual proficiency. SSN is used for positive identification purposes only.
DISCLOSURE: Voluntary. However, mass rating and scoring require some tracking method. Therefore, only those soldiers who provide their SSNs can receive scores or qualify on weapons or equipment.

1a. LAST NAME: DOE | 1b. FIRST NAME: JOHN | 1c. MI: A | 2. DATE *(YYYYMMDD)*: 20021201

3. SSN: 222-22-2222 | 4. UNIT: 2 / 29 TH INF | 5. RANGE: RUTH | 6. LANE: 4

EXAMPLE

7a. ENGAGEMENT	7b. STANDARDS	7c. AMMO	7d. TIME *(Minutes)*	7e. RANGE *(Meters)*	7f. GO	7g. NO GO
ZERO	1 BURST HIT	14	NA	400	NA	NA
2	1 BURST HIT	14	1.5	1,100 (±200)	X	
3	1 BURST HIT	14	1.5	1,500 (±200)	X	
4	1 BURST HIT	14	1.5	600 (±100)	X	
5	1 BURST HIT	14	1.5	800 (±100)	X	
6	1 BURST HIT	14	1.5	400	X	
8. MULTIPLE TARGETS						
7	1 BURST HIT	28	2.5	1,100 (±200)	X	
	1 BURST HIT			600 (±100)	X	
8	1 BURST HIT	28	2.5	800 (±100)	X	
	1 BURST HIT			1,500 (±200)	X	

9. SCORING *(Check one)*

EXPERT - 9 [X] SHARPSHOOTER - 8 [ ]

MARKSMAN - 6 [ ] UNQUALIFIED - 5 OR BELOW [ ]

10. GUNNER'S SIGNATURE: *PFC John A. Doe*

11a. GRADER'S PRINTED OR TYPED NAME: SSG DAVID R. JONES | 11b. GRADER'S SIGNATURE: *SSG David R. Jones*

DA FORM 7448-R, JAN 2002 USAPA V1.00

Figure G-8. Example completed DA Form 7448-R (Day Practice Scorecard).

M2 CALIBER .50 HEAVY BARREL MACHINE GUN FIRING TABLE II
DAY QUALIFICATION SCORECARD

For use of this form, see FM 23-65, Change 1; the proponent agency is TRADOC.

DATA REQUIRED BY THE PRIVACY ACT OF 1974

AUTHORITY: 10 USC 3012(g)/Executive order 9397.
PRINCIPAL PURPOSE: To aid individual training on targets at various ranges.
ROUTINE USES: To evaluate individual proficiency. SSN is used for positive identification purposes only.
DISCLOSURE: Voluntary. However, mass rating and scoring require some tracking method. Therefore, only those soldiers who provide their SSNs can receive scores or qualify on weapons or equipment.

1a. LAST NAME	1b. FIRST NAME	1c. MI	2. DATE *(YYYYMMDD)*
DOE	JOHN	A	20021201

3. SSN	4. UNIT	5. RANGE	6. LANE
222-22-2222	2 / 29 TH INF	RUTH	4

7a. ENGAGEMENT	7b. STANDARDS	7c. AMMO	7d. TIME *(Minutes)*	7e. RANGE *(Meters)*	7f. GO	7g. NO GO
ZERO	1 BURST HIT	14	NA	400	NA	NA
2	1 BURST HIT	14	1	1,100 (±200)		X
3	1 BURST HIT	14	1	1,500 (±200)	X	
4	1 BURST HIT	14	1	600 (±100)	X	
5	1 BURST HIT	14	1	800 (±100)	X	
6	1 BURST HIT	14		400	X	
8. MULTIPLE TARGETS						
7	1 BURST HIT	28	2	1,100 (±200)	X	
	1 BURST HIT			600 (±100)	X	
8	1 BURST HIT	28	2	800 (±100)	X	
	1 BURST HIT			1,500 (±200)	X	

EXAMPLE

9. SCORING *(Check one)*

EXPERT - 9 [ ] SHARPSHOOTER - 8 [X]

MARKSMAN - 6 [ ] UNQUALIFIED - 5 OR BELOW [ ]

10. GUNNER'S SIGNATURE

PFC John A. Doe

11a. GRADER'S PRINTED OR TYPED NAME	11b. GRADER'S SIGNATURE
SSG DAVID R. JONES	SSG David R. Jones

DA FORM 7449-R, JAN 2002 USAPA V1.00ES

Figure G-9. Example completed DA Form 7449-R (Day Qualification Scorecard).

M2 CALIBER .50 HEAVY BARREL MACHINE GUN FIRING TABLES III (A) AND III (B)
NIGHT PRACTICE SCORECARDS

For use of this form, see FM 23-65, Change 2; the proponent agency is TRADOC.

DATA REQUIRED BY THE PRIVACY ACT OF 1974

AUTHORITY: 10 USC 3012(g)/Executive order 9397.
PRINCIPAL PURPOSE: To aid individual training on targets at various ranges.
ROUTINE USES: To evaluate individual proficiency. SSN is used for positive identification purposes only.
DISCLOSURE: Voluntary. However, mass rating and scoring require some tracking method. Therefore, only those soldiers who provide their SSNs can receive scores or qualify on weapons or equipment.

1a. LAST NAME	1b. FIRST NAME	1c. MI	1d. RANK	2. SSN
Doe	John	A	PFC	222-22-2222

TABLE III (A). NIGHT PRACTICE SCORECARD

NOTE: Use this table if the M2 has the TWS, or if it has an AN/TVS-5 that has been upgraded with the third-generation tube.

a. RANGE	b. UNIT	c. LANE	d. DATE *(YYYYMMDD)*
Ruth	2 / 29 th	4	20021120

e. ENGAGEMENT	f. STANDARDS	g. AMMO	h. TIME *(Minutes)*	i. RANGE *(Meters)*	j. GO	k. NO GO
ZERO	1 BURST HIT	14	NA	400	NA	NA
2	1 BURST HIT	14	1.5	1,100 (±200)	X	
3	1 BURST HIT	14	1.5	1,500 (±200)	X	
4	1 BURST HIT	14	1.5	600 (±100)		X
5	1 BURST HIT	14	1.5	800 (±100)	X	
6	1 BURST HIT	14	1.5	400	X	
l. MULTIPLE TARGETS						
7	1 BURST HIT	28	2.5	1,100 (±200)	X	
	1 BURST HIT			600 (±100)	X	
8	1 BURST HIT	28	2.5	800 (±100)	X	
	1 BURST HIT			1,500 (±200)	X	

4. SCORING *(Check one)*
EXPERT - 9 [ ]
SHARPSHOOTER - 8 [X]
MARKSMAN - 6 [ ]
UNQUALIFIED - 5 OR BELOW [ ]

5. GUNNER'S SIGNATURE: PFC John A. Doe
6a. GRADER'S PRINTED OR TYPED NAME: SSG David R. Jones
6b. GRADER'S SIGNATURE: SSG David R. Jones

7. TABLE III (B). NIGHT PRACTICE SCORECARD

NOTE: Use this table if the M2 has an AN/TVS-5 that has not been upgraded.

a. RANGE	b. UNIT	c. LANE	d. DATE *(YYYYMMDD)*
Ruth	2 / 29 th	4	20021120

e. ENGAGEMENT	f. STANDARDS	g. AMMO	h. TIME *(Minutes)*	i. RANGE *(Meters)*	j. GO	k. NO GO
ZERO	1 BURST HIT	14	NA	400	NA	NA
2	1 BURST HIT	14	1.5	1,100 (±200)	X	
3	1 BURST HIT	14	1.5	800 (±100)	X	
4	1 BURST HIT	14	1.5	600 (±100)	X	
5	1 BURST HIT	14	1.5	800 (±100)	X	
6	1 BURST HIT	14	1.5	400	X	
l. MULTIPLE TARGETS						
7	1 BURST HIT	28	2.5	1,100 (±200)	X	
	1 BURST HIT			600 (±100)	X	
8	1 BURST HIT	28	2.5	800 (±100)	X	
	1 BURST HIT			600 (±100)	X	

8. SCORING *(Check one)*
EXPERT - 9 [X]
SHARPSHOOTER - 8 [ ]
MARKSMAN - 6 [ ]
UNQUALIFIED - 5 OR BELOW [ ]

9. GUNNER'S SIGNATURE: PFC John A. Doe
10a. GRADER'S PRINTED OR TYPED NAME: SSG David R. Jones
10b. GRADER'S SIGNATURE: SSG David R. Jones

EXAMPLE

DA FORM 7450-R, OCT 2002 — DA FORM 7450-R, JAN 2002, IS OBSOLETE. — USAPA V1.00ES

Figure G-10. Example completed DA Form 7450-R (Night Practice Scorecard).

M2 CALIBER .50 HEAVY BARREL MACHINE GUN FIRING TABLES IV (A) AND IV (B) NIGHT QUALIFICATION SCORECARDS

For use of this form, see FM 23-65, Change 2; the proponent agency is TRADOC.

DATA REQUIRED BY THE PRIVACY ACT OF 1974

AUTHORITY:	10 USC 3012(g)/Executive order 9397.
PRINCIPAL PURPOSE:	To aid individual training on targets at various ranges.
ROUTINE USES:	To evaluate individual proficiency. SSN is used for positive identification purposes only.
DISCLOSURE:	Voluntary. However, mass rating and scoring require some tracking method. Therefore, only those soldiers who provide their SSNs can receive scores or qualify on weapons or equipment.

1a. LAST NAME	1b. FIRST NAME	1c. MI	1d. RANK	2. SSN
Doe	John	A	PFC	222-22-2222

3. TABLE IV (A). NIGHT QUALIFICATION SCORECARD

NOTE: Use this table if the M2 has the TWS, or if it has an AN/TVS-5 that has been upgraded with the third-generation tube.

a. RANGE	b. UNIT	c. LANE	d. DATE *(YYYYMMDD)*
Ruth	2 / 29	4	20021204

e. ENGAGEMENT	f. STANDARDS	g. AMMO	h. TIME *(Minutes)*	i. RANGE *(Meters)*	j. GO	k. NO GO
ZERO	1 BURST HIT	14	NA	400	NA	NA
2	1 BURST HIT	14	1	1,100 (±200)	X	
3	1 BURST HIT	14	1	1,500 (±200)	X	
4	1 BURST HIT	14	1	600 (±100)		X
5	1 BURST HIT	14	1	800 (±100)	X	
6	1 BURST HIT	14	1	400	X	
l. MULTIPLE TARGETS						
7	1 BURST HIT	28	2	1,100 (±200)	X	
	1 BURST HIT			600 (±100)		X
8	1 BURST HIT	26	2	800 (±100)	X	
	1 BURST HIT			1,500 (±200)	X	

4. SCORING *(Check one)*
- EXPERT - 9 [ ]
- SHARPSHOOTER - 8 [ ]
- MARKSMAN - 6 [X]
- UNQUALIFIED - 5 OR BELOW [ ]

5. GUNNER'S SIGNATURE: PFC John A. Doe

6a. GRADER'S PRINTED OR TYPED NAME: SSG David R. Jones

6b. GRADER'S SIGNATURE: SSG David R. Jones

7. TABLE IV (B). NIGHT QUALIFICATION SCORECARD

NOTE: Use this table if the M2 has an AN/TVS-5 that has not been upgraded.

a. RANGE	b. UNIT	c. LANE	d. DATE *(YYYYMMDD)*
Ruth	2 / 29	4	20021204

e. ENGAGEMENT	f. STANDARDS	g. AMMO	h. TIME *(Minutes)*	i. RANGE *(Meters)*	j. GO	k. NO GO
ZERO	1 BURST HIT	14	NA	400	NA	NA
2	1 BURST HIT	14	1	1,100 (±200)	X	
3	1 BURST HIT	14	1	800 (±100)	X	
4	1 BURST HIT	14	1	600 (±100)	X	
5	1 BURST HIT	14	1	800 (±100)	X	
6	1 BURST HIT	14	1	400	X	
l. MULTIPLE TARGETS						
7	1 BURST HIT	28	2	1,100 (±200)		X
	1 BURST HIT			600 (±100)	X	
8	1 BURST HIT	28	2	800 (±100)	X	
	1 BURST HIT			600 (±100)	X	

8. SCORING *(Check one)*
- EXPERT - 9 [ ]
- SHARPSHOOTER - 8 [X]
- MARKSMAN - 6 [ ]
- UNQUALIFIED - 5 OR BELOW [ ]

9. GUNNER'S SIGNATURE: PFC John A. Doe

10a. GRADER'S PRINTED OR TYPED NAME: SSG David R. Jones

10b. GRADER'S SIGNATURE: SSG David R. Jones

EXAMPLE

DA FORM 7451-R, OCT 2002 — DA FORM 7451-R, JAN 2002, IS OBSOLETE. — USAPA V1.00ES

Figure G-11. Example completed DA Form 7451-R (Night Qualification Scorecard).

DANGER

BEFORE INSTALLING THE BRACKET, MAKE SURE THE WEAPON IS *UNLOADED* AND THE SAFETY IS SET ON *SAFE*. A LOADED WEAPON CAN DISCHARGE ACCIDENTALLY, INJURING OR KILLING SOMEONE.

NOTE: If the bracket sticks before reaching its correct position, rock the bracket up and down while sliding it.

b. **Dismounting Procedure**. Ensure the weapon is clear and the bolt is forward.

(1) Raise the top cover assembly to the upright position.

(2) Push the three locking cams forward until the bracket is free.

(3) Slide the mounting bracket forward over the breach until the mounting bracket clears the weapon.

Section IV. BORE-LIGHTING PROCESS AND BORESIGHT TARGET OFFSETS

The bore light allows gunners to accurately zero weapons and most aided vision equipment without the use of bullets. Table G-2 shows different combinations of weapon and aided-vision device that gunners can use to zero the bore light.

DANGER

- ***DO NOT STARE* INTO THE VISIBLE LASER BEAM.**
- ***DO NOT LOOK* INTO THE VISIBLE LASER BEAM THROUGH A TELESCOPE OR A PAIR OF BINOCULARS.**
- ***DO NOT POINT* THE VISIBLE LASER BEAM AT A MIRROR-LIKE SURFACE.**
- ***DO NOT SHINE* THE VISIBLE LASER BEAM INTO OTHER PEOPLE'S EYES.**

WARNING

- **Make sure the weapon is CLEAR and on SAFE before using the bore light.**
- **When rotating the bore light to zero it, ensure the mandrel turns. *Do not* tighten the bore light down on the mandrel; doing so could strip or break the mandrel's stud.**

	M16A2	M4/MWS	M249	M240B/M60	M2	M203	AT4
AN/PAQ-4C	X	X	X	X		X	X
AN/PEQ-2A	X	X	X	X	X	X	X
AN/PAS-13	X	X	X	X	X		X
M68 CCO	X	X					
AN/PVS-4	X	X	X	X		X	
AN/TVS-5					X		

Table G-2. Possible weapon-aided vision-device combinations.

G-10. BORESIGHTING PROCESS

Properly boresighting a weapon requires two gunners--a firer and a target-holder.

a. **Personnel.**

(1) *Firer*. The firer zeroes the bore light and adjusts the aided vision device. Before boresighting, he locks the M2 down in a cradle or some other device to keep the weapon stable during the boresighting process. In the absence of such a device, he assumes the most stable supported firing position possible and makes sure the weapon does not cant during boresighting.

(2) *Target holder*. The target holder keeps the target straight up and down at the appropriate distance from the firer and directs the firer in making necessary sight adjustments. The target holder *must* wear night vision goggles when viewing laser aiming light offsets.

b. **Steps**.

(1) *Align Bore Light*. Checking the alignment of the bore light requires the following actions:

(a) Place the appropriate mandrel, with bore light attached, in the muzzle of the weapon.

(b) Turn the bore light on so that the laser beam strikes the target (offset) 10 meters away.

(c) Slowly rotate the bore light one-half turn (180 degrees) while watching the beam on the target area, noting any circular patterns made.

(d) If the beam remains stationary, you have boresighted the M2. Move forward to the next paragraph and use the appropriate boresight target for the given weapon.

(e) If the beam rotates in a circle, adjust the windage, elevation, or both, until the beam remains stationary or rotates on itself no more than 1 centimeter.

(2) *Adjust Bore Light*. If necessary, the firer and target holder must adjust the bore light.

(a) Move the target 2 meters away.

(b) Mark the location of the laser beam.

(c) Slowly rotate the bore light one-half turn.

(d) Note the new location of the laser beam.

(e) Adjust the windage and elevation until the laser beam moves halfway back to its original location.

(3) *Adjust elevation adjustment screw*. One click at 25 meters equals 4 millimeters; clockwise equals down.

(4) *Adjust windage adjustment screw*. One click at 25 meters equals 4 millimeters; clockwise equals left.

(5) *Continue adjusting the bore light*. Keep adjusting it until the laser beam either stops moving or spins upon itself within 1 centimeter.

(6) *Move the target, and recheck the boresight*. Move it 10 meters away and recheck the boresight. Repeat this process every 10 meters if necessary.

NOTE: Because the M2 requires such a large offset, place the offset for the boresight target 5 meters away.

(7) *Boresight the Aided Vision Device to the Weapon*. Select the boresight target offset for the appropriate aided vision device. For laser aiming lights, position the weapon so the bore light strikes the black dot on the boresight target.

(a) Adjust the *laser* until it centers on the cross on the boresight target offset. For aided vision optics, position the weapon so the *reticle* centers on the cross on the boresight target offset (the firer must physically aim the weapon).

(b) Adjust the aided vision optic until the bore light strikes the black dot on the boresight target. Refer to specific instructions on each boresight-offset target.

(c) You have boresighted the weapon when both the laser bore light and the aiming point or laser move to the appropriate position on the boresight target offset.

G-11. TARGET OFFSETS

Figure G-12 (page G-23), Figure G-13 (page G-24), and Figure G-14 (page G-25) show example boresight target offsets. The gunner measures boresight offsets in 1-centimeter squares. For targets, he copies M16A2 25-meter zero targets, with the silhouette representing the point of aim.

a. **Aligning the M2 with the AN/PEQ-2A**. Using the 10-meter boresight target, align the bore light on the dot and adjust the AN/PEQ-2A to the cross. This gives an 800-meter zero (Figure G-12, page G-23).

b. **Aligning the M2 with the AN/TVS-5**. Using the 10-meter boresight target, place the target at a distance of 10 meters. Aim the 400-meter AN/TVS-5 reticle at the cross and adjust the sight so that the bore light strikes the dot (Figure G-13, page G-24).

c. **Aligning the M2 with the AN/PAS-13**. Using the 10-meter boresight target, place the tips of the index fingers on the gray circles. Aim between the hot spots provided by the fingers. Adjust the TWS so that the bore light strikes the dot when you aim between the hot spots at 10 meters. If using a spacer, increase the offset by the height of the spacer (Figure G-14, page G-25).

10-meter boresight target AN/PEQ-2A mounted on an AN/TVS-5 feed tray cover bracket and plastic spacer on M2 MG

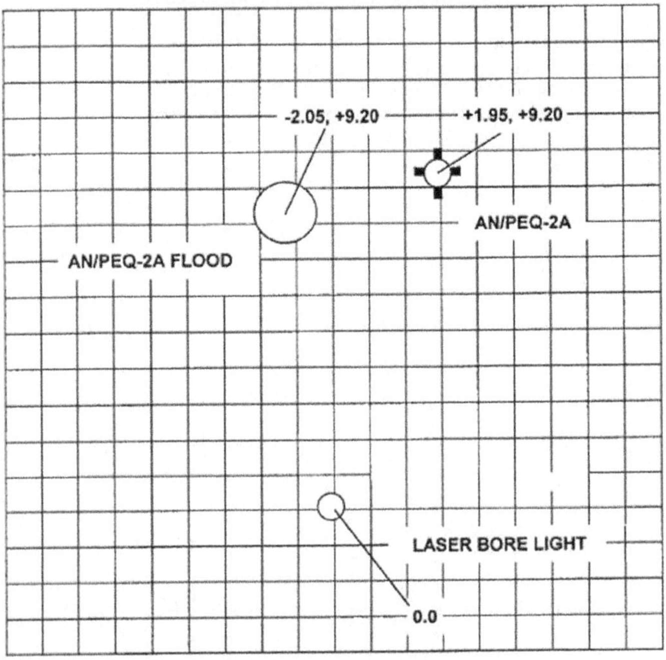

This is the 10-meter boresight target for 500-meter zero.
Grids are 1-centimeter square.
Units manufacture targets locally.

Zero target data for M2 MG:

1. **Stabilize the weapon.**
2. **Align laser borelight on its dot.**
3. **Adjust the AN/PEQ-2A until aiming laser centers on the dot cross hair.**

Figure G-12. Alignment of M2 with the AN/PEQ-2A.

10-meter boresight target
AN/TVS-5 mounted on AN/TVS-5 feed tray cover bracket on the M2 MG.

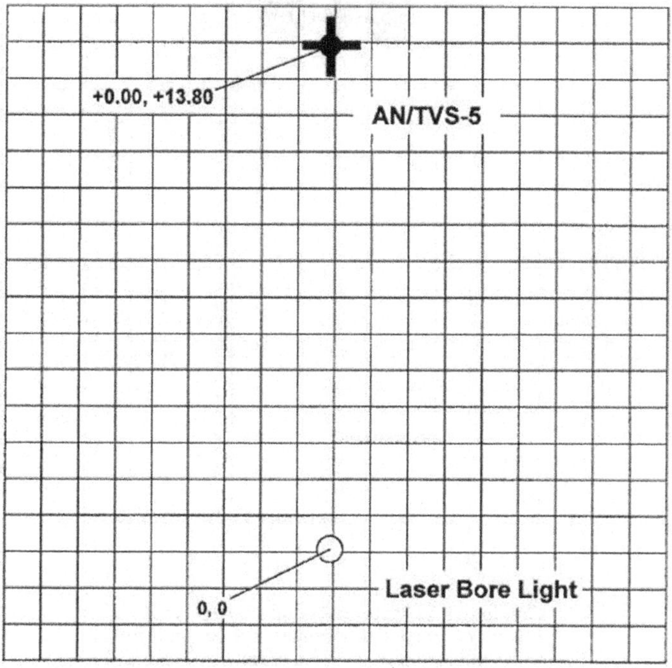

This is the 10-meter boresight target for the 500-meter zero. Grids measure 1 cm square. Units manufacture targets locally.

Zero target data for M2 MG:

1. Stabilize the weapon.
2. Aim the AN/TVS-5 at the cross hair and adjust until the borelight strikes the laser's borelight dot.

Figure G-13. Alignment of M2 with the AN/TVS-5.

10-meter boresight target AN/PAS-13 mounted on AN/PAS-13 feed tray cover bracket on the M2 MG.

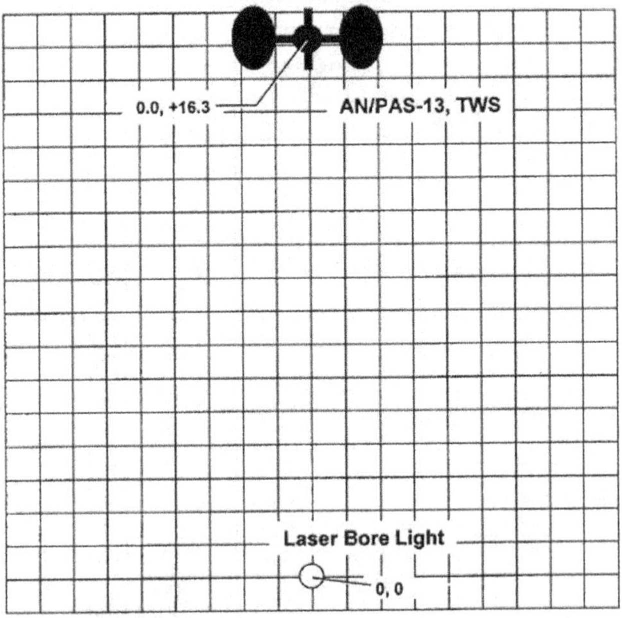

This is the 10-meter boresight target for the 500-meter zero. Grids measure 1 cm square. Units manufacture targets locally.

Zero target data for M2 MG:

1. Stabilize the weapon.
2. Place the thermal signature material or the tips of fingers on the gray circles.
3. Aim between the gray circles at the cross hairs.
4. Adjust the AN/PAS-13 until the borelight laser centers on its dot.
5. Zero the reticle in *both* wide and narrow fields of view.

Figure G-14. Alignment of M2 with the AN/PAS-13.

Section V. T&E MANIPULATION

The T&E mechanism allows engagement of preselected target areas at night or during limited visibility. The gunner records (in mils) the direction and elevation readings from the traversing bar and T&E mechanism.

G-12. ZEROING THE T&E MECHANISM

Before a gunner can effectively engage targets with the T&E, he must zero it to the weapon:

a. **Zero Traversing Handwheel.** Hold the T&E so the traversing handwheel sits on the left as you look at it.

(1) Turn the traversing handwheel toward you until it stops.

(2) Loosen the locking nut slightly.

(3) Align the zero on the scale with the zero on the elevating screw yoke.

(4) Hold the scale (with the zeroes aligned) and tighten the locking nut. Make sure the zeroes remain aligned.

(5) Turn the traversing handwheel two complete revolutions away from you. If doing this at night, count 50 "clicks" away from you.

b. **Zero Elevating Handwheel to Upper Elevating Screw.** Align the two zeroes.

(1) Rotate the elevating handwheel up or down until you can see a zero with a line below it on the upper elevating screw.

(2) Position the elevating handwheel so the indicator points to the zero on the handwheel.

c. **Zero Elevating Mechanism Sleeve to Lower Elevating Screw.**

(1) Rotate the elevating mechanism sleeve all the way up, and then rotate it down until it stops. Note the number of complete turns down.

(2) Rotate the elevating mechanism sleeve up half that many turns.

(3) Position the slide lock lever to face you.

(4) Attach the T&E to the tripod and gun.

G-13. LAYING THE GUN FOR DIRECTION

After the gunner receives an assigned sector of fire, he should--

a. Pick up the rear legs of the tripod.

b. Shift the tripod until the muzzle of the weapon points to the center of the sector of fire.

c. After laying the weapon for direction, firmly stamp in the tripod shoes and place sandbags on the legs. This aids stability and may prevent accidental movement.

d. Obtain and record directional readings to all targets within the sector and perform the following:

(1) Loosen the traversing slide lock lever and move the slide along the traversing bar until the weapon lays either on the center of a point target or on the flank of a linear target.

(2) Lock the traversing bar. Read the direction from the scale on the traversing bar. If the left edge of the traversing slide falls anywhere

outside the 5-mil tick mark, the gunner moves the left edge of the traversing bar slide back to the next smaller mil reading. Then, the gunner uses the traversing handwheel to complete the initial lay.

e. Obtain a reading to the target by the direction of the weapon barrel. If the barrel is moved to the right, the gunner records a right heading. He reads the number on the traversing bar from the left side of the traversing slide lock. If the barrel is moved to the left and the traversing slide lock is on the right side of the zero, he records a left reading.

f. After taking a directional reading for a target, measure the width of the target in mils. Then, he uses the traversing handwheel to move the barrel across the target--one click equals one mil.

g. Before moving to another target, reposition the traversing mechanism.

h. Obtain elevation readings. The gunner verifies that the weapon is laid on the center base of the target. Read the elevation from two scales.

i. Get the first, or major, part of the elevation from the elevating screw plate scale.

j. Get the second, or minor, part of the elevation from the elevating handwheel.

k. Separate the two parts of the elevation reading with a slash (/). For example, the gunner could write "-50/3." An elevation reading might apply only to the T&E where he reads it.

l. Note that, if the number of threads increases or decreases after he records the data, he cannot place accurate fire on the target. For example, rotating the base of the T&E mechanism to engage a secondary sector makes the data incorrect--unless the same number of threads appear both before and after the move.

* GLOSSARY

AP	armor piercing
APC	armored personnel carrier
approx	approximately
attn	attention
ba	battery
BFA	blank firing attachment
BFV	Bradley fighting vehicle
BMP	Soviet fighting vehicle
cal	caliber
cdr	commander
CLP	cleaner, lubricant, and preservative
CVKI	combat vehicle kill indicator
DA	Department of the Army
FEBA	forward edge of the battle area
FM	field manual
FOV	field of view
FPF	final protective fires
FPL	final protective line
FSN	federal stock number
FTX	field training exercise
GA	Georgia
HB	heavy barrel
HMMWV	high-mobility, multi-purpose, wheeled vehicle
HWTS	heavy weapon thermal sight
IAW	in accordance with
IN	infantry
IR	infrared
LED	light-emitting diode
LFX	live-fire exercise
LSA	semifluid lubricating oil
LTA	local training area
m	meter
METL	mission-essential task list
MG	machine gun
MILES	multiple integrated laser engagement system
mm	millimeter
MOPP	mission-oriented protection posture
mph	miles per hour
MOUT	military operations on urbanized terrain
MTA	major training area
MTP	mission training plan
NBC	nuclear, biological, and chemical
NCO	noncommissioned officer

NCOES	noncommissioned officer education system
NCOIC	noncommissioned officer in charge
NSN	national stock number
NVD	night vision device
OIC	officer in charge
PDF	principal direction of fire
PL-M	lubricating oil, general purpose
PL-S	lubricating oil, special purpose
POI	program of instruction
RABA	recoil amplifier barrel assembly
RBC	rifle bore cleaner
SM	soldiers manual
SMCT	soldiers manual of common tasks
SOP	standing operating procedure
SRTA	short-range training ammunition
SRTA-T	short-range training ammunition-tracer
STP	soldiers training publication
STRAC	Standards in Training Commission
STX	situational training exercise
T&E	traversing and elevating
TM	technical manual
TRADOC	Training and Doctrine Command
TRP	target reference point
US	United States
v	volt

* REFERENCES

SOURCES USED

These are the Army publications quoted or paraphrased in this manual.

DA Pam 350-38	Standards in Weapons Training. 3 July 1997
FM 3-5	NBC Decontamination. 17 November 1993.
FM 44-80	Visual Aircraft Recognition. 30 September 1996.
STP 21-24-SMCT	Soldier's Manual of Common Tasks, Skill Levels 2/3/4. 1 October 1992.
TC 25-8	Training Ranges. 25 February 1992, with Change 1, 16 November 1992.
TM 9-1005-213-10	Operator's Manual for Machine Guns, Caliber .50; Browning, M2, Heavy Barrel Flexible, 6 August 1992, with Changes 1 and 2, 1 August 1995 and 6 August 1996.
TM 9-1300-200	Ammunition General. 3 October 1969; with changes 1 through 5, dated 22 January 1971 through 30 September 1993.
TM 11-5855-214-10	Operator's Manual for Night Vision Sight, Crew Served Weapon, AN/TVS-5. 15 February 1989.
TM 11-5855-302-12&P	Operator's and Unit Maintenance Manual (including Repair Parts and Special Tools List), Sight, Thermal, AN/PAS-13(V)2, AN/PAS-13(V)3. To be published.
TM 11-5855-312-10	Operator's Manual, Sight, Thermal, AN/PAS-13B(V)2, AN/PAS-13B(V)3. To be published.

DOCUMENTS NEEDED

These Army publications must be available to the intended users of this manual.

Army Forms

These forms can be locally reproduced on 81/2 x 11-inch paper. The reproducible form is located at the back of this manual. The forms are also available on the Army Electronic Library (AEL) CD-ROM (EM0001) and at the USAPA website located at (www.usapa.army.mil):

DA Form 2404	Equipment Inspection and Maintenance Worksheet. 1 April 1979.
DA Form 5517-R	Standard Range Cards (LRA). February 1986.
DA Form 7007-R	Machine Gun Scorecard for M2. August 2001.
DA Form 7448-R	M2 Caliber .50 Heavy Barrel Machine Gun, Firing Table I, Day Practice Scorecard. January 2002.
DA Form 7449-R	M2 Caliber .50 Heavy Barrel Machine Gun, Firing Table II, Day Qualification Scorecard. January 2002.
* DA Form 7450-R	M2 Caliber .50 Heavy Barrel Machine Gun, Firing Tables III(A) and III(B), Night Practice Scorecard. October 2002.
* DA Form 7451-R	M2 Caliber .50 Heavy Barrel Machine Gun, Firing Tables IV(A), Night Qualification Scorecard; and IV(B), Day Practice Scorecard. October 2002.

Technical Manuals

TM 11-5855-306-10	Operator's Manual Monocular Night Vision Device (MNVD), AN/PVS-14. June 2000.
TM 11-5855-306-23&P	Unit and Direct Support Maintenance Manual (Including Repair Parts and Special Tools List), Monocular Night Vision Device (MNVD), AN/PVS-14. June 2000.
TM 11-5855-309-12&P	Operator's and Unit Maintenance Manual (including Repair Parts and Special Tools List), Sight, Thermal, AN/PAS-13A(V)2, AN/PAS-13A(V)3. To be published.

INTERNET WEB SITES

U.S. Army Publishing Agency (USAPA):
www.usapa.army.mil

Reimer Digital Library (RDL):
www.adtdl.army.mil

Air Force Publishing:
afpubs.hq.af.mil/

* INDEX

adjustment of fire methods
- aiming point, 5-18 (*illus*), 6-24
- mil relation, 5-17 (*illus*), 6-24
- observation, 5-16, 6-23

advanced gunnery
- mounted, 5-45, C-26 thru C-31
- mounted NBC, 5-45, C-26 thru C-31
- predetermined firing, 5-45, C-31 thru C-34
- tracking and leading, 5-42 thru 5-44

aerial defense
- active, F-1
- engagements, F-1, F-2
- passive, F-1
- techniques, F-3

aided vision devices, G-1

ammunition, 1-17 (*illus*)
- ballistic data, 1-18 (*illus*)
- care, 1-19
- classification, 1-17, 1-18
- penetration data, 1-19, 1-20

antiaircraft gunnery, 6-25, 6-26
- position, D-6 (*illus*)

AN/PAS-13 (V3), G-1
- aiming exercise, G-5
- aligning with the M2, G-22
- boresight targets, G-25
- boresighting, G-4, G-21
- components, G-1
- data, G-1, G-2 (illus)
- fundamentals of marksmanship, G-2
- nonfiring exercise, G-5
- practice qualification, G-6
- qualification, G-6
- qualifications, G-4
- target, G-25
- weapons that use, G-2 (illus)

AN/PEQ-2A
- aligning with the M2, G-22
- boresight targets, G-23
- target, G-23

AN/TVS-5 night sight, 5-38, G-6
- aligning with the M2, G-22
- boresight targets, G-24
- boresighting, G-10, G-21
- components, G-7 (illus)
- data, G-7 (illus)
- dismounting, 5-41, G-8
- fundamentals of marksmanship, G-2, G-9
- installing sight to mount, 5-39, G-9
- mounting, 5-38, G-8
- operation, G-8
- range estimation, G-10
- target, G-24
- zeroing, 5-40, G-11

application of fire, 6-22, 6-23 (*illus*)

assembly, general, 2-12
- backplate group, 2-16
- barrel, 2-17
- barrel buffer assembly, 2-13
- barrel buffer body group, 2-13
- barrel buffer group, 2-14
- barrel extension group, 2-14
- bolt, 2-14
- bolt stud, 2-15
- driving spring rod group, 2-16
- function check, 2-17

beaten zone, 6-3

bore lights, G-20

boresighting, G-4, G-10, G-21

classes of fire, with respect to--
- ground, 6-3, 6-4 (illus)
- gun, 6-6, 6-7 (illus)
- target, 6-4, 6-5 (illus)

cleaning the weapon, 2-9, 2-10

clearing the weapon, 2-1 (*illus*)

cone of fire, 6-2, 6-3 (*illus*)

crews
duties, 5-23, 5-32
equipment, 5-23
exercises, 5-22
gun placement, 5-26
inspection, 5-24
positions, 5-23
relocating gun, 5-32 thru 5-35 (*illus*)
CVKI (combat vehicle kill indicator), B-2
cycle of functioning
chambering, 3-6 (illus)
cocking, 3-11
ejecting, 3-10
extracting, 3-10
feeding, 3-4 (illus), 3-5 (illus), 3-6 (illus), 3-11
firing, 3-7, 3-8 (illus)
locking, 3-6, 3-7 (illus)
unlocking, 3-8, 3-9 (illus), 3-10 (illus)
DA Forms
7007-R, Machine Gun Scorecard for M2, C-22
7448-R, M2 Caliber .50 Heavy Barrel Machine Gun, Firing Table I, Day Practice Scorecard, G-16
7449-R, M2 Caliber .50 Heavy Barrel Machine Gun, Firing Table II, Day Qualification Scorecard, G-16
7450-R, M2 Caliber .50 Heavy Barrel Machine Gun, Firing Tables III(A) and III(B), Night Practice Scorecard, G-17
7451-R, M2 Caliber .50 Heavy Barrel Machine Gun, Firing Table IV Scorecard, G-18
defilade positions
full, 6-18
laying methods, 6-20, 6-21
defilade positions (continued)
partial, 6-18, 6-19
description, 1-5 (*illus*)
components, 1-8, 1-9 (illus)
front sight, 1-7 (illus)
general data, 1-7
rear sight, 1-6 (illus)
destruction procedures
ammunition, 4-5
antiaircraft mount, 4-5
burning, 4-6
disposal, 4-6
gun, 4-5
spare parts, 4-5
tripod mount, 4-5
devices, training (*see also* SRTA)
M19 BFA, B-1 (illus)
M3 RABA, B-8 (illus)
MILES, B-2 thru B-6
sighting bar, B-6, B-7 (illus)
differences between practice and qualification tables, G-12
dismounting procedures, G-8, G-19
disassembly, general
backplate group, 2-4
barrel buffer assembly, 2-9
barrel buffer body group, 2-8
barrel extension group, 2-8
barrel group, 2-2
bolt group, 2-7
bolt stud, 2-6
driving spring rod assembly, 2-5
receiver group, 2-7 (illus)
dry fire
range setting and laying, 5-8
sighting and aiming, 5-6, 5-7
traversing and elevating, 5-8 thru 5-12
elevation, setting, 5-37
fighting positions, construction
antiaircraft, D-6 (illus)
three-man, D-1 (*illus*)

fighting positions, construction
(continued)
tripod-mounted, D-2 thru D-5 (*illus*)
final protective fires, 6-21
fire commands
elements of initial command
alert, 5-19
command to open fire, 5-21
description, 5-20
direction, 5-19
method of fire, 5-21
range, 5-21
subsequent commands, 5-22
fire control methods, 6-8, 6-28
firing exercises
mounted, C-26 thru C-31
NBC, 5-16, 5-41, 5-45, C-14 thru C-17
night-fire, 5-41, 5-45, C-17 thru C-20, G-11
predetermined, C-31 thru C-34
qualifications, C-21
scorecard, C-22, C-23 (illus)
ten-meter, 5-16, C-5 thru C-11
transition, 5-41, C-11 thru C-14
firing positions
antiaircraft, 5-5, 5-6 (illus)
prone, 5-3 (illus)
sitting, 5-3 (illus)
standing, 5-4 (illus)
vehicular, 5-5 (illus)
firing tables
I, C-5, C-6
II, C-13, C-14
III, C-16, C-17
IV, C-20
V, C-26
VI, C-30, C-31
VII, C-33, C-34
forms, example completed
DA Form 7007-R, C-22
DA Form 7448-R, G-16
forms, example completed
(continued)
DA Form 7449-R, G-16
DA Form 7450-R, G-17
DA Form 7451-R, G-18
FPF (final protective fires), 6-21
FPL (final protective line), 6-21
FTX (field training exercise), B-1
headspace, 3-12 thru 3-16 (*illus*)
heavy weapon thermal sight, G-1
immediate action, 6-29
inspection, 2-10, 2-11
laying methods, 6-20, 6-21
limited visibility
difficulties, 6-26
fire control, 6-28
NBC considerations, 6-29, 6-30
preplanned fires, 6-28
target engagement, 6-27
terminology, 6-26, 6-27
loading procedures, 3-1, 3-3 (*illus*)
automatic mode, 3-2
single-shot mode, 3-2
lubrication, 2-11
maintenance, 2-10
NBC conditions, 2-12
procedures, 2-12 (illus)
malfunctions, 4-1
marksmanship training
advanced, 5-41 thru 5-45
basic, 5-36 thru 5-41
dry fire, 5-6
fundamentals, 5-3 thru 5-36
phases, 5-2
planning, 5-1
remedial, 5-2
sustainment, 5-2
MILES (multiple integrated laser engagement system), B-2
mounting bracket assembly, G-9 (*illus*)
mounting procedures, G-8, G-15

mounts, ground
accessories, 1-12, 1-13, 1-14
antiaircraft (M63), 1-9, 1-11 (illus)
tripod (M3), 1-9, 1-10 (illus)
mounts, vehicular
armored vehicle cupola, 1-14, 1-16 (illus)
gun cradle (MK 64), 1-16 (illus)
pedestal truck (M31C and M24A2), 1-14, 1-15 (illus)
truck (M36), 1-14, 1-15 (illus)
MOUT, B-8
MRC (multipurpose range complex), C-24, C-25 (*illus*)
multipurpose MG transition, C-1, C-2 (*illus*)
multipurpose range complex, C-24, C-25 (*illus*)
NBC
considerations, 2-, 5-, 6-29, C-1, C-11, C-17
firing, 5-41
maintenance, 2-12
qualification
standards, C-21
table III, C-16 thru C-17 (*illus*)
night
course of fire qualification tables, G-11
firing exercises, 5-41, 5-45
optics, G-1
overhead fire
conditions, 6-16
gunner's rule, 6-16 (illus)
leader's rule, 6-17 (illus), 6-18
minimum clearance, 6-15 (illus)
precautions, 6-18
safety angles, 6-15
terrain, 6-16
practice and qualification tables, differences between, G-12
preplanned fires, 6-28
qualification
standards, night, C-21
tables, night course of fire, C-16 thru C-17 (*illus*), G-11
RABA (recoil amplifier barrel assembly), B-8
range card preparation, E-1 thru E-3 (*illus*)
range determination
estimating by eye, 5-13, 5-14, 5-15
firing the gun, 5-15
range safety, A-1, A-2, A-3
range setting and laying, 5-8
ranges, ten-meter layout, C-3 (*illus*), C-4
scorecard, C-22, C-23 (*illus*)
qualifications, C-21
reproducible form, back of manual
sighting and aiming, 5-6, 5-7 (illus)
sight picture, 5-37
SRTA (short-range training ammunition), B-7
SRTA-T (short-range training ammunition-tracer), B-8
stoppages, 4-1, 4-2 (illus)
immediate action, 4-2, 4-3, 6-29
remedial action, 4-32, 4-4 (illus)
STRAC (Standards in Training Commission) table, G-11 (*illus*)
tables, practice and qualification, differences between, G-12
target detection, 6-29
target engagements
area targets, 6-13, 6-14 (illus)
deep targets, 6-12
limited visibility, 6-27
linear targets, 6-9
linear targets with depth, 6-10
NBC considerations, 6-29

target engagements (continued)
point targets, 6-9
target offsets, G-22
test, fundamental skills, 5-36
thermal sight, G-1, G-2 (illus)
timing
field expedient methods, 3-18, 3-19
gauge, 3-12 (illus)
procedures, 3-16, 3-17, 3-18 (illus)
training
collective, 1-2
devices, B-1
initial, 1-1, 1-3
marksmanship, 5-1 thru 5-45
safety, 7-1, 7-2, A-1
sustainment, 1-2, 1-3, 1-4 (illus), 5-2
train the trainer, 1-3, 7-1, 7-2, 7-3
training strategy, 1-1, 1-4 (*illus*)
trajectory, 6-1, 6-2 (*illus*)
traversing and elevating, 5-8 thru 5-12
direction reading, 5-10
elevation reading, 5-11
manipulation exercise, 5-11, 5-12, G-26
zeroing, 5-10
unloading procedures, 3-3
vision devices, aided, G-1
windage, set, 5-37
zeroing/targeting
AN/TVS-5, 5-38, 5-39, 5-40
bore lights, G-20
field, 5-38
procedures, G-11
ten-meter, 5-36

MACHINE GUN SCORECARD FOR M2

For use of this form, see FM 23-65, Change 1, or the back of this form. The proponent agency is TRADOC.

DATA REQUIRED BY THE PRIVACY ACT OF 1974

AUTHORITY: 10 USC 3012(g)/Executive order 9397.

PRINCIPAL PURPOSE: To aid individual training on targets at various ranges.

ROUTINE USE: To evaluate individual proficiency. SSN is used for positive identification purposes only.

DISCLOSURE: Voluntary. However, mass rating and scoring require some tracking method. Therefore, only those soldiers who provide their SSNs can receive scores or qualify on weapons or equipment.

1a. LAST NAME	1b. FIRST NAME	1c. MI	2. DATE *(YYYYMMDD)*

3. SSN	4. UNIT	5. LANE

	6a. FIRING TABLE I			7a. FIRING TABLE II					8a. FIRING TABLE III					9a. FIRING TABLE IV				
TSK	RANGE (M)	HIT	PTS	RANGE (M)	TIME	HIT	PTS	BON	RANGE (M)	TIME	HIT	PTS	BON	RANGE (M)	TIME	HIT	PTS	BON
1	10	NA	NA	550	None	NA	NA	NA	550	None	NA	NA	NA	550	None	NA	NA	NA
2	10	NA	NA	800	20 Sec				800	20 Sec				800	20 Sec			
3	10	NA	NA	400	20 Sec				400	20 Sec				400	20 Sec			
4	10	NA	NA	700	25 Sec				700	25 Sec				700	25 Sec			
5	10			1,000	25 Sec				1,000	25 Sec				1,000	25 Sec			
6	10			400	35 Sec				400	35 Sec				400	35 Sec			
				700					700					700				
7	10			550	35 Sec				550	35 Sec				550	35 Sec			
				800					800					800				
8	NA	NA	NA	400	45 Sec				400	45 Sec				400	45 Sec			
				550					550					550				
				1,000					1,000					1,000				

6b. SUBTOTAL FIRING TABLE I	7b. SUBTOTAL FIRING TABLE II	8b. SUBTOTAL FIRING TABLE III	9b. SUBTOTAL FIRING TABLE IV

10. TOTAL SCORE	11. RATING
12a. OIC's PRINTED OR TYPED NAME	13a. GRADER'S PRINTED OR TYPED NAME
12b. OIC'S SIGNATURE	13b. GRADER'S SIGNATURE

DA FORM 7007-R, JAN 2002 DA FORM 7007-R, JUN 1991, IS OBSOLETE. USAPA V1.00

COMPLETE THE BLOCKS ON THIS SCORECARD AS FOLLOWS:

1. NAME. Enter last name, first name, middle initial.

2. DATE. Enter today's date in YYYYMMDD format.

3. SSN. Enter gunner's social security number.

4. UNIT. Enter gunner's unit.

5. LANE. Enter gunner's firing lane.

6a - 9a. FIRING TABLE I thru IV

HIT column. As applicable, mark block with an "X" if gunner hit target.

POINTS column. As applicable, enter points earned for each target hit.

BON *(Bonus)* column *(Firing Tables I thru III only)*. As applicable, enter bonus points earned for each target hit.

6b-9b. SUBTOTALS. Subtotal all of the points earned for each firing table.

10. TOTAL SCORE. Add firing table subtotals *(Blocks 6b - 9b)* to obtain this score.

11. RATING. Compare total score *(Block 10)* to rating chart to determine the soldier's rating.

12a. OIC's PRINTED OR TYPED NAME. Print or type OIC's name.
12b. OIC's SIGNATURE. *(OIC must sign the actual form).*

13a. GRADER's PRINTED OR TYPED NAME. Print or type grader's name.
13b. GRADER's SIGNATURE. *(Grader must sign the actual form).*

COMPUTE RATING:

Maximum................................ 218 points

Expert.................................... 196 to 217 points

Gunner first class...................... 174 to 195 points

Gunner second class................. 153 to 173 points

Unqualified.............................. 152 points or less

M2 CALIBER .50 HEAVY BARREL MACHINE GUN FIRING TABLE I
DAY PRACTICE SCORECARD

For use of this form, see FM 23-65, Change 1; the proponent agency is TRADOC.

DATA REQUIRED BY THE PRIVACY ACT OF 1974

AUTHORITY: 10 USC 3012(g)/Executive order 9397.

PRINCIPAL PURPOSE: To aid individual training on targets at various ranges.

ROUTINE USES: To evaluate individual proficiency. SSN is used for positive identification purposes only.

DISCLOSURE: Voluntary. However, mass rating and scoring require some tracking method. Therefore, only those soldiers who provide their SSNs can receive scores or qualify on weapons or equipment.

1a. LAST NAME	1b. FIRST NAME	1c. MI	2. DATE *(YYYYMMDD)*

3. SSN	4. UNIT	5. RANGE	6. LANE

7a. ENGAGEMENT	7b. STANDARDS	7c. AMMO	7d. TIME *(Minutes)*	7e. RANGE *(Meters)*	7f. GO	7g. NO GO
ZERO	1 BURST HIT	14	NA	400	NA	NA
2	1 BURST HIT	14	1.5	1,100 (±200)		
3	1 BURST HIT	14	1.5	1,500 (±200)		
4	1 BURST HIT	14	1.5	600 (±100)		
5	1 BURST HIT	14	1.5	800 (±100)		
6	1 BURST HIT	14	1.5	400		
8. MULTIPLE TARGETS						
7	1 BURST HIT	28	2.5	1,100 (±200)		
	1 BURST HIT			600 (±100)		
8	1 BURST HIT	28	2.5	800 (±100)		
	1 BURST HIT			1,500 (±200)		

9. SCORING *(Check one)*

EXPERT - 9 ☐ SHARPSHOOTER - 8 ☐

MARKSMAN - 6 ☐ UNQUALIFIED - 5 OR BELOW ☐

10. GUNNER'S SIGNATURE

11a. GRADER'S PRINTED OR TYPED NAME	11b. GRADER'S SIGNATURE

M2 CALIBER .50 HEAVY BARREL MACHINE GUN FIRING TABLE II
DAY QUALIFICATION SCORECARD

For use of this form, see FM 23-65, Change 1; the proponent agency is TRADOC.

DATA REQUIRED BY THE PRIVACY ACT OF 1974

AUTHORITY: 10 USC 3012(g)/Executive order 9397.

PRINCIPAL PURPOSE: To aid individual training on targets at various ranges.

ROUTINE USES: To evaluate individual proficiency. SSN is used for positive identification purposes only.

DISCLOSURE: Voluntary. However, mass rating and scoring require some tracking method. Therefore, only those soldiers who provide their SSNs can receive scores or qualify on weapons or equipment.

1a. LAST NAME	1b. FIRST NAME	1c. MI	2. DATE *(YYYYMMDD)*

3. SSN	4. UNIT	5. RANGE	6. LANE

7a. ENGAGEMENT	7b. STANDARDS	7c. AMMO	7d. TIME *(Minutes)*	7e. RANGE *(Meters)*	7f. GO	7g. NO GO
ZERO	1 BURST HIT	14	NA	400	NA	NA
2	1 BURST HIT	14	1	1,100 (±200)		
3	1 BURST HIT	14	1	1,500 (±200)		
4	1 BURST HIT	14	1	600 (±100)		
5	1 BURST HIT	14	1	800 (±100)		
6	1 BURST HIT	14	1	400		
8. MULTIPLE TARGETS						
7	1 BURST HIT	28	2	1,100 (±200)		
	1 BURST HIT			600 (±100)		
8	1 BURST HIT	28	2	800 (±100)		
	1 BURST HIT			1,500 (±200)		

9. SCORING *(Check one)*

EXPERT - 9 ☐ SHARPSHOOTER - 8 ☐

MARKSMAN - 6 ☐ UNQUALIFIED - 5 OR BELOW ☐

10. GUNNER's SIGNATURE

11a. GRADER'S PRINTED OR TYPED NAME	11b. GRADER'S SIGNATURE

DA FORM 7449-R, JAN 2002 USAPA V1.00

M2 CALIBER .50 HEAVY BARREL MACHINE GUN FIRING TABLES III (A) AND III (B) NIGHT PRACTICE SCORECARDS

For use of this form, see FM 23-65, Change 2; the proponent agency is TRADOC.

DATA REQUIRED BY THE PRIVACY ACT OF 1974

AUTHORITY: 10 USC 3012(g)/Executive order 9397.
PRINCIPAL PURPOSE: To aid individual training on targets at various ranges.
ROUTINE USES: To evaluate individual proficiency. SSN is used for positive identification purposes only.
DISCLOSURE: Voluntary. However, mass rating and scoring require some tracking method. Therefore, only those soldiers who provide their SSNs can receive scores or qualify on weapons or equipment.

1a. LAST NAME	1b. FIRST NAME	1c. MI	1d. RANK	2. SSN

TABLE III (A). NIGHT PRACTICE SCORECARD

NOTE: Use this table if the M2 has the TWS, or if it has an AN/TVS-5 that has been upgraded with the third-generation tube.

a. RANGE	b. UNIT	c. LANE	d. DATE *(YYYYMMDD)*

e. ENGAGEMENT	f. STANDARDS	g. AMMO	h. TIME *(Minutes)*	i. RANGE *(Meters)*	j. GO	k. NO GO
ZERO	1 BURST HIT	14	NA	400	NA	NA
2	1 BURST HIT	14	1.5	1,100 (±200)		
3	1 BURST HIT	14	1.5	1,500 (±200)		
4	1 BURST HIT	14	1.5	600 (±100)		
5	1 BURST HIT	14	1.5	800 (±100)		
6	1 BURST HIT	14	1.5	400		
l. MULTIPLE TARGETS						
7	1 BURST HIT	28	2.5	1,100 (±200)		
	1 BURST HIT			600 (±100)		
8	1 BURST HIT	28	2.5	800 (±100)		
	1 BURST HIT			1,500 (±200)		

4. SCORING *(Check one)*	
EXPERT - 9 ☐	5. GUNNER'S SIGNATURE
SHARPSHOOTER - 8 ☐	6a. GRADER'S PRINTED OR TYPED NAME
MARKSMAN - 6 ☐	
UNQUALIFIED - 5 OR BELOW ☐	6b. GRADER'S SIGNATURE

7. TABLE III (B). NIGHT PRACTICE SCORECARD

NOTE: Use this table if the M2 has an AN/TVS-5 that has not been upgraded.

a. RANGE	b. UNIT	c. LANE	d. DATE *(YYYYMMDD)*

e. ENGAGEMENT	f. STANDARDS	g. AMMO	h. TIME *(Minutes)*	i. RANGE *(Meters)*	j. GO	k. NO GO
ZERO	1 BURST HIT	14	NA	400	NA	NA
2	1 BURST HIT	14	1.5	1,100 (±200)		
3	1 BURST HIT	14	1.5	800 (±100)		
4	1 BURST HIT	14	1.5	600 (±100)		
5	1 BURST HIT	14	1.5	800 (±100)		
6	1 BURST HIT	14	1.5	400		
l. MULTIPLE TARGETS						
7	1 BURST HIT	28	2.5	1,100 (±200)		
	1 BURST HIT			600 (±100)		
8	1 BURST HIT	28	2.5	800 (±100)		
	1 BURST HIT			600 (±100)		

8. SCORING *(Check one)*	
EXPERT - 9 ☐	9. GUNNER'S SIGNATURE
SHARPSHOOTER - 8 ☐	10a. GRADER'S PRINTED OR TYPED NAME
MARKSMAN - 6 ☐	
UNQUALIFIED - 5 OR BELOW ☐	10b. GRADER'S SIGNATURE

M2 CALIBER .50 HEAVY BARREL MACHINE GUN FIRING TABLES IV (A) AND IV (B) NIGHT QUALIFICATION SCORECARDS

For use of this form, see FM 23-65, Change 2; the proponent agency is TRADOC.

DATA REQUIRED BY THE PRIVACY ACT OF 1974

AUTHORITY: 10 USC 3012(g)/Executive order 9397.
PRINCIPAL PURPOSE: To aid individual training on targets at various ranges.
ROUTINE USES: To evaluate individual proficiency. SSN is used for positive identification purposes only.
DISCLOSURE: Voluntary. However, mass rating and scoring require some tracking method. Therefore, only those soldiers who provide their SSNs can receive scores or qualify on weapons or equipment.

1a. LAST NAME	1b. FIRST NAME	1c. MI	1d. RANK	2. SSN

3. TABLE IV (A). NIGHT QUALIFICATION SCORECARD

NOTE: Use this table if the M2 has the TWS, or if it has an AN/TVS-5 that has been upgraded with the third-generation tube.

a. RANGE	b. UNIT	c. LANE	d. DATE *(YYYYMMDD)*

e. ENGAGEMENT	f. STANDARDS	g. AMMO	h. TIME *(Minutes)*	i. RANGE *(Meters)*	j. GO	k. NO GO
ZERO	1 BURST HIT	14	NA	400	NA	NA
2	1 BURST HIT	14	1	1,100 (±200)		
3	1 BURST HIT	14	1	1,500 (±200)		
4	1 BURST HIT	14	1	600 (±100)		
5	1 BURST HIT	14	1	800 (±100)		
6	1 BURST HIT	14	1	400		
l. MULTIPLE TARGETS						
7	1 BURST HIT	28	2	1,100 (±200)		
	1 BURST HIT			600 (±100)		
8	1 BURST HIT	28	2	800 (±100)		
	1 BURST HIT			1,500 (±200)		

4. SCORING *(Check one)*
- EXPERT - 9 ☐
- SHARPSHOOTER - 8 ☐
- MARKSMAN - 6 ☐
- UNQUALIFIED - 5 OR BELOW ☐

5. GUNNER'S SIGNATURE

6a. GRADER'S PRINTED OR TYPED NAME

6b. GRADER'S SIGNATURE

7. TABLE IV (B). NIGHT QUALIFICATION SCORECARD

NOTE: Use this table if the M2 has an AN/TVS-5 that has not been upgraded.

a. RANGE	b. UNIT	c. LANE	d. DATE *(YYYYMMDD)*

e. ENGAGEMENT	f. STANDARDS	g. AMMO	h. TIME *(Minutes)*	i. RANGE *(Meters)*	j. GO	k. NO GO
ZERO	1 BURST HIT	14	NA	400	NA	NA
2	1 BURST HIT	14	1	1,100 (±200)		
3	1 BURST HIT	14	1	800 (±100)		
4	1 BURST HIT	14	1	600 (±100)		
5	1 BURST HIT	14	1	800 (±100)		
6	1 BURST HIT	14	1	400		
l. MULTIPLE TARGETS						
7	1 BURST HIT	28	2	1,100 (±200)		
	1 BURST HIT			600 (±100)		
8	1 BURST HIT	28	2	800 (±100)		
	1 BURST HIT			600 (±100)		

8. SCORING *(Check one)*
- EXPERT - 9 ☐
- SHARPSHOOTER - 8 ☐
- MARKSMAN - 6 ☐
- UNQUALIFIED - 5 OR BELOW ☐

9. GUNNER'S SIGNATURE

10a. GRADER'S PRINTED OR TYPED NAME

10b. GRADER'S SIGNATURE

Made in United States
Orlando, FL
04 November 2026

90206685R00331